The impact of nutrition on the health and welfare of horses

EAAP – European Federation of Animal Science

EWEN – European Workshop on Equine Nutrition

BSAS – british society of animal science

The European Association for Animal Production wishes to express its appreciation to the *Ministero per le Politiche Agricole e Forestali* and the *Associazione Italiana Allevatori* for their valuable support of its activities

The impact of nutrition on the health and welfare of horses

5th European Workshop Equine Nutrition
Cirencester, United Kingdom
19-22 September 2010

EAAP publication No. 128

edited by:

A.D. Ellis, A.C. Longland, M. Coenen and N. Miraglia

Wageningen Academic
Publishers

ISBN: 978-90-8686-155-2
e-ISBN: 978-90-8686-711-0
DOI: 10.3920/978-90-8686-711-0

ISSN 0071-2477

First published, 2010

Wageningen Academic Publishers
P.O. Box 220
6700 AE Wageningen
The Netherlands
www.WageningenAcademic.com
copyright@WageningenAcademic.com

Table of contents

Part 3. Promoting health and preventing disease

Part 4. Nutrition and gastro-intestinal health

Part 5. Nutrition and metabolic disease

Foreword

It is with great pleasure that we have compiled this volume representing both invited and open papers presented in the Scientific Sessions of the 5th European Workshop on Equine Nutrition, held at the Royal Agricultural College, Cirencester, UK from 19-22nd September 2010. The focus of the meeting was The Impact of Nutrition on Equine Behaviour and Welfare.

This biannual meeting is held under the umbrella of the Horse commission of the European association for Animal Production (EAAP) as a satellite of the annual meeting of EAAP. EWEN UK 2010 was also organised in conjunction with the British Society for Animal Science (BSAS, Member of EAAP).

Each year, the Horse commission sets up six scientific sessions, within the scope of the EAAP meeting. These sessions are devoted to specific disciplinary areas in order to initiate discussion between scientists, education and industry in order to provide a multidisciplinary approach to issues raised by the equine industry. Within this meeting only limited time is available for in-depth discussions during the sessions. Hence, European working groups (EWG) have been set up, at the initiative of the Horse commission, to allow for in depth exploration of topic areas and to communicate latest research to the public. The European Workshop of Equine Nutrition (EWEN) is a biannual workshop which was first held in Dijon in 2002. It came under the umbrella of the EAAP 'EWG Nutrition' from its second meeting in 2004.

The founding aims of the workshop are to:
1. Facilitate discussion between equine nutrition scientists in Europe.
2. Create a bridge between scientists, practitioners and the horse industry.

Key objectives have been identified as:
- To evaluate differences and applications of different nutritional systems for rationing horses, comparing feeding practices throughout Europe, with a view to giving clear guidelines to practitioners.
- To discuss the increased understanding of interactions between health and nutrition with a focus on performance and welfare of horses.

EWEN UK 2010 held both Scientific and Applied Sessions. Scientific papers were delivered on two days with applied sessions being held on the second. The conference aimed to provide up-to-date reviews and findings of the latest research of interest to academics, researchers, veterinarians, equine nutritionists and students. The Applied Sessions focused on how this information could be used in a practical manner by both professional practitioners within the equine industry and by lay horse owners.

The conference sessions included: Nutritional requirements and physiology, Nutrition, behaviour and welfare, Promoting health and preventing disease, Ration formulation for horses, Nutrition and gastro-intestinal health, Nutrition and metabolic disease, Functional nutritional ingredients and finally Nutrition, health and performance. There was also a well-supported Poster Session, with prizes for those young scientists judged to have given the best oral and poster presentations.

This conference could not have been held without the inspiration and hard work of the UK Organising Committee (P.A. Harris & A.D. Ellis [Co-Chairs] and the UK team members: A.C. Longland, J. Murray, C. Dunnett, C. Hale, J. Newbold, M. Steele, C. Argo, D. Goodwin, B. Young, G. Crossman, A. Hemmings and C. Williams) and the substantial efforts delivered by the Editorial team (A.C. Longland [Chief Editor], A.D. Ellis, M. Coenen and N. Miraglia,). We thank Pat Harris in particular for bringing the organising committee together and her unwavering, open minded support throughout.

We also thank all of the Referees (which in addition to those acknowledged in their other roles within EWEN 2010, included: I. Vervuert, V. Julliand, A. Durham, G. Janssen, A. Goncalves, D. Bergero, A. Zeyner. R. Geor, N. Lutherson) and the input of the Scientific Committee (M. Coenen [Chair], N. Miraglia [President of the EAAP Horse Commission], M. Saastamoinen, W. Martin-Rosset, J.E. Lindberg,. P.A. Harris, A.D. Ellis, A.S. Santos).

Catherine Dunnet from *Independent Equine Nutrition* deserves a special mention for attracting and looking after our Sponsors that have been so very generous in their support of this conference (Alltech Inc., Equine Health & Nutrition Conference & CAVALOR jointly, Lanes Health (LitoVET), Saracen Horse Feed, SPILLERS Horse Feeds, Thoroughbred Remedies Ireland (TRM), Probiotics International Ltd. (Protexin), WALTHAM).

Grateful thanks are due to all of the Invited Speakers who provided a high standard of informative review papers as well as to scientists who presented their recent research findings.

We also thank Jo-Anne Murray and the Edinburgh University graphics laboratory for designing and updating the web-site and providing scientific flyers. Finally last but not least the UK Organising committee and The Scientific committee of EWEN would like to thank BSAS for their support and hard work in helping to ensure the success of the 5th European Workshop in Equine Nutrition (particularly: M. Steele, B. Cooke and B. Hilton).

All of the above gave up their time freely; we are deeply indebted to them.

Annette C. Longland, *Director of Research, Equine and Livestock Nutrition Services, UK*
Andrea D. Ellis, *Nottingham Trent University, UK*
Manfred Coenen, *Institute of Animal Nutrition, Leipzig, Germany*
Nicoletta Miraglia, *Molise University, Italy*

Part 1. Tailoring nutrition to the physiology of the horse

Digestive strategy and flexibility in horses with reference to dietary carbohydrates

R.J. Geor
Department of Large Animal Clinical Sciences, College of Veterinary Medicine, Michigan State University, East Lansing, Michigan 48824, USA

Abstract

Nutritional management of horses is often in conflict with the gastrointestinal physiology of horses. Whereas horses evolved as grazing and browsing animals with a gastrointestinal tract adapted for the 'trickle feeding' of feedstuffs with high structural fibre content, today's horse is often fed a ration contain substantial quantities of starch and other nonstructural carbohydrates. This review discusses selected aspects of digestive function in horses, with special reference to mechanisms of carbohydrate digestion in the small intestine. Adaptive responses to an increase in dietary non-structural carbohydrates and current knowledge concerning upper limits or tolerances for dietary intake are also discussed. A central thesis is that horses have limited digestive flexibility in the face of abrupt dietary change, and that this constraint at least in part explains the high risk for colic and other manifestations of intestinal dysfunction following recent changes in feeding.

Keywords: nutrition, physiology, dietary carbohydrates

Introduction

The term phenotypic flexibility refers to reversible adjustments in traits of organisms that occur in response to alterations in the internal or external environment (Piersma and Drent, 2003). From a nutritional perspective, the rate and magnitude of changes in digestive and metabolic function in response to, for example, a change in diet composition will reflect phenotypic flexibility. Successful adjustment to alterations in diet will assure continued nourishment and health, whereas failure to adapt may result in nutritional compromise and/or adverse health events. The concept of phenotypic flexibility is relevant to an understanding of the impact of diet and feeding strategy on gastrointestinal and metabolic physiology in horses – knowledge of factors that regulate the digestion and utilization of dietary constituents should enable the development of feeding recommendations that are commensurate with physiologic mechanisms and tolerances, thereby lessening risk of diet-associated problems.

Modern nutritional management is often in conflict with the gastrointestinal and, potentially, metabolic physiology of horses. Over millions of years, equids have evolved as grazing and browsing animals with a gastrointestinal tract adapted for 'trickle feeding' and the assimilation of nutrients and energy from a diet that is predominantly plant fibre. Feral equids consume a variety of forages with grasses, sedges and rushes accounting for 90% of dietary intake (McInnes and Vavra, 1987). These animals spend up to 20 hours per day engaged in grazing activity, with sustained periods of feeding separated by short (30 to 60 min) rest periods. Similarly, domesticated ponies kept at pasture will graze for 12 to 14 hours per day, with most feed intake occurring during daytime and early evening (Crowell-Davis *et al.*, 1985; Husted *et al.*, 2009; Prache *et al.*, 1998). When feed is freely available periods of inanition rarely extend longer than 3 to 5 hours (Ralston *et al.*, 1984).

These circumstances are in marked contrast with the feeding management of many domesticated horses, particularly those kept in confinement. The ration often contains substantial quantities of cereal grain or grain by-products (starch rich ingredients) with a much lower proportion of dry matter coming from structural carbohydrates when compared to horses kept in a grazing habitat. Second, the ration is often provided as two meals (e.g. 7:00 and 16:00). When compared to forages, energy-dense concentrate feeds are consumed quickly resulting in lengthy periods of feed withholding throughout a 24-hour period. Additionally, the dynamics of feed and preserved forage availability and horse owner preferences may result in fairly abrupt dietary changes that challenge gastrointestinal tract health and function. The potential for these feeding practices to adversely affect gastrointestinal health is highlighted by the results of several epidemiological studies that have identified cereal grain (or starch) feeding as a risk factor for gastric ulcer syndrome (Luthersson *et al.*, 2009) or colic (Tinker *et al.*, 1997). Similarly, sudden changes in feeding (forage or cereal grain/concentrate) appear to markedly increase risk of colic during the 7- to 14-day period following the dietary change (Hillyer *et al.*, 2002; Hudson *et al.*, 2001). The alterations in the equine colon ecosystem that accompany abrupt changes in diet (Julliand *et al.*, 2001; Muhonen *et al.*, 2009) may contribute to increased risk of colic under these circumstances.

Nutritionists and veterinarians have also been interested in the impact of diet and feeding management on the metabolic function and health of horses. For example, it has been argued that the pronounced glycaemic and insulinaemic responses that occur following ingestion of starch-rich meals may result in a decrease in insulin sensitivity, a concern given the recently recognized associations between insulin resistance, hyperinsulinemia and risk of laminitis (Treiber *et al.*, 2006a). Consequently, a number of recent studies have examined the effects of dietary energy source on insulin sensitivity and other aspects of metabolism (Treiber *et al.*, 2006b; Pratt *et al.*, 2006; Treiber *et al.*, 2008). The apparent negative effects of high-starch feeding have spurned interest in the utility of vegetable oils in rations for horses, including research on the digestibility (Kronfeld *et al.*, 2004) and metabolic effects (Hoffman *et al.*, 2003; Treiber *et al.*, 2008) of oil-supplemented diets. However, despite the widespread use of vegetable oils in equine diets, little is known concerning the mechanisms of fat digestion and absorption in the horse. Similarly, there is only limited information on the regulation of fat metabolism.

This paper reviews selected aspects of digestive function in horses, with special reference to mechanisms of carbohydrate digestion in the small intestine. Discussion on adaptive responses to an increase in dietary nonstructural carbohydrates and current knowledge concerning upper limits or tolerances for dietary intake are also included. A central thesis is that horses have limited digestive flexibility in the face of abrupt dietary change, and that this constraint at least in part explains the high risk for colic and other manifestations of intestinal dysfunction following recent changes in feeding.

Aspects of carbohydrate digestion

Structural and nonstructural carbohydrates are primary constituents of feedstuffs for horses (Hoffman *et al.*, 2001). Nonstructural carbohydrates (NSC) may be hydrolysed to monosaccharides in the small intestine, mostly by pancreatic and brush-border enzymes, while structural carbohydrates are subjected to bacterial fermentation, mostly in the caecum and large colon. Carbohydrates with α-1,4 or α-1,6 bonds between sugar molecules are subject to enzymatic hydrolysis; these include hexoses, disaccharides, some oligosaccharides and starches (although some starches are resistant to enzymatic hydrolysis). Carbohydrates with β-1,4 linked molecules cannot be digested by mammalian enzymes and must be fermented. The fermentable carbohydrates include soluble fibres (e.g. gums, pectins), some oligosaccharides and polysaccharides such as fructans, galactans and starches resistant to enzymatic hydrolysis, hemicelluloses and cellulose. Fructans and soluble fibres (as well as starch) are subject to rapid fermentation, whereas insoluble fibres (e.g. cellulose) are fermented at a slower rate. The delivery of excessive loads of rapidly fermentable carbohydrates to the large intestine can result in destabilisation of the large intestinal microbiome and the precipitation of colic and/or laminitis (Milinovich *et al.*, 2006, 2010).

In common with other monogastric species, the small intestine is the major site of host enzymatic carbohydrate hydrolysis. Equine saliva contains little or no α-amylase activity and probably has minimal impact on starch digestion. There may be some hydrolysis in the stomach via the action of gastric acid but intragastric fermentative activity is likely to have a more substantial impact on the breakdown of ingested starch. It has been reported that the concentration of total anaerobic bacteria in gastric content is at least equivalent to that of the caecum or colon (De Fombelle *et al*., 2003). Additionally, the composition of the intragastic microbiome appears to change rapidly following meal ingestion (Varloud *et al*., 2007). Varloud *et al.* (2007) demonstrated marked increases in total anaerobes and lactate-utilising bacteria during a 210-min postprandial period, in association with linear increases in the L-lactate, D-glucose and volatile fatty acids (VFA) concentrations of gastric content. Further studies are needed to examine the quantitative impact of intragastric fermentation on precaecal starch disappearance.

In non-ruminant species there are three primary steps in the digestion of starch within the small intestine. First, there is hydrolysis of α-1,4 linkages by pancreatic α-amylase and intestinal glucoamylase, yielding primarily maltose and maltotriose but also the oligosaccharides amylose, amylopectin and α-dextrins. Disaccharidases present on the brush-border membrane, especially maltase and sucrase, hydrolyze these products to monosacharides that can be absorbed via specialised transport mechanisms. Sucrase hydrolyses sucrose and α-(1,6) links of residual α-dextrins to α-D-glucose and α-D-fructose. Maltose is hydrolysed by maltase yielding α-D-glucose. Dyer *et al.* (2002) reported that sucrase and maltase activities in the small intestine of pasture fed horses were higher in the jejunum and ileum than in the duodenum. Maltase activity in equine small intestine is very high compared to other mammalian species, suggesting that the digestion of maltose is not a limiting factor in the precaecal digestion of starch.

Glucose is transported across the enterocyte brush-border membrane by the sodium-glucose co-transporter protein isoform 1 (SGLT1), with subsequent movement of glucose across the basolateral membrane into the interstitium facilitated by the GLUT2 transport protein (Shirazi-Beechey *et al.*, 1991; Dyer *et al.*, 2002). In horses maintained on pasture, there was wide individual variation in the expression of SGLT1 but in general levels of expression decreased in an oral-aboral (proximal to distal) fashion (duodenum > jejunum > ileum) (Dyer *et al.*, 2002). A similar oral-aboral pattern of mRNA and protein expression has been described for GLUT2, the glucose transporter on the basolateral membrane (Salmon *et al.*, 2002).

A specific fructose transporter protein, equine GLUT5, also has been identified on the brush-border membrane (Fernandez-Castaño Merediz *et al.*, 2004). Studies in horses maintained at grass without grain supplementation demonstrated that the GLUT5 protein is highly expressed in duodenal and jejunal enterocytes, with lower level expression in the ileum (Fernandez-Castaño Merediz *et al.*, 2004). These molecular findings are consistent with observations of a glycemic/insulinemic response in horses following fructose administration (Bullimore *et al.*, 2000). There are no published data on adaptive responses to diet in relation to equine GLUT5 expression or the capacity for fructose absorption.

In recent years there has been intense interest in fructans, driven by associations between fructan accumulation in pasture forage and the development of laminitis in pasture-kept equids (Longland and Byrd, 2006) and the observation that administration of large doses (>7.5 g/kg BW) of a commercial fructan (oligofructose derived from the roots of chicory) induces laminitis in healthy horses (Van Eps and Pollitt, 2006). Fructans may comprise between 5 and 50% of the dry matter in temperate species such as perennial ryegrass (Longland *et al.*, 1999). Under some circumstances, when the fructan content of grass reaches >20-30% dry matter (e.g. during periods of intense photosynthesis and carbohydrate storage), it is conceivable that grazing horses could ingest more than 5 kg fructans per day – a load equivalent to that used in experimental models of laminitis, albeit ingested over a 24-h period vs. bolus administration (Longland and Byrd, 2006).

Fructans are water soluble polymers of fructose with either β-2,1 or β-2,6 linkages, all bonded to a terminal glucose moiety. The glycosidic bonds in fructan are not hydrolysed by host enzymes but may be susceptible to partial acid hydrolysis in the stomach and/or microbial fermentation in the small intestine (Coenen *et al.*, 2006). The breakdown of fructan in the foregut could result in release of fructose that would be absorbed in the small intestine via equine GLUT5. Such a mechanism could, at least in part, explain the circadian and seasonal patterns in plasma glucose and insulin noted in grazing horses that were correlated with changes in pasture forage water soluble carbohydrate content (Byrd *et al.*, 2002). The bulk of ingested fructan, however, is thought to reach the hindgut where it is rapidly fermented. Indeed, *in vitro* studies have demonstrated that commercial fructan elicits a more rapid fall in caecal pH than an equal amount of corn starch (Bailey *et al.*, 2002), and studies during oligofructose-induced laminitis have demonstrated marked hindgut acidosis and alterations in microbial populations consistent with carbohydrate overload of the caecum and large intestine (Milinovich *et al.*, 2006). However, numerous knowledge gaps exist regarding the impact of naturally-occurring fructans on the hindgut environment. Most information has been derived from studies utilizing commercially-available fructan (oligofructose) that has a different biochemical structure than levans, the predominant fructan in grasses. Oligofructose polymers are β-2,1-linked whereas levans are β-2,6-linked. Additionally, the degree of polymerisation (DP) of oligofructose (≤10) is lower when compared to levans (DP 30-50 or higher), which may contribute to a more rapid rate of fermentation and also impact the type of bacteria that proliferate (Milinovich *et al.*, 2010).

Recent studies have also examined the effects of dietary short-chain fructo-oligosaccharides (scFOS) on the composition and activity of hindgut microbiota in horses. The feeding of small amounts (30-40 g/day to a mature horse) of scFOS is associated with an increase in the quantity of lactate-utilizing bacteria in the large colon and an increase in total VFA (Berg *et al.*, 2005; Respondek *et al.*, 2007). Additionally, scFOS supplementation mitigated changes the bacterial composition of caecal and colonic contents in horses subjected to an abrupt increase in dietary starch (Respondek *et al.*, 2008).

Limitations to starch digestion in the small intestine

There is evidence that horses have a limited capacity for complete digestion of starch in the small intestine (Kienzle *et al.*, 1997). Moreover, as mentioned above, it is apparent that overload of small intestinal capacity for starch hydrolysis increases risk for colic and laminitis due to changes in the hindgut environment. Accordingly, there has been research directed toward the determination of the critical upper limit of small intestinal starch digestion. The results of early work suggested an upper limit in the range of 3.5 to 4.0 g starch per kg BW (Potter *et al.*, 1992) but several other studies have reported significant starch 'by-pass' of the small intestine at lower levels of starch intake depending on the source of starch (grain type) and method of grain processing (Radicke *et al.*, 1991; Meyer *et al.*, 1995; McLean *et al.*, 2000). Potter *et al.* (1992) observed that the precaecal digestion of corn starch increased between intakes of 1.0 to 3.5 g/kg BW, with no further increment in starch digestion when intake was increased above 4.0 g/kg BW. These authors also observed that the starch content of ileal chyme remained low and unchanged between corn starch intakes between 1.0 and 2.5 g/kg BW but increased exponentially at higher intakes. Oats have higher precaecal starch digestibility than equivalent doses of unprocessed corn or barley (Radicke *et al.*, 1991; Meyer *et al.*, 1995), in part due to differences in starch structure (Kienzle *et al.*, 1997). At intakes between 1.0 and 4.0 g starch/kg BW, corn starch resulted in lower caecal pH when compared to equivalent amounts of oat starch. Furthermore, this difference in pH between starch sources increased in proportion to starch intake (Radicke *et al.*, 1991). Mechanical processing of oats, corn or barley modestly increased starch digestibility, while more substantial increases in apparent precaecal starch digestibility were observed when corn or barley were subjected to thermo-mechanical processing techniques such as popping or steam flaking (Potter *et al.*, 1992; Meyer *et al.*, 1995; McLean *et al.*, 2000; Vervuert *et al.*, 2008).

Based on the above studies, it has been widely recommended that individual meals contain no more than 2.0 g starch/kg BW for optimization of starch digestion within the small intestine and minimization of gastrointestinal disturbances associated with the flow of undigested starch to the large intestine. Nonetheless, findings from recent experimental and epidemiological studies suggest that this recommendation may require re-examination – with some authors recommending a starch intake of ≤1.1 g/kg BW/meal for the purposes of minimizing digestive disturbances and mitigating postprandial increases in blood glucose and insulin concentrations (Vervuert *et al.*, 2009). Willing *et al.* (2009) compared the effects of a high-energy forage only diet (F; 10.4 MJ/kg DM) vs. a forage-concentrate diet (C) on faecal microbiota in six Standardbred trotters in race training. The concentrate was 35.8% starch (predominantly from oats) and provided 50% of DM intake in the C diet. Importantly, starch intake in C was considered to be safe at less than 2 g/kg BW/meal (approximately 5 g/kg BW/day, divided into 3 equal meals). Faecal samples, collected on days 7, 14, 21 and 29 in each experimental period, were cultured and analyzed for bacterial 16S rRNA genes by use of terminal-restriction fragment length methodology. There were major differences between diets for faecal microbiota. The C diet resulted in higher counts and relative abundance of lactic acid bacteria (LAB) than in the F diet, specifically members of the *Streptococcus bovis/equinus* complex that have been associated with the induction of laminitis. Additionally, diet C resulted in an increase in the members of *Clostridiaceae* cluster III and a concomitant reduction in an unknown group of *Bacteroidales*. Although there was wide variation between horses with respect to stability of faecal micobiota and response to diet, there was greater stability in the microbial composition between sampling periods on the F diet. The authors concluded that 'feeding recommendations for safe starch inclusion levels in the diet of horses may have to be re-evaluated and possibly changed.' The wide variability between horses in faecal microbial diversity and instability in response to the C diet implied that some horses may be more or less susceptible to adverse effects associated with dietary change (Willing *et al.*, 2009).

An epidemiological study of risk factors for equine gastric ulceration syndrome (EGUS) in Danish horses found that a diet that provided more than 2 g/kg BW of starch intake per day was associated with a two-fold increase in the likelihood of EGUS (severity grade of ≥2 on a 5-point scale), and that feeding more than 1 g/kg BW of starch per meal was associated with a 2.6-fold increase in the likelihood of an EGUS score ≥2 (Luthersson *et al.*, 2009). This study cannot confirm a cause-and-effect relationship between starch feeding and EGUS, and it is possible that other dietary factors, such as low structural fibre intake, contributed to this apparent association. However, it is known that higher starch meals result in more extensive intra-gastric fermentation and production of VFA, which have been shown to decrease squamous gastric mucosal integrity (Andrews *et al.*, 2006; Al Jassim and Andrews, 2009).

Finally, it should be noted that the data on pre-caecal starch digestibility were derived from studies in which horses or ponies were adapted to grain feeding for a 3-4 week period. The upper limit of pre-caecal starch digestibility may be lower in horses not accommodated to grain or concentrate feeding.

Mechanistic basis for limitations in precaecal starch digestion

Factors that may constrain starch hydrolysis in the small intestine include the availability and/or activity of α-amylase or brush-border maltase and the effect of meal size and/or composition on ingesta transit through proximal intestine tract, and therefore the time available for complete starch hydrolysis (Kienzle *et al.*, 1997).

The concentration of α-amylase within pancreatic juice and intestinal chyme of horses is very low in comparison to other species and also highly variable between horses (Alexander and Hickson, 1970; Roberts, 1974; Kienzle *et al.*, 1994). Similarly, α-amylase activity of pancreatic tissue is very low relative to pig, rat and calf (Lorenzo-Figueras *et al.*, 2007; Table 1). It is therefore quite

Table 1. Specific activities of pancreatic enzymes in different species. Data adapted from Lorenzo-Figueras et al. *(2007).*

Enzyme activity (units per mg pancreatic protein)	Horse	Pig	Rat	Calf
Amylase	2.3 (1.00)	107 (1.00)	56 (1.00)	2.3 (1.00)
Elastase	0.07 (0.03)	0.22 (<0.01)	0.02 (<0.01)	0.03 (0.01)
Chymotrypsin	0.36 (0.16)	2.26 (0.02)	1.34 (0.02)	1.51 (0.66)
Lipase	41.5 (18.04)	49 (0.46)	39 (0.69)	11 (4.78)

Values in parentheses are ratio to amylase value

possible that the availability and/or activity of α-amylase is one factor that limits precaecal starch hydrolysis, thereby contributing to the flow of undigested starch into the large intestine in horses provided starch-rich meals (Richards *et al.,* 2004).

On the other hand, intestinal brush-border maltase activity is unlikely to be a limiting factor in starch hydrolysis in the horse (Roberts, 1975; Kienzle *et al.,* 1994; Dyer *et al.,* 2002). Maltase activity is fairly consistent along the length of the small intestine and is very high compared with activity measured in other species (Dyer *et al.,* 2002). The horse also has high activity of sucrase, the disaccharidase that hydrolyses sucrose to glucose and fructose, with activity being highest in the proximal when compared to distal small intestine (Kienzle *et al.,* 1993; Dyer *et al.,* 2002).

Intestinal transit time also may affect the rate and extent of precaecal starch hydrolysis. However, only a small number of studies have examined the effects of diet composition on direct measures of small intestinal transit or retention time. In a recent review, Julliand *et al.* (2008) concluded that 'taking into account gastric emptying, it seemed that transit in the small intestine was more rapid when the intake of concentrate is increased but, for equal intakes, the retention time was not modified by the composition of the concentrate.' It should be noted that interpretation of studies performed to date has been confounded by simultaneous alterations in starch and structural fibre intake (i.e. an increase in meal starch content is most often accompanied by a decrease in fibre content, and it is difficult to determine which factor has contributed to observed changes in transit). Further studies are needed to determine the extent to which variation in small intestinal transit affect the efficiency of precaecal starch digestion.

Adaptive responses to increased dietary starch

Another important question in relation to precaecal carbohydrate digestion is whether (and to what extent) there are alterations in the capacity for starch digestion and/or glucose absorption in response to changes in dietary intake. Studies in many herbivorous and omnivorous species have demonstrated a substantial increase in the ability to absorb monosaccharides in response to increased dietary carbohydrates (Ferraris and Diamond, 1989; Buddington *et al.,* 1991; Shirazi-Beechey *et al.,* 1991; Dyer *et al.,* 2003). In mice, there is very rapid up-regulation with a significant increase in small intestinal SGLT1 as soon as 12 h after an abrupt increase in dietary nonstructural carbohydrates (Ferraris, 2001). In addition, enhanced levels of pancreatic α-amylase protein abundance and activity have been detected following an increase in dietary hydrolysable carbohydrate in several species (Brannon, 1990; Swanson *et al.,* 2000). In horses, it has been proposed that the small intestine may have a slower or inadequate adaptive response to increased dietary load (Buddington and Rashmir-Raven, 2002), which may be a contributing factor to the increased risk of colic attributed to recent changes in grain or concentrate feeding (Clarke *et al.,* 1990; Hudson *et al.,* 2001; Hillyer *et al.,* 2002).

Recent studies from the laboratory of Professor Soraya Shirazi-Beechey have examined the impact of dietary nonstructural carbohydrate load on digestive mechanisms in the equine small intestine (Dyer *et al.*, 2009). In one study, the activity of brush-border disaccharidase enzymes and the expression of SGLT1 along the length of the small intestine were measured in horses that were either kept at pasture (forage only diet) or fed hay plus grain-concentrate. Horses on the grain-concentrate supplemented diet exhibited 2-fold higher jejunal SGLT1 mRNA and protein expression when compared to pasture-fed animals, and a 4-fold higher expression of SGLT1 in the ileum. This increase in SGLT1 abundance was associated with proportional increases in the rates of sodium-dependent glucose transport, measured in brush-border membrane vesicle preparations. The grain-concentrate fed horses also demonstrated increased GLUT2 mRNA and protein abundance and glucose transport function in the basolateral membrane (Salmon *et al.*, 2002). The activity of brush-border disaccharidases was similar between groups suggesting no effect of diet composition on disaccharidase activity (Dyer *et al.*, 2009). The design of this study did not allow for assessment of the time course of changes in small intestinal glucose absorptive capacity. Furthermore, precise information on diet composition and intake was not available. Nonetheless, this study provided evidence of an adaptive increase in small intestinal glucose absorptive capacity in response to increased dietary nonstructural carbohydrates.

A subsequent trial examined the effects of a controlled increase in starch feeding on the magnitude and time course of any changes in SGLT1 expression (Dyer *et al.*, 2009). In this study, 9 horses were fed timothy hay (with vitamin-mineral supplement) for 3 months prior to collection of full-thickness biopsies (B1) from the proximal duodenum and distal jejunum via a laparoscopic technique (Bracamonte *et al.*, 2008). Six of the horses were then fed a diet of 60% hay and 40% grain (2/3 oats and 1/3 corn) that provided about 3 g starch/kg BW/day, while the other horses remained on the base diet. Further intestinal biopsies were obtained after 1 week (B2) and 1 month (B3) on this diet. Subsequently, the diet was changed to 40% hay, 60% grain (6.0 g starch/kg BW/day) for a further month after which a final set (B4) of duodenal and jejunal biopsies were collected. After one week of grain feeding, there was minimal change in in SGLT1 mRNA expression in the duodenum but a 2-fold increase in SGLT1 expression in the distal jejunum. With a further 3 weeks of grain feeding, duodenal SGLT1 mRNA expression was 2-fold higher when compared to levels measured on the base diet but there was no further change in distal jejunal expression. A doubling of starch intake resulted in a further increase in distal jejunal SGLT1 mRNA expression (3-fold higher when compared to the base diet) but no change in duodenal SGLT1 mRNA (Dyer *et al.*, 2009). Notably, the rate of change in SGLT1 expression was slow relative to observations in other species (Figure 1).

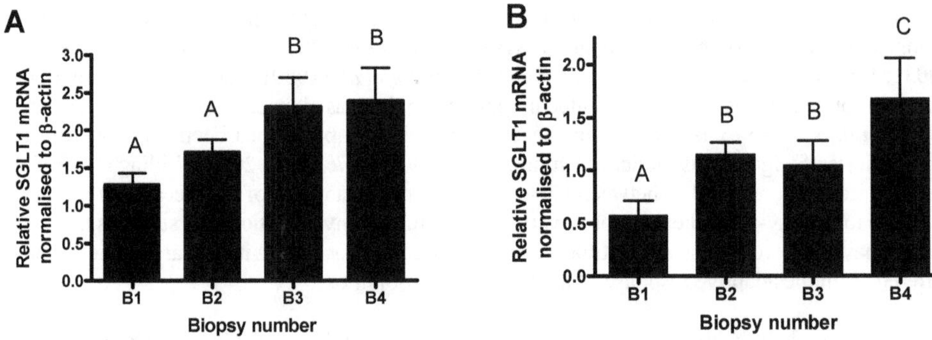

*Figure 1. Effect of dietary starch on SGLT1 expression in equine small intestine. Steady-state levels of SGLT1 mRNA were determined by qPCR in RNA isolated from duodenal (A) and distal jejunal (B) biopsies; data were normalized to β-actin mRNA within each sample. All values are expressed relative to SGLT1 in the duodenum of horses on hay-only diets (B1) as means ± SEM (with n=6 in each group). Statistically significant results (*P*<0.05*) are indicated by different letters above error bars (ABC). The data are adapted from Dyer* et al. *(2009).*

Studies in several species have shown that small intestinal SGLT1 is upregulated by monosaccharides but not by starch (Ferraris, 2001). Accordingly, the observation that SGLT1 expression was enhanced in response to increased starch feeding implies a concomitant increase in capacity for starch hydrolysis with release of monosaccharides (glucose) that signalled upregulation of SGLT1 (Dyer *et al.*, 2007). Dyer *et al.* (2009) proposed that the slow rate of increase in SGLT1 reflected a lag in upregulation of mechanisms for starch hydrolysis, in particular α-amylase activity. It was further suggested that the primary rate limiting step for an increase in small intestinal glucose absorptive capacity is the inability of the horse to hydrolyse starch to glucose rapidly. Although technically challenging, studies on the effect of diet on pancreatic α-amylase expression, secretion and activity are needed to further explore these issues. Kienzle *et al.* (1994) reported that α-amylase activity in samples of jejunal chyme from horses fed a grain diet for 24 h was approximately double that of horses maintained on an-all hay diet but the mechanism of this change in enzyme activity was not determined.

Tolerance to acute dietary change

Several epidemiological studies have identified a recent, abrupt change in diet (change in hay *or* grain feeding) as a risk factor for an episode of colic (Proudman, 1991; Coenen *et al.*, 1999; Hudson *et al.*, 2001; Hillyer *et al.*, 2002). Risk of colic is highest during the initial 7- to 14-day period following dietary change. These observations together with clinical experience suggest that horses have low tolerance to abrupt dietary changes; or, viewed another way *limited digestive flexibility*, particularly with respect to increased loads of starch or water-soluble carbohydrates. Although specific mechanisms linking diet change to development of colic have not been elucidated, it is reasonable to hypothesise that an abrupt increase in the quantity of rapidly fermentable carbohydrate reaching the hindgut plays a role (Clarke *et al.*, 1990; Geor and Harris, 2007). This may occur with: (1) a sudden introduction to grain feeding or an abrupt increase in the amount of grain-concentrate; (2) the feeding of large grain meals that, even in horses adapted to such feeds, overwhelm the hydrolytic and/or absorptive capacity of the small intestine; and (3) an abrupt change in forage composition such as introduction to 'lush' pasture with a high water-soluble carbohydrate content (or even circadian fluctuations in pasture forage composition). Several studies have reported marked changes in large intestinal microbial profiles and activities after an abrupt increase in grain feeding (e.g. Goodson *et al.*, 1988; De Fombelle *et al.*, 2001; Julliand *et al.*, 2001). In these circumstances, there likely is disturbance to the hindgut microbiome and clinical manifestations of colic may reflect the inability of the microbiome to appropriately adapt to the change in dietary substrate.

The recent application of modern microbiological tools has provided new opportunities to study changes in a highly complex microbial ecosystem (Daly *et al.*, 2001; Daly and Shirazi-Beechey, 2003; Milinovich *et al.*, 2010). As detailed above, Willing *et al.* (2009) observed that adaptation to a conventional hay plus grain diet resulted in a significant increase in members of the *Streptococcus bovis/equinus* complex isolated from faeces, a species group that has been implicated in the pathogenesis of oligofructose-induced laminitis (Milinovich *et al.*, 2006, 2010). Willing *et al.* (2009) also reported wide variability between horses with respect to stability of the faecal microbiome in response to the hay-grain diet; this observation merits further investigation as it suggests that some horses may be more or less susceptible to adverse effects associated with dietary changes due to differences in the adaptive response of the hindgut microbiome.

There is relatively little information on the effects of a change in forage feeding, including circadian and seasonal alterations in pasture forage, on the hindgut ecosystem and other aspects of large intestinal physiology. In this regard, Muhonen and colleagues (2009) reported that an abrupt change from grass hay to grass silage or grass haylage (all forages harvested from the same swath on the same day) does not induce major alterations in the colon ecosystem during the first 24 h with a decrease in colon and faecal dry matter and some alterations in bacterial populations during the subsequent 3-week period. More studies of this type are needed to expand knowledge of the effects of 'real-world' diet changes on in the hindgut microbial community and environment.

Conclusions

A detailed understanding of digestive and metabolic physiology, including the speed and extent of adaptive mechanisms (both in the host and the intestinal microbiome), is crucial to the development of sound feeding recommendations, i.e. *tailoring nutrition to the physiology of the horse*. The equine digestive tract is well adapted to hindgut fermentation of structural carbohydrates, favoured by a fairly stable microbial community and only minor fluctuations in the substrate delivered to the caecum and large intestine. Whereas the horse has all of the 'machinery' for starch digestion in the small intestine, there may be limited flexibility for upregulation of capacity (e.g. at the level of pancreatic α-amylase) and/or there is a relatively slow rate of adaptation in response to increased dietary load. Potentially, low dietary starch during evolution of the horse did not favour development of a flexible system for small intestinal carbohydrate digestion. Regardless, these constraints demand a cautious approach to the feeding of starch-rich feeds to horses, limiting the amount of starch per meal (e.g. <1-1.5 g/kg BW) and ensuring a slow increase in the rate of starch feeding.

The concept of digestive flexibility also applies to the complex microbial ecosystem in the hindgut. Diets and feeding patterns that favour delivery of rapidly fermentable carbohydrates to the hindgut can result in dramatic changes in the composition and activity of the microbial community, including the proliferation of streptococcal species that have been associated with development of laminitis. An inability of the microbiome to appropriately adapt to the change in dietary substrate will not only compromise digestive efficiency but also increases risk of adverse health events such as colic. A more detailed examination of the interactions between diet, hindgut microbial community and intestinal function/dysfunction represents a growth area for equine nutrition research.

References

Alexander, F. and J.C.D. Hickson, 1970. The salivary and pancreatic secretions of the horse. In: Phillipson, A.T., ed. Physiology of digestion and metabolism in the ruminant. Oriel Press, Newcastle Upon Tyne, United Kingdom, p. 375-389.

Al Jassim, R.A. and F.M. Andrews, 2009. The bacterial community of the horse gastrointestinal tract and its relation to fermentative acidosis, laminitis, colic, and stomach ulcers. Vet. Clin. North Am. Equine Pract. 25, 199-215.

Andrews, F.M., B.R. Buchanan, S.H. Smith, S.B. Elliott and A.M Saxton, 2006. *In vitro* effects of hydrochloric acid and various concentrations of acetic, propionic, butyric, or valeric acids on bioelectric properties of equine gastric squamous mucosa. Am. J. Vet. Res. 67, 1873-1882.

Bailey, S.R., A. Rycroft and J. Elliot, 2002. Production of amines in equine cecal contents in an *in vitro* model of carbohydrate overload. J. anim. Sci. 80, 2656-2662.

Berg, E.L., C.J. Fu, J.H. Porter and M.S. Kerley, 2005. Fructooligosaccharide supplementation in the yearling horse: Effects on fecal pH, microbial content and volatile fatty acid concentrations. J. anim. Sci. 83, 1549-1553.

Bracamonte, J.L., L.P. Bouré, R.J. Geor, J.R. Runciman, S.G. Nykamp, A.M. Cruz, M.G. Teeter and H.L. Waterfall, 2008. Evaluation of a laparoscopic technique for collection of serial full-thickness small intestinal biopsy specimens in standing sedated horses. Am. J. vet. Res. 69, 431-439.

Brannon, R.M., 1990. Adaptation of the exocrine pancreas to diet. Annu. Rev. Nutr. 10, 85-105.

Buddington, R.K., J.W. Chen and J.M. Diamond, 1991. Dietary regulation of intestinal brush-border sugar and amino acid transport in carnivores. Am. J. Physiol. 261, R794-R801.

Buddington, R.K. and A.M. Rashmir-Raven, 2002. Carbohydrate digestion by the horse: is there a limiting factor? Equine vet. J. 34, 326-327.

Bullimore, S.R., J.D. Pagan, P.A. Harris, K.E. Hoekstra, K.A. Roose, S.C. Gardner and R.J. Geor, 2000. Carbohydrate supplementation of horses during endurance exercise: comparison of fructose and glucose. J. Nutr. 130, 1760-1765.

Byrd, B., K. Treiber, D. Kronfeld, W. Staniar, R. Geor and P.A. Harris, 2006. Circadian variation of pasture NSC and insulin concentration in horses. Proc. Am. Soc. Anim. Sci. 84, 330-331 (abstract).

Clarke, L.L., M.C. Roberts and R.A. Argenzio, 1990. Feeding and digestive problems in horses. Physiologic responses to a concentrated meal. Vet. Clin. N. Am.: Equine Pract. 6, 433-450.

Coenen, M., A. Mösseler and I. Vervuert, 2006. Fermentative gases in breath indicate that inulin and starch start to be degraded by microbial fermentation in the stomach and small intestine of the horse in contrast to pectin and cellulose. J. Nutr. 136, 2108S-2110S.

Cohen, N.D., P.G. Gibbs and A.M. Woods, 1999. Dietary and other management factors associated with colic in horses. J. Am. vet. med. Ass. 215, 53-60.

Crowell-Davis, S.L, K.A. Houpt and J. Carnevale, 1985. Feeding and drinking behaviour of mares and foals with free access to pasture and water. J. Anim. Sci. 60, 883-889.

Daly, K. and S.P. Shirazi-Beechey, 2003. Design and evaluation of group-specific oligonucletide probes for quantitative analysis of intestinal ecosystems: Their application to assessment of equine colonic microflora. FEMS Microbiol. Ecol. 44, 243-252.

Daly, K., C. Stewart, H.J. Flint and S.P. Shirazi-Beechey, 2001. Bacterial diversity within the equine large intestine as revealed by molecular analysis of cloned 16S rRNA genes. FEMS Microbiol. Ecol. 38, 141-151.

De Fombelle, A., V. Julliand, C. Drogoul and E. Jacotot, 2001. Feeding and microbial disorders in horses: 1. Effects of an abrupt incorporation of two levels of barley in a hay diet on microbial profile and activities. J. equine vet. Sci. 21, 439-445.

De Fombelle, A., M. Varloud, A.-G. Goachet, E. Jacotot, C. Philippeau, C. Drogoul and V. Julliand, 2003. Characterisation of the microbial and biochemical profile of the different segments of the digestive tract of horses fed two distinct diets. Anim. Sci. 77, 293-304.

Dyer, J., S. Vayro, T.P. King and S.P. Shirazi-Beechey, 2003. Glucose sensing in the intestinal epithelium. Eur. J. Biochem. 270, 3377-3388.

Dyer, J., E. Fernandez-Castaño Merediz, K.S. Salmon, C.J. Proudman, G.B. Edwards and S.P. Shirazi-Beechey, 2002. Molecular characterisation of carbohydrate digestion and absorption in the equine small intestine. Equine vet. J. 34, 349-358.

Dyer, J., K. Daly, K.S. Salmon, D.K. Arora, Z. Kokrashvili, R.F. Margolskee and S.P. Shirazi-Beechey, 2007. Intestinal glucose sensing and regulation of intestinal glucose absorption. Biochem. Soc. Trans. 35, 1191-1194.

Dyer, J., M. Al-Rammahi, L. Waterfall, K.S.H. Salmon, R.J. Geor, L. Boure, G.B. Edwards, C.J. Proudman and S.P. Shirazi-Beechey, 2009. Adaptive response of equine intestinal Na^+/glucose co-transporter (SGLT1) to an increase in dietary soluble carbohydrate. Pflugers Arch. Eur. J. Physiol. 458, 419-430.

Fernandez-Castaño Merediz, E., J. Dyer, K.S.H. Salmon and S.P. Shirazi-Beechey, 2004. Molecular characterisation of fructose transport in the equine small intestine. Equine Vet. J. 36, 532-538.

Ferraris, R.P., 2001. Dietary and developmental regulation of intestinal sugar transport. Biochem. J. 360, 265-276.

Ferraris, R.P. and J.M. Diamond, 1989. Specific regulation of intestinal nutrient transporters by their dietary substrates. Annu. Rev. Physiol. 51, 125-141.

Geor, R.J. and P.A. Harris, 2007. How to minimize gastrointestinal disease associated with carbohydrate nutrition in horses. In: Proceedings of the American Association of Equine Practitioners, Orlando, Florida, December 1-5. pp. 178-185.

Goodson, J., W.J. Tyznik, J.H. Cline and B.A. Dehority, 1988. Effects of an abrupt diet change from hay to concentrate on microbial numbers and physical environment in the caecum of the pony. Appl. Env. Microbiol. 54, 1946-1950.

Hillyer, M., F.G. Taylor, C.J. Proudman, G.B. Edwards, J.E. Smith and N.P. French, 2002. Case control study to identify risk factors for simple colonic obstruction and distension colic in horses. Equine vet. J. 34, 455-463.

Hoffman, R.M., J.A. Wilson, D.S. Kronfeld, W.L. Cooper, L.A. Lawrence, D. Sklan and P.A. Harris, 2001. Hydrolyzable carbohydrates in pasture, hay, and horse feed: direct assay and seasonal variation. J. anim. Sci. 79, 500-506.

Hoffman, R.M., R.C. Boston, D. Stefanovski, D.S. Kronfeld and P.A. Harris, 2003. Obesity and diet affect glucose dynamics and insulin sensitivity in Thoroughbred geldings. J. anim. Sci. 81, 2333-2342.

Hudson, J.M., N.D. Cohen, P.G. Gibbs and J.A. Thompson, 2001. Feeding practices associated with colic in horses. J. Am. vet. med. Ass. 219, 1419-1425.

Husted, L., L.C. Sanchez, K.E. Baptiste and J. Scherzer, 2009. Effect of a feed/fast protocol on pH in the proximal equine stomach. Equine Vet. J. 41, 658-662.

Julliand, V., A. de Fombelle, C. Drogoul and E. Jacotot, 2001. Feeding and microbial disorders in horses: 3. Effects of three hay:grain rations on microbial profiles and activities. J. Equine Vet. Sci. 21, 543-546.

Julliand, V., C. Philippeau, A.-G. Goachet and S. Ralston, 2008. Physiology of intake and digestion in equine animals. In: 4th European Workshop on Equine Nutrition. pp. 1-23.

Kienzle, E. and S. Radicke, 1993. Effect of diet on maltase, sucrase and lactase in the small intestinal mucosa of the horse. J. Anim. Physiol. Anim. Nutr. 70, 97-103.

Kienzle, E., S. Radicke, E. Landes, D. Kleffken, M. Illenseer and H. Meyer, 1994. Activity of amylase in the gastrointestinal tract of the horse. J. Anim. Physiol. Anim. Nutr. 72, 234-241.

Kienzle, E., J. Pohlenz and S. Radicke, 1997. Morphology of starch digestion in the horse. J. Vet. Med. A 44, 207-221.

Kronfeld, D.S., J.L. Holland, G.A. Rich, T.N. Meacham, J.P. Fontenot, D.J. Sklan and P.A. Harris, 2004. Fat digestibility in Equus caballus follows increasing first-order kinetics. J anim. Sci. 82, 1773-1780.

Longland, A.C., A.J. Cairns and M.O. Humphreys, 1999. Seasonal and diurnal changes in fructan concentration in Lolium perenne: Implications for the grazing management of equine pre-disposed to laminitis. In: Proceedings of the 16th Equine Nutrition Physiology Society Symposium, Raleigh. Pp. 258-259.

Longland, A.C. and B.M. Byrd, 2006. Pasture nonstructural carbohydrates and equine laminitis. J. Nutr. 136, 2099S-2102S.

Lorenzo-Figueras, M., S.M. Morisset, J. Morisset, J. Lainé and A.M. Merritt, 2007. Digestive enzyme concentrations and activities in healthy pancreatic tissue of horses. Am. J. Vet. Res. 68, 1070-1072.

Luthersson, N., K. Hou Nielsen, P.A. Harris and T. Parkin, 2009. Risk factors associated with equine gastric ulceration syndrome in 201 horses in Denmark. Equine Vet. J. 41, 625-630.

McInnes, M.L. and M. Vavra, 1987. Dietary relationships among feral horses, cattle and Pronghorn sheep in southeastern Oregon. J. Range Management 40, 60-66.

McLean, B.M., J.J. Hyslop, A.C. Longland, D. Cuddeford and T. Hollands, 2000. Physical processing of barley and its effects on intracaecal fermentation patterns in ponies. Anim. Feed Sci. Tech. 85, 79-87.

Meyer, H., S. Radicke, E. Kienzle, S. Wilke, D. Kleffken and M. Illenseer, 1995. Investigations on preileal digestion of starch grain, potato and manioc in horses. J. Vet Med. A 42, 371-381.

Milinovich, G.J., D.J. Trott, P.C. Burrell, A.W. van Eps, M.B. Thoefner, R.A. Blackall, R.A.M. Al Jassim, J.M. Morton and C.C. Pollitt, 2006. Changes in equine hindgut bacterial populations during oligofructose-induced laminitis. Environ. Microbiol. 8, 885-898.

Milinovich, G.J., A.V. Klieve, C.C. Pollitt and D.J. Trott, 2010. Microbial events in the hindgut during carbohydrate-induced equine laminitis. Vet. Clin. Nth. Am. Eq. Pract. 26, 79-94.

Muhonen, S., V. Julliand, J.E. Lindberg, J. Bertilsson and A. Jansson, 2009. Effects on the equine colon ecosystem of grass silage and haylage diets after an abrupt change from hay. J. anim. Sci. 87, 2291-2298.

Piersma, T. and J. Drent, 2003. Phenotypic flexibility and the evolution of organismal design. Trends Ecol. Evol. 18, 228-233.

Potter, G.D., F.F. Arnold, D.D. Householder, D.H. Hansen and K.M. Brown, 1992. Digestion of starch in the small and large intestine of the equine. Pferdeheilkunde 1, 107-111.

Prache, S., I.J. Gorson and A.J. Rook, 1998. Foraging behaviour and diet selection in domestic herbivores. Annales de Zootechnie 47, 335-345.

Pratt, S.E., R.J. Geor and L.J. McCutcheon, 2006. Effects of dietary energy source and physical conditioning on insulin sensitivity and glucose tolerance in standardbred horses. Equine vet. J. Suppl. 36, 579-584.

Proudman, C.J., 1991. A two year, prospective survey of equine colic in general practice. Equine vet. J. 24, 90-93.

Radicke, S., E. Kienzle and H. Meyer, 1991. Preileal apparent digestibility of oats and corn starch and consequences for cecal metabolism. In: Proceedings of the 12th Equine Nutrition and Physiology Symposium, Clagary, Canada. pp. 43-48.

Ralston, S.L., 1984. Controls of feeding in horses. J. anim. Sci. 59, 1354-1361.

Respondek, F., A.G. Goachet and V. Julliand, 2008. Effects of dietary short-chain fructooligosaccharides on the intestinal microflora of horses subjected to a sudden change in diet. J. anim. Sci. 86, 316-323.

Respondek, F., A.G. Goachet, F. Rudeaux and V. Julliand, 2007. Effects of dietary short-chain fructooligosaccharides on the microbial and biochemical profiles of different segments of the gastrointestinal tract of horses. Pferdeheilkunde 22, 146-150.

Richards, N., M. Choct, G.N. Hinch and J.B. Rowe, 2004. Examination of the use of exogenous α-amylase and amyloglucoside to enhance starch digestion in the small intestine of horse. Anim. Feed Sci. Tech. 114, 295-305.

Roberts, M.C., 1974. Amylase activity in the small intestine of the horse. Res. vet. Sci. UK, 17, 400-401.

Roberts, M.C., 1975. Carbohydrate digestion and absorption in the equine small intestine. J. S. Afr. vet. Ass. 46, 19-27.

Salmon, K.S.H., J. Dyer, E. Fernandez-Castaño Merediz and S.P. Shirazi-Beechey, 2002. Absorption of monosaccharides in equine small intestine; adaptive response to dietary change. Gastroenterol. 122, T1316.

Shirazi-Beechey, S.P., 2008. Molecular insights into dietary induced colic in the horse. Equine Vet. J. 40, 414-421.

Shirazi-Beechey, S.P., B.A. Hirayama, Y. Wang, D. Scott, M.W. Smith and E.M. Wright, 1991. Ontogenic development of lamb intestinal sodium-glucose cotransporter is regulated by diet. J. Physiol. 437, 699-708.

Swanson, K.C., J.C. Matthews, A.D. Matthews, J.A. Howell, C.J. Richards and D.L. Harmon, 2000. Dietary carbohydrate source and energy intake influence the expression of pancreatic α-amylase in lambs. J. Nutr. 130, 2157-2165.

Tinker, M.K., N.A. White, P. Lessard, C.D. Thatcher, K.D. Pelzer, B. Davis and D.K. Carmel, 1997. Prospective study of equine colic risk factors. Equine Vet. J. 29, 454-458.

Treiber, K.H, D.S. Kronfeld, T.M. Hess, B.M. Byrd, R.K. Splan and W.B. Staniar, 2006a. Evaluation of genetic and metabolic predispositions and nutritional risk factors for pasture-associated laminitis in ponies. J. Am. Vet. Med. Assoc. 228, 1538-1545.

Treiber, K.H, T.M. Hess, D.S. Kronfeld, R.C. Boston, R.J. Geor, M. Friere, A.M. Silva and P.A. Harris, 2006b. Glucose dynamics during exercise: dietary energy sources affect minimal model parameters in trained Arabian geldings during endurance exercise. Equine vet. J. Suppl. 36, 631-636.

Treiber, K.H, R.J. Geor, R.C. Boston, T.M. Hess, P.A. Harris and D.S. Kronfeld, 2008. Dietary energy source affects glucose kinetics in trained Arabian geldings at rest and during endurance exercise. J. Nutr. 138, 964-970.

Van Eps, A.W. and C.C. Pollitt, 2006. Equine laminitis induced with oligofructose. Equine Vet. J. 38, 203-208.

Varloud, M., G. Fonty, A. Roussel, A. Guyonvarch and V. Julliand, 2007. Postprandial kinetics of some biotic and abiotic characteristics of the gastric ecosystem of horses fed a pelleted concentrate meal. J. anim. Sci. 85, 2508-2516.

Vervuert, I., K. Voight, T. Hollands, D. Cuddeford and M. Coenen, 2008. Effects of processing barley on its digestion by horses. Vet. Rec. 162, 684-688.

Vervuert, I., K. Voight, T. Hollands, D. Cuddeford and M. Coenen, 2009. Effects of feeding increasing quantities of starch on glycaemic and insulinaemic responses in healthy horses. Vet. Rec. 182, 67-72.

Willing, B., A. Vörös, S. Roos, C. Jones, A. Jansson and J.E. Lindberg, 2009. Changes in faecal bacteria associated with concentrate and forage-only diets fed to horses in training. Equine Vet. J. 41, 908-914.

Plasma pancreatic α-amylase profile does not reflect starch digestion in healthy horses

I. Vervuert and M. Coenen
Institute of Animal Nutrition, Nutrition Diseases and Dietetics, Faculty of Veterinary Medicine, University of Leipzig, Gustav-Kührstrasse 8, 04159 Leipzig, Germany

Abstract

In laboratory diagnosis, equine postprandial pancreatic α-amylase activities were partly used to reflect starch digestion capacity in the equine, however, the role of plasma pancreatic α-amylase activity is unknown in horses. The aim of the study was to monitor postprandial plasma pancreatic α-amylase profiles in healthy horses. Twelve horses, all geldings, mean age 13 years, mean body mass (± SD) 601 (±64 kg), mean body condition score 6 (on a scale of 1 to 9) were fed cracked corn to supply 2 g starch/kg BW/meal. Each dietary treatment consisted of a 7 day feeding period of the respective test diet. At day 7, blood samples for serum glucose, serum insulin and plasma pancreatic α-amylase determination were collected 30 minutes before feeding the test meal, immediately after finishing the test meal, and at 30 min intervals until eight hours after finishing feed intake. A significant increase in serum glucose and serum insulin was measured 30 to 45 min after feeding cracked corn, reflecting starch digestion and glucose absorption in the small intestine. In contrast, plasma pancreatic α-amylase showed no significant postprandial changes after cracked corn intake. In conclusion, plasma pancreatic α-amylase activity is not a tool to assess starch digestion capacity in healthy horses.

Keywords: plasma pancreatic α-amylase, starch, digestion

Introduction

The rate of starch digestion in the small intestine primarily determines the glycaemic and pancreatic insulin responses after a meal. In general, starch is hydrolysed enzymatically in the small intestine by secreted pancreatic α-amylase to a disaccharide (maltose), a trisaccharide (maltotriose), and branched α-dextrins (Gray, 1992). However, pancreatic α-amylase levels are significantly lower in horses than those observed in carnivores and omnivores (Roberts, 1974, Radicke *et al.*, 1992, Kienzle *et al.*, 1994, Lorenzo-Figueras *et al.*, 2007). Plasma pancreatic α-amylase activity is known to reflect exocrine pancreatic function in cats (Steiner *et al.*, 2004) and dogs (Simpson *et al.*, 1991). However, the role of plasma pancreatic α-amylase activity is unknown in horses. The aim of the study was to monitor postprandial plasma pancreatic α-amylase profiles following a starch-based meal in healthy horses.

Materials and methods

Animals

Twelve horses, all geldings, mean age 13 years, mean body mass (± SD) 601 (± 64 kg), mean body condition score 6 (on a scale of 1 to 9) were individually housed in box stalls, bedded on wood shavings and turned out onto a dry lot for 8 h/day. The horses had free access to water at all times. The horses were familiarised with the experimental procedure for several months.

Diet

Cracked corn was fed to supply 2 g starch/kg BW/meal. Horses were adapted to the corn diet at least for 7 days. Except on blood sampling days horses were given additionally 1.2 kg hay/100 kg BW per day. Test diet was fed once a day (08:00 a.m.) and hay was divided into two portions which were offered at 09:00 a.m. and 07:00 p.m. A commercial mineral mixture was added at 07:00 p.m.

Blood collection

On day 7 of the experiment horses were fed their respective test diet without any hay. One hour before feeding (07:00), an indwelling jugular catheter (1.8 x 2.35 mm/14 G, Braun Melsungen AG, Melsungen, Germany) was inserted. Blood samples for serum glucose, serum insulin and plasma pancreatic α-amylase concentrations were collected 30 minutes before feeding the test meal, immediately after finishing the test meal, at 30 min intervals until eight hours after finishing the feed. Blood samples were centrifuged within 5 minutes of collection. Serum was stored at -20 °C until analysis.

Analytical methods

Diets were analysed for starch polarimetrically (Polartronic® E, Schmidt & Haensch, Berlin, Germany) and for other dietary components using the Weende system (VDLUFA, 1997); sugar was determined using the LUFF-SCHOORL method (1975). Serum glucose concentrations were determined by the glucose oxidase assay (Unimate 7 GLUC GDH®, Roche Diagnostics GmbH, Mannheim, Germany), serum insulin concentrations were determined by radioimmunoassay (Insulin RIA, Coat-A-Count® (^{125}I), DPC Biermann, Bad Nauheim, Germany) and plasma pancreatic α-amylase concentrations were performed enzymatically by EPS assay (ethylene protected substrate, Fa. Roche Diagnostics, Mannheim, Germany).

Statistical methods

Data were subjected to an analysis of one-way variance, factoring the effect of time. Statistical significance was accepted at $P<0.05$.

Results

The rate of cracked corn intake varied around 10.0±3.4 min/kg DM. Significant ($P<0.05$) increases in mean serum glucose and insulin concentrations were measured 30 to 45 min after feeding cracked corn (Figure 1). Mean basal and peak values for serum glucose and insulin concentrations are shown in Table 1. Mean resting plasma pancreatic α-amylase concentrations varied around 4.3±1.3 U/L without any significant changes in the postprandial phase after cracked corn intake (Figure 1, Table 1).

Discussion

The rate of starch digestion in the small intestine primarily determines the glycaemic and hormonal responses after a starchy meal. However, considerable differences in precaecal starch digestibility were found between the different starch sources. In general, the precaecal digestibility of oat starch is greater than that of corn starch or of barley starch. In the present study, the postprandial glycaemic and insulinaemic responses after cracked corn intake were comparable to results obtained by other experiments in horses fed a starchy meal (Vervuert, 2009). Furthermore, Jose-Cunilleras *et al.* (2004)

Table 1. Mean (±SD) basal and mean (±SD) peak values for serum glucose (mmol/l), serum insulin (µU/ml) and mean (±SD) plasma pancreatic α-amylase (U/L) before and after cracked corn intake (2 g starch/kg BW) in healthy horses (n=12).

Parameter	Basal value	Peak value
Serum glucose, mmol/l	5.2±0.4	8.2±1.5
Serum insulin, µU/ml	5.9±3.4	148±115
Plasma pancreatic α-amylase (U/L)	4.3±1.3	5.7±1.6

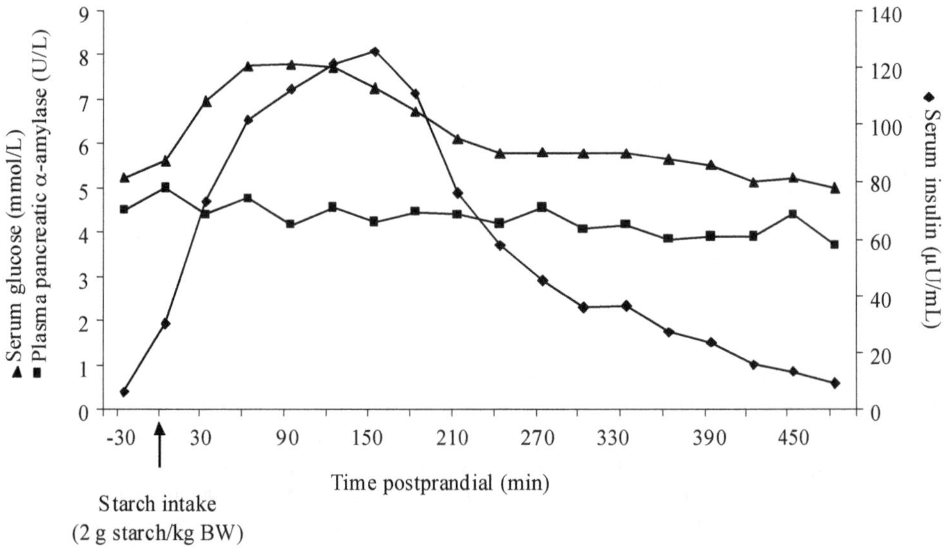

Figure 1. Mean serum glucose (mmol/l), mean serum insulin (µU/ml) and mean plasma pancreatic α-amylase (U/L) before and after cracked corn intake (2 g starch/kg BW)in healthy horses (n=12).

found no differences in the glycaemic response of thoroughbred horses to cracked corn, oat groats and rolled barley when fed 2 g available carbohydrates/kg BW. Resting plasma pancreatic α-amylase concentrations were lower in comparison to cats (1380±410 U/L) and dogs (1600±370 U/L) and starch intake was not accompanied by a respective postprandial response. In general, pancreatic α-amylase levels are significantly lower in horses than those observed in carnivores and omnivores (Roberts, 1974, Radicke *et al.*, 1992, Kienzle *et al.*, 1994, Lorenzo-Figueras *et al.*, 2007), thereby limiting starch degradation. In laboratory diagnosis, equine postprandial pancreatic α-amylase activities were partly used to reflect starch digestion capacity in the equine. However, our data clearly show no link between the typical postprandial pattern in glycaemic and insulinaemic responses, and the non-responsive postprandial reaction of plasma pancreatic α-amylase activity after starch intake. In consequence, plasma pancreatic α-amylase activity is not a tool to assess starch digestion capacity in healthy horses.

References

Gray, G.M., 1992. Starch digestion and absorption in nonruminants. J. Nutr. 122, 172-177.

Jose-Cunilleras, E., L.E. Taylor and K.W. Hinchcliff, 2004. Glycemic index of cracked corn, oat groats and rolled barley in horses. J. Anim. Sci. 82, 2623-2629.

Kienzle, E., S. Radicke, E. Landes, D. Kleffken, M. Illenseer and H. Meyer, 1994. Activity of amylase in the gastrointestinal tract of the horse. J. Anim. Physiol. Anim. Nutr. 72, 234-241.

Lorenzo-Figueras, M., S.M. Morisset, J. Morisset, J. Lainé and A.M. Merritt, 2007. Digestive enzyme concentrations and activities in healthy pancreatic tissue of horses. Am. J. Vet. Res. 68, 1070-1072.

Radicke, S., E. Landes, E. Kienzle and H. Meyer, 1992. Aktivität der Amylase im Darmkanal des Pferdes in Abhängigkeit von der Futterart. Pferdeheilkunde Suppl., 1. Eur. Conf. Horse Nutr. Hannover, 99-102.

Roberts, M.C., 1974. Amylase activity in the small intestine of the horse. Res. Vet. Sci. 17, 400-401.

Simpson, K.W., J.W. Simpson, S. Lake, D.B. Morton and R.M. Batt, 1991. Effect of pancreatectomy on plasma activities of amylase, isoamylase, lipase and trypsin-like immunoreactivity in dogs. Res. Vet. Sci. 51, 78-82.

Steiner, J.M, B.G. Wilson and D.A. Williams, 2004. Development and analytical validation of a radioimmunoassay for the measurement of feline pancreatic lipase immunoreactivity in serum. Can. J. Vet. Res. 68, 309-314.

Vervuert, I., 2009. Food properties affecting starch digestion by healthy horses as measured by glycaemic and insulinaemic responses. Habilitationsschrift, University of Leipzig, Germany.

Revision of the energy and nutrient requirements of mares; I. Gestation

M. Coenen[1], S. Schmidt[1], A. Zeyner[2] and E. Kienzle[3]
[1]Institute of Animal Nutrition, Nutrition Diseases & Dietetics, Gustav-Kühn-Straße 8, 04159 Leipzig, Germany
[2]Chair for Nutrition Physiology and Animal Nutrition, University of Rostock, Justus-von-Liebig-Weg 8, 18059 Rostock, Germany
[3]Chair for Animal Nutrition and Dietetics, Ludwig-Maximilians-University Munich, Schönleutner Str. 8, 85764 Oberschleißheim, Germany

Abstract

The latest version of the German feeding standards for horses were published 1994 (Gesellschaft für Ernährungsphysiologie der Haustiere (GfE) 1994). In between the literature provides more data on factors which are needed for the calculation of energy and nutrient requirements in pregnant mares. In addition, equine specific data on oxygen consumption in the pregnant uterus are challenging to to revise the current requirement data. The expanded data set on intrauterine weight indicates a very intensive growth at the end of gestation. This results in a higher estimate of energy and nutrient accretion in conspectuses than calculated formerly. Statistical data processing yields exponential models for daily energy and nutrient accretion. Conflicting data need to be discussed regarding energy. The data for oxygen consumption of fetus and pregnant uterus results in a high energy expenditure. Remarkably is the level itself but also the enormous contribution of the adenxes and the uterine tissue to total oxygen consumption and energy expenditure. These data end up with a higher energy requirement in comparison to the traditional factorial approach.

Keywords: gestation, growth, foetus, energy, nutrient, requirement

Introduction

Objective of the present literature review was to revise the current requirement data for pregnant mares. Data on oxygen consumption of the foetus, the adnexa and the uterus need to be considered. These data show a high energy expenditure (EE) in the pregnant uterus and a much higher contribution of the adnexa to total EE.

Material and methods

The literature includes data on intrauterine growth for 242 foetuses (Coenen *et al.*, 2010). The composition of the foetus at different gestational stages is reported by Meyer and Ahlswede (1976). Data for placenta and uterine tissue are taken from pigs (Noblet *et al.*, 1985) and ruminants (Leo *et al.*, 1995). Data on the amino acid profile in foal muscle and plasma (Manso Filho *et al.*, 2009) were used for amino acid accretion. Calculation on energy requirements use literature data on oxygen consumption of the pregnant uterus (Schmidt *et al.*, 2010, this Volume)

Results and discussion

The foetal weight and subsequently growth can be described as a fraction of birth weight (BiW); setting BiW = 100 we receive a universal term which can be applied to any individual foal: y (% of BiW)=$e^{0.013638x}$; x= day (d) of gestation, r=0.95. The foetus contributes 51.6 and 63.7% of total weight of pregnant uterus at day 253 and 338 of parturition respectively (foetal portion of uterus y=0.1424x + 15,586; partition of total weight of conceptuses: uterine tissue, adnexa, foetal fluids: 15.6, 14.1, 18.7% at d 250 and 11.7, 13.0, 11.6% at d 340). But for modelling tissue accretion it is necessary to exclude the fetal fluids.

The model for weight of pregnant uterus without foetal fluids (y) at particular days (x) is: y (kg/kg BiW)=$0.0375*e^{0.011x}$. The energy stored in conceptuses (foetal fluids excluded) are expressed by the equation y (MJ ME/kg BiW)=$0.033*e^{0.01544x}$ (x=day of gestation). The adnexa are considered by 1.99 kJ/g (adapted from Noblet *et al.*, 1985). Data processing using the chemical body composition of foetuses and adnexa produces models for nutrient deposition. The basic data for chemical composition of the fetus and the model parameters representing the deposited quantities per kg BiW at a given day of gestation are summarized in Table 1.

The first derivation of the models gives the daily accretion and the daily net requirements respectively. The use of foetal chemical composition for adnexa as well results probably in an overestimation of protein and mineral deposition in the pregnant uterus.

In contrast to former calculations, the energy expenditure of the pregnant mare is deduced by the daily energy accretion in the pregnant uterus plus the caloric equivalent of the oxygen consumption of the entire pregnant uterus. The daily heat production of conceptuses (foetus plus adnexa and uterus) amounts 404.8 kJ per kg accordingly the experimental data in equine foetus from Fowden and Silver (1995) and Fowden *et al.* (2000, 2000a). This approach defines energy expenditure universally for late gestation and can be projected to any specific mare.

The usage of oxygen consumption results in higher energy requirements of pregnant mares in comparison to the common factorial approach. The reasons for these discrepancies may be explained by the basic differences between both methods. The nature of oxygen consumption data offers a superior access to energy turnover in the pregnant uterus as a dynamic biological phenomenon.

Table 1. Energy concentration, chemical composition of fetuses in dependence on day (x) of gestation (day 240 up to 340; modified accordingly Meyer and Ahlswede, 1976) and estimated quantities deposited in the conteptuses (foetal fluids excluded)[1].

Parameter	Concentration per kg foetus depending on day (x)	Deposited quantity per kg BiW by the general model $y=a*e^{b*x}$; x=day	
		a	b
Energy, MJ	= 0.3836+0.0145x	0.0330	0.01544
Protein, g	= -51.77+0.64x	0.7658	0.01688
Calcium, g	= -11.572+0.09x	0.0560	0.01820
Phosphorus, g	= -5.228+0.045x	0.0337	0.01783
Magnesium,g	= -0.148+0.002x	0.0014	0.01770
Sodium,g	= 2.0	0.0461	0.01213
Chloride, g	= 1.2	0.0438	0.01213
Potassium, g	= 1.9	0.0277	0.01213
Copper, mg	= -0.258+0.016x	0.0352	0.01577
Zinc, mg	= -35.5+0.23x	0.0858	0.01936
Iron, mg	= 80 up to day 304	0.0033	0.01854
	= 120 for day >304[a]	1.8721	0.01207
Manganese mg	= -0.818+0.006x	2.7	0.01220

[1] Using foetal data for adnexa as well.

[a] Assuming an increase by reason of haemoglobin and myoglobin synthesis.

References

Fowden, A.L. and M. Silver, 1995. Glucose and oxygen metabolism in the fetal foal during late gestation. Am J Physiol 269, R1455-1461.

Fowden, A.L., A.J. Forhead, K.L. White and P.M. Taylor, 2000a. Equine uteroplacental metabolism at mid- and late gestation. Exp. Physiol. 85, 539-545.

Fowden, A.L., P.M. Taylor, K.L. White and A.J. Forhead, 2000b. Ontogenic and nutritionally induced changes in fetal metabolism in the horse. J. Physiol. 528 Pt 1, 209-219.

Gesellschaft für Ernährungsphysiologie der Hautiere (GfE), 1994. Energie- und Nährstoffbedarf landwirtschaftlicher Nutztiere; Nr. 2, Empfehlungen zur Energie- und Nährstoffversorgung der Pferde. DLG Verlag, Frankfurt (Main), Germany.

Leo, M., E. Kolb, J. Erices and K. Nestler, 1995. The content of Na, K, Ca, Mg, total P, Fe, Cu and Zn in different tissues and in the placenta of sheep fetuses and newborn lambs. Berl. Münch. Tierärztl. Wochenschr. 108, 179-184.

Manso Filho, H.C., K.H. McKeever, M.E. Gordon, H.E. Manso, W.S. Lagakos, G. Wu and M. Watford, 2009. Developmental changes in the concentrations of glutamine and other amino acids in plasma and skeletal muscle of the Standardbred foal. J. Anim. Sci. 87, 2528-2535.

Meyer, H. and L. Ahlswede, 1976. Über das intrauterine Wachstum und die Körperzusammensetzung von Fohlen sowie den Nährstoffbedarf tragender Stuten. Übersichten Tierernährung 4, 263-292.

Noblet, J., W.H. Close, R.P. Heavens and D. Brown, 1985. Studies on the energy metabolism of the pregnant sow. 1. Uterus and mammary tissue development. Br. J. Nutr.53, 251-265.

Revision of the energy and nutrient requirements of mares; II. Lactation

M. Coenen[1], *S. Kirchhof*[2], *E. Kienzle*[3] *and A. Zeyner*[4]
[1]*Institute of Animal Nutrition, Nutrition Diseases & Dietetics, Gustav-Kühn-Straße 8, Leipzig, Germany*
[2]*Institute of Animal Nutrition, University of Hohenheim, Emil-Wolff-Straße 8 u. 10, Stuttgart Germany*
[3]*Chair for Animal Nutrition and Dietetics, Ludwig-Maximilians-University Munich, Schönleutner Str. 8, 85764 Oberschleißheim, Germany*
[4]*Chair for Nutrition Physiology and Animal Nutrition, University of Rostock, Justus-von-Liebig-Weg 8, 18059 Rostock, Germany*

Abstract

The literature on composition of equine milk and the milk secretion in mares was reviewed in order to update the energy and nutrient requirements of lactating mares. The data on milk volume are limited and less precise in contrast to those on milk composition. As a compromise a model for a lactation curve is developed based on the common model of Wood. However the estimated milk yield for an average mare covers the needs of the nursing foal.

Keywords: mare, lactation, requirement, milk yield, milk composition

Introduction

Objective of the present literature review was to revise the current requirement data for lactating mares. Conflicting data are published on milk yield and initiate the discussion for an update.

Material and methods

The literature over >100 years was reviewed, beginning with Cameron (1875) up to Salamon *et al.*, (2009). A detailed list of references is published by Coenen *et al.*, 2010) Data on fat, protein, lactose and energy represent 2,997, 3,250, 2,847 and 2,567 individual measurements. Data on macro elements are obtained from >1000 horses except Na, Cl, S and trace elements. Included are the fatty acids, amino acids and vitamins and finally Li, Mo, Si, Sr, Ti, Al, B and Ba. The relationship between milk constituents and stage of lactation were processed by regression analysis (Statistica 7.1®). Milk yield data are taken from 21 papers representing 219 milk volumes at specific days of lactation produced by 1191 measurements. The relationship between milk yield and foal growth in these data is discussed by Schmidt *et al.* (2010).

Results and discussion

There was no significant change in gross energy concentration in milk over time? (GE, kJ/kg =2,138±294, colostrum excluded) while a decrease for protein, Ca, P and Mg could be described by the following models: Protein, g/kg =34-7.19logx, Ca, mg/kg= 1,620-426logx, P, mg/kg= 948-283logx, Mg, mg/kg= 126-35.5logx; x is the day of lactation. Constant values can be taken for other elements (mg/kg, Na 174, K 659, Cl 278, S 241, Cu 0.27, Zn 2.04, Fe 0.65; µg/kg, Mn 30, Se 45). A major problem is the description of milk yield; data transformed to g/kg BW or $BW^{0.75}$ does not support a simple model for a lactation curve. The literature data are adjusted to a daily intake of 883 kJ gross energy per kg body $weight^{0.82}$ via milk at peak of lactation; this data is obtained by an interspezies analysis (Riek, 2007), Table 1 summarizes the model for milk yield and milk energy output of mares as well as the models for lactogenic nutrient secretion. The model for a lactation curve of Wood (1967) is applied: $a * d^b * c^{cd}$, where d is day of lactation. The parameter a of the equation multiplicated by concentration of energy or nutrients in the milk serves the model for daily net requirements. The corresponding data are summarized in Table 1.

Table 1. Daily net requirement of mares for milk production (per kg $BW^{0.82}$); d=day of lactation.

Parameter	Per $kg^{0,82}d^{-1}$	Net-requirement ($y = a*d^{b}*e^{cd}$)	
1. Milk volume	g	$73*d^{0.144}*e^{-0.0059d}$	
2. Energy[1]	kJ	$156*d^{0.144}*e^{-0.0059d}$ [1]	
3. Protein	g	$2.51*d^{0.035}*e^{-0.0064*d}$	
3. nn amino acids	g	(AA (g/g protein) $*2.51*d^{0.035}*e^{-0.0064*d}$	
		AA, /g*2.51=	
3. a Lysine	g	0.199	$*d^{0.035}*e^{-0.0064d}$
3. b Met+Cys	g	0.082	
3. c Threonine	g	0.114	
3. d Leucine	g	0.235	
3. e Isoleucine	g	0.113	
3. f Phe+Tyr	g	0.224	
3. g Tryptophan	g	0.060	
3. h Valin	g	0.138	
4. Calcium	g	$0.1199*d^{0.0043}*e^{-0.007d}$	
5. Phosphorus	g	$0.0703*d^{-0.0167}*e^{-0.0077d}$	
6. Magnesium	g	$0.0093*d^{-0.0069}*e^{-0.0073d}$	
7. Sodium	g	0.0127	$*d^{0,144}*e^{-0.0059d}$
8. Potassium	g	0.0481	
9. Chloride	g	0.0203	
10. Sulfur	g	0.0176	
11. Copper	mg	0.0197	
12. Zink	mg	0.1486	
13. Iron	mg	0.0477	
14. Manganese	mg	0.0022	
15. Selenium	mg	0.0033	
16. Jodine	mg	0.0073	
17. Cobalt	mg	0.0073	

[1]The constant parameter (a) in the equation is the product of 73 (a from equation 1) and the concentration of energy or nutrients per g milk if there is no change in dependence on day of lactation; for amino acids it is 2.51 * individual amino acid in g/g milk protein.

References

Cameron, C.A., 1875. Some point on the chemistry of milk. Chemical News & J. industrial Sci 31, 54-55.

Coenen, M., S. Kirchhof, E. Kienzle and A. Zeyner, 2010. Zum Energie- und Nährstoffbedarf von Stuten – Aktualisierung und Neuerung der Bedarfsableitung und Versorgungsempfehlungen, I. Laktation (Energy and nutrient requirements of mares – Update and new aspects; I. Lactation. Übersichten Tierernährung, submitted.

Riek, A., 2007. Relationship between milk energy intake and growth rate in suckling mammalian young at peak lactation: an updated meta-analysis. J. Zoology 274, 160-170.

Salamon, R.V., Sz. Salamon, Z. Csapó-Kiss and J. Csapó, 2009. Composition of mare´s colostrum and milk I. Fat content, fatty acid composition and vitamin contents. Acta Univ. Sapientiae, Alimentaria 2, 119-131.

Schmidt, B., Zeyner, A., Kienzle, E., Kirchhof, S. and Coenen, M., 2010. Milk intake and foal growth-a review of literature, In: Ellis, A.D., A.C. Longland, M. Coenen and N. Miragli (eds.), The impact of nutrition on the health and welfare of horses. EAAP publication no. 128, Wageningen Academic Publishers, Wageningen, the Netherlands, p. 43-46.

Wood, P.D.P., 1967. Algebraic Model of the Lactation Curve in Cattle, Nature 216, 164-165.

Metabolizable energy for horses: development of a simple and flexible feed evaluation system

E. Kienzle[1] *and A. Zeyner*[2]
[1]*Chair for Animal Nutrition and Dietetics, Ludwig-Maximilians-University Munich, Schönleutner Str. 8, 85764 Oberschleißheim, Germany*
[2]*Chair for Nutrition Physiology and Animal Nutrition, University of Rostock, Justus-von-Liebig-Weg 8, 18059 Rostock, Germany*

Abstract

To develop a ME-system for horses the relationship between feed protein content and renal energy losses and the relationship between feed fibre content and methane energy losses were analysed in a literature review. Renal energy losses were correlated to protein intake, methane energy losses were correlated to crude fibre intake. The renal energy losses amount to 0.008 MJ/g crude protein, the methane energy losses amount to 0.002 MJ ME/g of crude fibre. These factors can be used to transform DE into ME: ME MJ/kg = DE MJ/kg - 0.008 MJ/g crude protein - 0.002 MJ/g crude fibre. The equation of Zeyner and Kienzle (2002) to predict DE was adapted as mentioned above to predict ME: ME (MJ/kg dry matter) = -3.54 + 0.0129 crude protein + 0.0420 crude fat - 0.0019 crude fibre + 0.0185 N-free extract (crude nutrients in g/kg dry matter).

Keywords: metabolizable energy, predictive equation

Introduction

A digestible energy (DE) system considers faecal energy losses. A metabolizable energy (ME) system considers also methane and renal energy losses. For practical use of ME it is necessary to predict these losses from chemical feed analysis, which was the aim of this study.

Materials and methods

Data from Fingerling (1931-1939), Burlacu *et al.* (1993), Vermorel *et al.* (1997a,b), Ragnarsson (2009), Kienzle *et al.* (2009) was analysed (Kienzle and Zeyner, 2010). The relationships between crude protein intake, digestibility and renal energy losses (113 observations) and crude fibre intake and methane losses (91 observations) were analysed.

Results and discussion

Renal energy losses were correlated to protein intake (Figure 1; $r^2 = 0.77$) better than to digestible protein intake, even in horses which were not at maintenance. Phenolic acids from forage cell walls contribute to urinary energy losses being metabolized via phenylpropanoic acid to benzoic acid, which is conjugated with glycine and excreted in the urine as hippurate. Renal energy losses by hippurate amount to 303.8 kJ per gram of nitrogen compared to losses by urea of 22.9 kJ/g of nitrogen. Crude protein digestibility in forage is usually lower than in concentrates. The net result is a rather constant urinary energy loss of 0.008 MJ per gram of crude protein in the feed.

Methane energy losses were closely related to crude fibre intake (Figure 2; $r^2 = 0.79$). They amounted to 0.002 MJ ME/g of crude fibre. A subtraction of 0.008 MJ/g crude protein and of 0.002 MJ/g crude fibre can be made to transform DE into ME. This can be done with any predictive equation for DE.

The equation of Zeyner and Kienzle (2002) to predict DE was adapted to predict ME:

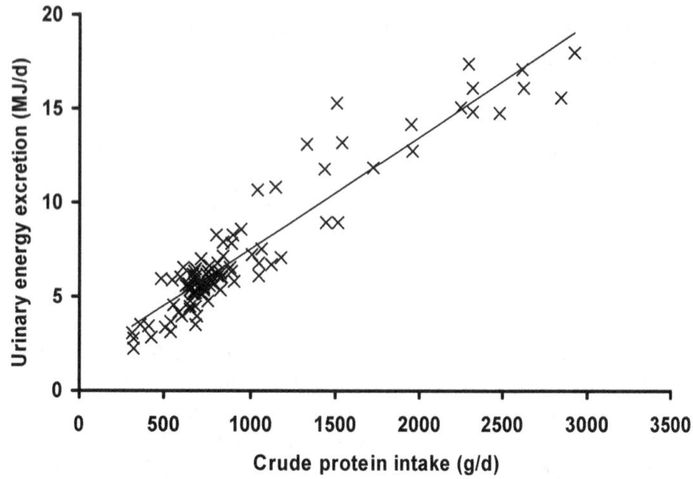

Figure 1. Relationship between protein intake and renal energy excretion in horses.

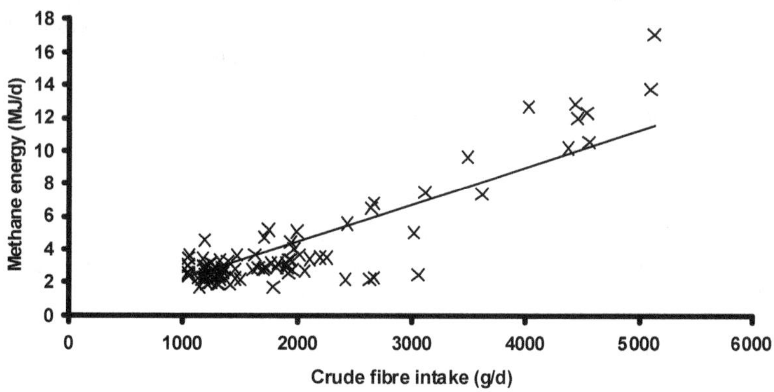

Figure 2. Relationship between crude fibre intake and methane energy losses in horses.

Equation 1 for DE:

DE (MJ/kg dry matter) = -3.54 + 0.0209 crude protein + 0.0420 crude fat + 0.0001 crude fibre + 0.0185 N-free extractives (1)
(crude nutrients in g/kg dry matter).

Equation 2 for ME:

ME (MJ/kg dry matter) = -3.54 + 0.0129 crude protein + 0.0420 crude fat - 0.0019 crude fibre + 0.0185 N-free extractives (2)
(crude nutrients in g/kg dry matter).

The following limitations (in dry matter) for validity are recommended for use: <35% crude fibre and <8% fat in the whole ration. Feedstuffs exceeding these limitation can be evaluated by the equation provided they are not used in rations which are outside of these limitations

Especially feedstuffs with a high protein content have lower ME content in relation to feedstuffs with medium or low protein content when compared to the DE values. However, feedstuffs rich in high quality protein could be appreciated otherwise by a more sophisticated protein evaluation system as has been proposed by Zeyner *et al.* (2010).

References

Fingerling, G., 1931-1939. Cited from: Festschrift anlässlich des 100jährigen Bestehens der Landwirtschaftlichen Versuchsstation Leipzig-Möckern. Volume I pp 327-334, Volume III pp 441-447.

Kienzle, E., L. Berchtold and A. Zeyner, 2009. Effects of hay versus concentrate on urinary energy excretion in horses. Proc. Soc. Nutr. Physiol. 18, 118.

Kienzle, E. and A. Zeyner, 2010. The development of a ME-system for energy evaluation in horses. J. Animal Physiology and Animal Nutrition, in press.

Ragnarsson, S., 2009. Digestibility and metabolism in Icelandic horses fed forage-only diets. Doctoral thesis No. 2009/92, Swedish University of Uppsala, Sweden.

Vermorel, M., W. Martin-Rosset and J. Vernet, 1997a. Energy utilization of twelve forages or mixed diets for maintenance by sport horses. Livestock Prod. Sci. 47, 157-167.

Vermorel, M., J. Vernet and W. Martin-Rosset, 1997b. Digestive and energy utilisation of two diets by ponies and horses. Livestock Prod. Sci. 51, 13.

Zeyner, A. and E. Kienzle, 2002. A method to estimate digestible energy in horse feed. J. Nutr. 132, 1771S-1773S.

Zeyner, A., S. Kirchhof, A. Susenbeth, K.-H. Südekum and E. Kienzle, 2010. Protein evaluation of horse feed – a novel concept. Proc. Soc. Nutr. Physiol. 19, 20.

Protein evaluation of horse feed: a novel concept

A. Zeyner[1], S. Kirchhof[2], A. Susenbeth[3], K.-H. Südekum[4] and E. Kienzle[5]
[1]Chair for Nutrition Physiology and Animal Nutrition, University of Rostock, Justus-von-Liebig-Weg 8, 18059 Rostock, Germany
[2]Institute of Animal Nutrition, University of Hohenheim, Emil-Wolff-Str. 8, 70599 Stuttgart, Germany
[3]Institute of Animal Nutrition and Physiology, Christian-Albrechts-University Kiel, Hermann-Rodewald-Str. 9, 24118 Kiel, Germany
[4]Institute of Animal Science, Rheinische Friedrich-Wilhelms-University Bonn, Endenicher Allee 15, 53115 Bonn, Germany
[5]Chair for Animal Nutrition and Dietetics, Ludwig-Maximilians-University Munich, Schönleutner Str. 8, 85764 Oberschleißheim, Germany

Abstract

In equines, amino acids can be absorbed exclusively from the small intestine. However, only few experimental results on small intestinal digestible (sid) crude protein (CP) and amino acids (AA) are available. Therefore the analytical fraction of neutral detergent insoluble protein (NDICP) and the corresponding neutral detergent soluble protein (NDSCP) have been used to develop a protein evaluation system based on estimated sidCP and sidAA. There was a strong positive correlation of the intake of sidCP reported in the literature to that of NDSCP. The slope of the linear regression line revealed the small intestinal digestibility of NDSCP being 90%. Results indicate that the analytical fraction of NDICP can be used indirectly to estimate the part of protein available for auto-enzymatic digestion. Because literature refers to a similar AA profile in NDICP and NDSCP within a given feed the NDICP/NDSCP concept may be transferred to AA. The proposed system is open to be adjusted depending on the scientific progress without altering its structure.

Keywords: protein, amino acids, small intestinal digestibility, Cornell System

Introduction

In equines, amino acids (AA) can only be absorbed from the small intestine. AA from microbial protein turnover in the hindgut can not be absorbed (Schubert, 1995). Consequently protein bound in cell walls is unlikely to be available. Therefore a concept of protein evaluation that allows estimating the amount of auto-enzymatic digestion would give a more accurate idea of the protein supply for monogastric animals that consume high fibre diets such as horses than concepts that are based on crude protein (CP) or digestible CP (DCP). However, there is limited data on small intestinal digestibility of protein or AA in horses. The 'Cornell Net Carbohydrate and Protein System' for cattle, which utilises (1) neutral detergent insoluble CP (NDICP; insoluble protein) and (2) the corresponding soluble CP (NDSCP; soluble protein), discriminates between cell wall protein and cell content protein. Data on NDICP are available. The idea behind this fractionation is that CP bound in neutral detergent fibre (i.e. mostly in cell walls) cannot be broken down by auto-enzymatic digestion, whereas the soluble protein which is not cell wall-bound can be digested by body own enzymes. AA profiles of both, NDICP and the corresponding fraction NDSCP, appear to be similar within a given feed (Rebolé *et al.*, 2001, Tedeschi *et al.*, 2001). Thus, based on the present knowledge it seems to be justified to transfer the AA profile of the whole feedstuff to both protein fractions, NDSCP and NDICP. The aim of this study was to investigate whether a concept of protein evaluation on the basis of soluble and insoluble protein is applicable to horse feed.

Material and methods

The following literature was used to investigate whether soluble/insoluble protein can be applied to estimate small intestinal digestible crude protein (sidCP) and small intestinal digestible amino acids (sidAA):

- NDICP and NDSCP in feedstuffs: DLG (1995), NRC (2001), Gruber *et al.* (2005) and Kirchhof (2007);
- sidCP: Reitnour *et al.* (1969), Hintz *et al.* (1970), Wootton and Argenzio (1975), Haley *et al.* (1979), Klendshoj *et al.* (1979), Muuß (1980), Krull (1984), Gibbs *et al.* (1988), Farley *et al.* (1995) and Gibbs *et al.* (1996);
- sidAA: Almeida *et al.* (1999) and Coleman *et al.* (2001).

Results and discussion

Taking the relevant data into account the intake of NDICP was positively correlated with the intake of experimentally determined small intestinal undigested protein ($R^2 = 0.691$, $P<0.01$). Although the relationship bases on limited data, results represent a wide spectrum of what is typical in horse feeding (grass hay, alfalfa hay, meadow grass, oats, barley, maize, mixed rations with hay to cereal grain being 3:2 and 1:4). There was also a strong positive dependency of sidCP from NDSCP intake ($R^2 = 0.693$). The intercept of the linear regression line revealed a small intestinal digestibility of NDSCP of 90% ($P<0.01$). The results indicate that NDICP in the feed can be used indirectly to estimate the part of CP that is available for auto-enzymatic digestion. Because of the fairly equal distribution of AA profiles in NDSCP and NDICP the proposed concept could be supplied on an AA basis. The proposed concept is open to be adjusted depending on the scientific progress without altering its structure.

References

Almeida, F.C., S.C. Valadares Filho, A.C. Queiroz, M.I. Leão, J.F.C. Silva and P.R. Cecon, 1998. Digestibilidade aparente e verdadeira pré-cecal e total da proteína em dietas com diferentes níveis protéicos em eqüinos. R. Bras. Zootec. 27, 521-529.

Coleman, R.J., G.W. Mathison, R.T. Hardin and J.D. Milligan, 2001. Effect of dietary forage and protein concentration on total tract, prececal and postilealprotein and lysine digestibilities of forage based diets fed to mature ponies. Proc. 17th Equine Nutr. Physiol. Symp. Lexington, Kentucky, May 31- June 2, 2001, 461-463.

DLG, 1995. Futterwerttabelle für Pferde. 3rd revised edition. DLG-Verlag, Frankfurt (Main), Germany.

Farley, E.B., G.D. Potter, P.G. Gibbs, J. Schuhmacher and M. Murray-Gerzik, 1995. Digestion of soybean meal protein in the equine small and large intestine. Proc. 14th Equine Nutr. Physiol. Soc., p. 24-29.

Gibbs, P.G., G.D. Potter, G.T. Schelling, J.L. Kreider and C.L. Boyd, 1988. Digestion of hay protein in different segments of the equine digestive tract. J. Anim. Sci. 66, 400-406.

Gibbs, P.G., G.D. Potter, G.T. Schelling, J.L. Kreider and C.L. Boyd, 1996. The significance of small vs large intestinal digestion of cereal grain and oilseed protein in the equine. J. Equine Vet. Sci. 16(2), 60-65.

Gruber, L., G. Stögmüller, K. Taferner, L. Haberl, G. Maierhofer, B. Steiner, A. Steinwidder, A. Schauer and W. Knaus, 2005. Protein- und Kohlenhydrat-Fraktionen nach dem Cornell System sowie ruminaler Trockenmasseabbau *in situ* von energie- und proteinreichen Kraftfuttermitteln. Übers. Tierernährg. 33, 129-143.

Haley, R.G., G.D. Potter and R.E. Lichtenwalner, 1979. Digestion of soybean and cottonseed protein in the equine small intestine. Proc. 6th Equine Nutr. Physiol. Symp., 85-98.

Hintz, H.F., D.E. Hogue, E.F. Walker, J.E. Lowe and H.F. Schryver, 1971. Apparent digestion in various segments of the digestive tract of ponies fed diets with varying roughage-grain ratios. J. Anim. Sci. 32, 245-248.

Kirchhof, S., 2007. Kinetik des ruminalen *in situ*-Nährstoffabbaus von Grünlandaufwüchsen des Alpenraumes unterschiedlicher Vegetationsstadien sowie von Maissilagen und Heu – ein Beitrag zur Weiterentwicklung der Rationsgestaltung für Milchkühe. Christian-Albrechts-University Kiel, Thesis.

Klendshoj, C., G.D. Potter, R.E. Lichtenwalner and D.D. Householder, 1979. Nitrogen digestion in the small intestine of horses fed crimped or micronized sorghum grain or oats. Proc. 6th Equine Nutr. Physiol. Symp., 91-94.

Krull, H.D., 1984. Untersuchungen über die Aufnahme und Verdaulichkeit von Grünfutter beim Pferd. Tierärztl. Hochschule Hannover, Doctoral thesis.

Muuß, H., 1980. Untersuchungen über Entleerung sowie Zusammensetzung des Ileumchymus beim Pferd. Tierärztl. Hochschule Hannover, Doctoral thesis.

NRC, 2001. Nutrient Requirements of dairy cattle. 7th revised edition. National Academy Press, Washington, DC.

Rebolè, A., J. Treviño, R. Caballero and C. Alzueta, 2001. Effect of maturity on the amino acid profiles of total and nitrogen fractions in common vetch forage. J. Sci. Food Agric. 81, 455-461.

Reitnour, C.M., J.P. Baker, G.E. Mitchell and C.O. Little, 1969. Nitrogen digestion in different segments of the equine digestive tract. J. Anim. Sci. 29, 332-334.

Schubert, R., 1995. Untersuchungen zum Stickstoff- und Aminosäuren-Stoffwechsel laktierender Stuten am Modelltier Shetland-Pony unter Verwendung oraler Gaben von 15N-Harnstoff. Habilitation thesis, Martin-Luther-University Halle, Germany.

Tedeschi, L.O., A.N. Pell, D.G. Fox and C.R. Llames, 2001. The amino acid profiles of the whole plant and of four plant residues from temperate and tropical forages. J. Anim. Sci. 79, 525-532.

Wootton, J.F. and R.A. Argenzio, 1975. Nitrogen utilization within equine large intestine. Am. J. Physiol. 229, 1062-1067.

Milk intake and foal growth-a review of literature

B. Schmidt[1]*, A. Zeyner*[2]*, E. Kienzle*[3]*, S. Kirchhof*[4] *and M. Coenen*[1]
[1]*Institute of Animal Nutrition, Nutrition Diseases & Dietetics, University of Leipzig, Gustav-Kühn-Str.8, 04159 Leipzig, Germany*
[2]*Chair for Nutrition Physiology and Animal Nutrition, University of Rostock, Justus-von-Liebig-Weg 8, 18059 Rostock, Germany*
[3]*Chair for Animal Nutrition and Dietetics, Ludwig-Maximilians-University Munich, Schönleutner Str. 8, 85764 Oberschleißheim, Germany*
[4]*Institute for Animal Nutrition, University of Hohenheim, Emil-Wolff-Str. 8/10, 70599 Stuttgart, Germany*

Abstract

Milk yield and composition are used to determine the energy and nutrient requirements of lactating mares and growing foals. The present study is a literature-based investigation as to whether growth dynamics of foals used for measuring milk intake differ from normal nursing foals not used for the determination of milk intake, conceivably due to methodology influences. Milk intake was measured using the weigh-suckle-weigh, the isotope dilution or the milking and feeding method. Growth dynamics of foals with ascertainment of milk intake did not considerably differ from normal nursing foals. Doubling of birth weight was realised within day 45 (studies growth data) respectively within day 47 (reference growth data: Hois, 2004; Vervuert *et al*., 2005). Until day 120, milk intake was dependent on day of life: kg milk intake/kg $BW^{0.82}$=0.4664-0.0025*d. At day 45 foals consume 354 g milk/kg $BW^{0.82}$. Consequently a foal with 100 kg BW ingests 15.5 kg milk/d 45 or 33.1 MJ GE via milk per day (equal to 760 kJ GE via milk/kg $BW^{0.82}$) and gain 1,300 g/d of live weight.

Keywords: foals, milk and gross energy intake, growth dynamics, milk yield

Introduction

The growth of foals is reliant on numerous genetic and environmental influences (Pool-Anderson *et al*., 1996). During the first eight to ten weeks of life mare's milk is confirmed to be the vital source for nutrient and energy supply for foals, which have very high growth rates (Doreau *et al*., 1986). Milk energy intake of about 1.5 MJ per kg $BW^{0.75}$ per day is considered to meet the maintenance and growth energy requirements of foals during the early days of life (Ousey *et al*., 1996). When birth weight doubles, after about eight weeks, milk energy intake decreases to 0.8 MJ per kg $BW^{0.75}$ per day, remaining at this level due to an increasing intake of solid feeds. Milk production and composition are used to define the energy and nutrient requirements of mares and foals (Neseni *et al*., 1958; Oftedal *et al*., 1983). Nevertheless, variations in the conduct of studies to determine milk yields of mares lead to uncertainties with regard to the correct determination of the energy and nutrient requirements of lactating mares and of the amount of milk required for physiologic growth of foals.

The present study is a literature-based investigation examining whether growth dynamics of foals from mares used for determination of milk production and foals' milk intake differ from normal nursing foals not used for determination of milk intake, conceivably due to methodology influences. Within the discussed studies milk intake and foal growth were measured from day 2 to 150 by different methods. For the weigh-suckle-weigh method, foals were separated from mares for 1 to 3 h, weighed, allowed to nurse approximately 15 to 20 min, reweighed and removed from the mares for the next period. This procedure was repeated to determine the 12- or 24 h milk intake (Pool-Anderson *et al*., 1996). By administration of a single isotope (^2H or ^3H) by stomach tube or intramuscular injection to foals, milk intake was estimated from water kinetics. Isotope concentration was corrected for body weight changes in evaluation of turnover rate and body water fraction. The relationship of water intake to milk intake was considered to be dependent on the free water content of milk and the fractions of milk contents that are catabolised to metabolic water (Martin *et al.*, 1992).

Material and methods

Literature data of 111 foals (Table 1) were analysed regarding growth dynamics of nursing foals, milk intake of the foals and mare's milk production, all measured using either the weigh-suckle-weigh (n=6), the isotope dilution (n=3) or the milking and separate feeding method (n=1). The data were used for statistical computation.

The relationship between gross energy (GE) intake via milk and the age of the foals were described with the conventional metabolic weight $BW^{0.75}$ compared with the preferred metabolic weight of suckling mammalian young at peak lactation ($BW^{0.82}$), when milk yield is comparable across species (Oftedal, 1984; Riek, 2008).

Results and discussion

Birth weights ranged from 54.9 kg (sem=4.8) among the Warmblood or Thoroughbred foals to 67 kg (sem=6.7) among the foals of the heavy breeds. Milk intake per kg $BW^{0.75}$ was highest in the first two weeks of life (590 g milk per kg $BW^{0.75}$; sem=0.13). In the first month of life foals required 7.1 to 14.0 kg milk for each kg of weight gain (excluding Burlacu *et al.*, 1993: 18.7-20.6 kg milk/kg weight gain) and 8.1 to 18.1 kg milk/kg weight gain from 4 to 8 weeks of age (Burlacu *et al.*, 1993: 17.8-25 kg milk/kg weight gain), when weight gain was no longer closely related to milk energy intake (Doreau *et al.*, 1986). Peak lactation of horses occurs within the 6th week (Riek, 2008). Until day 120 milk intake was dependent on day (d) of life (Figure 1): kg milk intake/kg $BW^{0.75}$=0.6351-0.0036*d (r=0.82; n=427, weighed by number of animals; outliers – Burlacu *et al.* (1993) – were excluded) or alternatively on kg milk intake/kg $BW^{0.82}$=0.4664-0.0025*d (r=0.83; n=427, weighed by number of animals). Milk intake per kg $BW^{0.82}$ declined steadily until weaning. At day 45 foals consume 491 g milk/kg $BW^{0.75}$ or 354 g milk/kg $BW^{0.82}$. Consequently a foal with 100 kg BW ingests 15.5 kg milk/d 45 or 33.1 MJ GE via milk/d (equal to 1050 kJ GE via milk/kg $BW^{0.75}$ or 760 kJ GE via milk/kg $BW^{0.82}$) and gains 1,300 g of live weight per day. At that time the daily milk energy intake is nearly one third lower in relation to other suckling mammalian young (46.2 MJ GE; Riek, 2008).

Table 1. Literature sources of study data, illustrated with applied material and methods.

Authors	Ascertainment of milk production (methods)	Breeds of the mares	Number of mares/foals
Bouwman and van der Schee, 1978	Weigh-suckle-weigh	Dutch warmblooded saddle horse	19
Burlacu *et al.*, 1993	Weigh-suckle-weigh	Romanian light draught horse	4
Doreau *et al.*, 1986	Single isotope dilution	Heavy French breeds	21
Glade, 1991	Weigh-suckle-weigh	Thoroughbred	10
Martin *et al.*, 1992	Single isotope dilution	Australian Stock Horse	10
Neseni *et al.*, 1958	Weigh-suckle-weigh	Heavy German breeds	9
		German warmblooded saddle horses, Arabians	12
Oftedal *et al.*, 1983	Single isotope dilution	Thoroughbred	5
Ousey *et al.*, 1996	Milking	Thoroughbred	5
Pool-Anderson *et al.*, 1994	Weigh-suckle-weigh	Quarter Horse	12
Zorn *et al.*, 1938	Weigh-suckle-weigh	Silesian heavy breeds	4

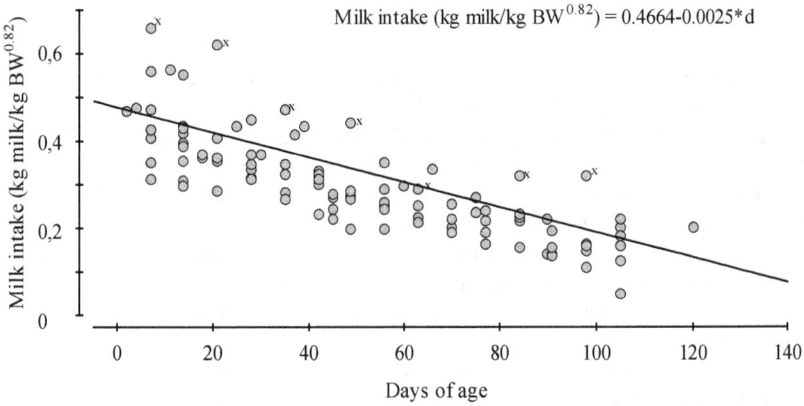

Figure 1. Milk intake (kg) of foals per kg $BW^{0.82}$ in the first four months of age (n=427, weighed by number of animals; x=Burlacu et al., *1993).*

Growth dynamics of foals used for determination of milk intake did not considerably differ from nursing foals not used for determination of milk intake. Doubling of birth weight (Figure 2) was realised within day 45 (studies growth data; weight as multiple of birth weight [wmb]=1+0.022265*d) respectively within day 47 (wmb=1+0.021689*d; reference growth data: Hois, 2004; Vervuert *et al.*, 2005).

Mare's milk production ranged from 44.4 to 7.0 g milk/kg BW (d 0-150) without a significant relationship to the time of lactation. In contrast Santos and Silvestre (2007) described lactation curves of nursing mares using Wood's model.

Different methods for the determination of milk production had no significant effect on mares' milk yield (g/kg BW) between lactation days 7 to 120, whereas the individual differences were considerable.

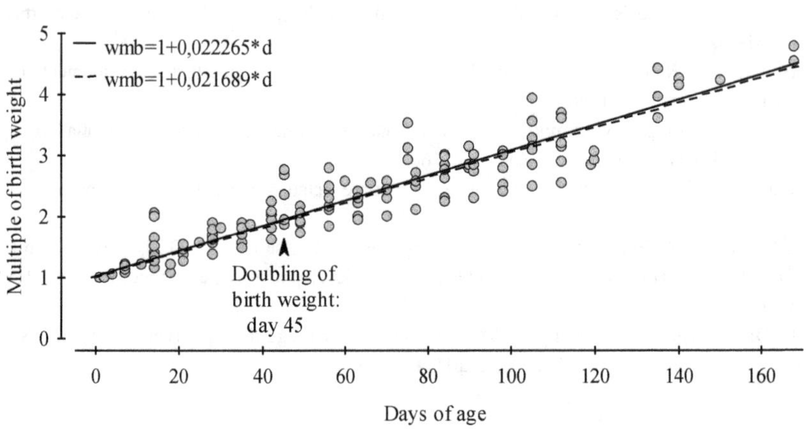

*Figure 2. Growth dynamics of foals used for determination of milk intake (wmb= 1+0.022265*d, full line) in relation to normal nursing foals(wmb=1+0.021689*d, broken line).*

References

Bouwman, H. and W. Van Der Schee, 1978. Composition and production of milk from Dutch warmblooded saddle horse mares. Z. Tierphysiol. Tierernährg. Futtermittelk. 40, 39-53.

Burlacu, G.H., D. Voicu, I. Voicu, M. Nicolae, E. Petrache, G.H. Georgescu and S. Balan, 1993. Study on the energy and protein metabolism in horses. Arch. Tierernahr. 45, 173-185.

Doreau, M., S. Boulot, W. Martin-Rosset and J. Robelin, 1986. Relationship between nutrient intake, growth and body composition of the nursing foal. Reprod. Nutr. Dev. 6, 683-690.

Doreau, M., S. Boulot, W. Martin-Rosset and H. Doubroeucq, 1988. Yield and composition of milk from lactating mares: effect of lactation stage and individual differences. J. Dairy Res. 57, 449-454.

Doreau, M. and S. Boulot, 1989. Methods of measurement of milk yield and composition in nursing mares: A review. Lait 69, 159-171.

Flade, E., 1955. Milchleistung und Milchqualität bei Stuten. Arch. Tierz. 9, 381-383.

Gibbs, P.G., G.D. Potter, R.W. Blake and W.C. McMullan, 1982. Milk production of Quarter Horse Mares during 150 days of lactation. J. Anim. Sci. 54, 496-499.

Glade, M.J., 1991. Dietary yeast culture supplementation of mares during late gestation and early lactation: Effects on milk production, milk composition, weight gain and linear growth of nursing foals. J. Equine Vet. Sci. 11, 89-95.

Hois, C., 2004. Feldstudie zur Gewichtsentwicklung und Gewichtsschätzung beim wachsenden Pferd. Ludwig-Maximilians-Universität München, 287 pp.

Martin, R.G., N.P. Mc Meniman and K.F. Dowsett, 1991. Effects of a proteindeficient diet and urea supplementation on lactating mares. J. Reprod. Fert. Suppl. 44, 543-550.

Martin, R.G., N.P. Mc Meniman and K.F. Dowsett, 1992. Milk and water intakes of foals sucking grazing mares. Equine Vet. J. 24, 295-299.

Neseni, R., E. Flade, G. Heidler and H. Steger, 1958. Milchleistung und Milchzusammen-setzung von Stuten im Verlauf der Laktation. Arch. Tierzucht. 1, 91-129.

Oftedal, O.T., H.F. Hintz and H.F. Schryver, 1983. Lactation in the horse: Milk composition and intake by foals. J. Nutr. 113, 2096-2206.

Oftedal, O.T., 1984. Milk composition, milk yield and energy output at peak lactation: a comparative review. Symp. Zool. Soc. Lond. 55, 33-85.

Ousey, J.C., N. Holdstock, P.D. Rosdale and A.J. McArthur, 1996. How much energy do sick neonatal foals require compared with healthy foals. Pferdeheilkunde 12.Sonderheft, 2.Europäische Konferenz über die Ernährung des Pferdes, 231-237.

Pool-Anderson, K., R.H. Raub and J.A. Warren, 1994. Maternal influences on growth and development of fill-sibling foals. J. Anim. Sci. 72, 1661-1666.

Riek, A., 2008. Relationship between milk energy intake and growth rate in suckling mammalian young at peak lactation: an updated meta-analysis. J. Zool. 274, 160-170.

Santos, A.S. and A.M. Silvestre, 2008. A study of Lusitano mare lactation curve with Wood's model. J. Dairy Sci. 91, 760-766.

Vervuert, I., M. Coenen, A. Borchers, M. Granel, S. Winkelsett, L. Christmann, O. Distl, E. Bruns and B.Hertsch, 2005. Growth rates in Hanoverian Warmblood foals and the development of osteochondrosis. EAAP Publication 114, 85-87.

Zorn, W., H. Brüggemann and W. Bünger, 1938. Ein Beitrag zur Frage der Futterverwertung von Saugfohlen. Biedermanns Zentralblatt Abt.B: Tierernährung 11, 1-18.

Energy requirements of pregnant mares according to oxygen consumption data of foetus and annexes: a meta analysis

S. Schmidt and M. Coenen
Institute of Animal Nutrition, Nutrition Diseases & Dietetics, Gustav-Kühn-Straße 8, 04159, Leipzig, Germany

Abstract

Current recommendations for energy requirements of pregnant mares are based on the factorial method using the daily energy accretion in the uterus. This procedure neglects available data on oxygen consumption of the pregnant uterus which can be taken as the indirect calorimetry of a defined fraction of tissue *in vivo*. To incorporate the energy expenditure of the foetus and its adnexes to the calculation, data on O_2 consumption of these tissues are reviewed. According to literature data, surrounding tissues represent a vast portion of energy expenditure and may be underestimated in former calculations.

Keywords: energy requirements, gestation, O_2 consumption

Introduction

Actual energy requirements of pregnant mares are estimated by factorial derivation based upon the composition of the foetus and fetal growth. But data on the O_2 consumption of the pregnant uterus (including foetus and adnexes) indicate a remarkable contribution of uterus and adnexes to energy expenditure. Therefore, these data require a revision of the energy requirements for pregnancy.

Material and methods

To calculate energy expenditure, literature data of O_2 consumption in the different fractions of the pregnant uterus acquired by arteriovenous differences (Comline and Silver, 1975; Silver and Comline, 1975; Fowden and Silver, 1995; Fowden *et al.*, 2000a,b) and information about the caloric equivalent of oxygen (Kleiber, 1961) were considered.

Results and discussion

Actual recommendations for an average mare of 500 kg, pregnant on day 335 (foetus 50 kg) amount to 15.9 MJ DE (including 20% for adnexes, GfE) or 19.7 MJ DE (NRC).

This information is based on the foetal growth and its composition. From that, one can calculate the energy content and therefore the amount of daily energy accretion. Actual data presented in literature allows definition of the weight of the foetus as % of birth weight = exp (0.013636*x) (x=day of gestation, R=0.94, weighted n=351) (Dusek, 1966; Douglas and Ginther, 1975; Meyer and Ahlswede, 1976; Platt, 1978; Cottrill *et al.*, 1991; Fowden and Silver, 1995; Fowden *et al.*, 2000a,b; Guissani *et al.*, 2005). Considering surrounding tissues and fluids, of the foetus plus adnexes arises the equation extramaternal accretion in % of birth weight = $0.0375*e^{0.011x}$ (x= day of gestation) (Coenen and Meyer, 1986). Therefore, the daily accretion of foetus and adnexes in g per kg foal's birth weight = $0.413*e^{0.011x}$ (this model assumes a portion of the foetus on d 250 of 51.8% and on d 340 of 63.7% from the total weight of the conceptus; the rest is contributed by the adnexes and uterine tissue).

The energy content of the foetus (as from day 240 of gestation) is 0.3836 + 0.0145x MJ GE/kg (Meyer and Ahlswede, 1976) and that of appended tissues can be assumed to be 1.99 MJ GE/kg according to data on pigs (Noblet *et al.*, 1985). Development of tissue mass and the inclusion of the energy in the foetus as well as adnexes can be summarized as a figure for daily energy accretion the model: MJ d^{-1} kg^{-1} birth weight = $0.000681*e^{0.0145x}$, where x=day of gestation.

Consequently daily energy accretion would be 4.7 MJ GE for a 500 kg mare at d 335 of gestation (birth weight of foal 52 kg). To obtain the resulting energy requirement of the mare, a conversion factor for energy of about 0.2 can be assumed.

According to Battaglia *et al.*, (1978), energy consumption by the products of conception consist of tissue energy accretion plus heat production. Considering O_2 consumption data by Fowden *et al.*, (2000b), it can be assumed that VO_2 ml/min = $6.48\ x^{0.9663}$ (x=foetal body weight). Because the exponent equals almost 1, we can simplify average O_2 consumption of the equine foetus to 6.13±1.25 ml VO_2 BW kg^{-1} min^{-1} (n=43). With regard to the respiratory quotient of about 0.9 and the associated oxidative equivalent of 20.91 kJ/l, there is a heat production of 184.6 kJ d^{-1} foetus kg^{-1}.

In late gestation, only 45.6% of energy expenditure is due to the foetus (Fowden *et al.*, 2000a) while surrounding tissues amount to 54.4% (Table 1). Therefore, the whole uterus accounts for a heat production of 404.8 kJ d^{-1} foetus kg^{-1}, resulting in 20.2 MJ d^{-1} for the given mare of 500 kg at an gestational age of 335 days (foetus 50 kg).

In addition to heat production, energy requirements are caused by energy accretion in the pregnant uterus of 4.7 MJ d^{-1}. This results in an overall requirement for gestation of 24.9 MJ ME d^{-1} (with assumed energy utilisation of 84% this would result in a requirement of 29.6 MJ DE d^{-1}).

The results obtained by the currently applied calculation based on O_2 consumption, indicate that pregnant mares may have higher energy requirements than previous estimations suggested. This difference may particularly reflect underestimation of the high proportion of heat production represented by the uteroplacental tissue. However, current feeding practice is orientated towards present recommendations without visible energy deficit. This tremendous contradiction requires scientific clarification and a re-inspection of requirement analysis should be performed with possible integration of the calorimetric findings.

References

Table 1. Distribution of foetus and adnexes to total intrauterine O_2 consumption (Fowden et al.*, 2000a).*

	O_2 consumption in ml/min^{-1}		
	Uterus	Foetus	Utero-placental tissues
Mid gestation (n=5)	74,8	30.3	45.6
	100%	40.5%	61%
Late gestation (n=4)	221.8	101.2	120.6
	100%	45.6%	54.4%

Battaglia, F.C. and G. Meschia, 1978. Principal substrates of fetal metabolism. Physiol Rev 58: 499-527.

Coenen, M. and H. Meyer, 1986. Energy and nutrient requirements of the pregnat mare. Proc. EAAP confenrence 1986, Budapest.

Comline, R.S. and M. Silver, 1975. Placental transfer of blood gases. Br Med Bull 31: 25-31.

Cottrill, C.M., J. Jeffers-Lo, J.C. Ousey, A.J. McGladdery, S.W. Ricketts, M. Silver and P.D. Rossdale, 1991. The placenta as a determinant of fetal well-being in normal and abnormal equine pregnancies. J Reprod Fertil Suppl 44: 591-601.

Douglas, R.H. and O.J. Ginther, 1975. Development of the equine fetus and placenta. J Reprod Fertil Suppl(23): 503-5.

Dusek, J., 1966. Notes on the problem of the prenatal developement of horse. Vedecke Prace Vysk. Stanice. Chov. Keni. Slatinany 2: 1-25.

Fowden, A.L. and M. Silver, 1995. Glucose and oxygen metabolism in the fetal foal during late gestation. Am J Physiol 269(6 Pt 2): R1455-61.

Fowden, A.L., A.J. Forhead, K.L. White and P.M. Taylor, 2000a. Equine uteroplacental metabolism at mid- and late gestation. Exp Physiol 85(5): 539-45.

Fowden, A.L., P.M. Taylor, A.J. Forhead and Kl White, 2000b. Ontogenic and nutritionally induced changes in fetal metabolism in the horse. J Physiol 528 Pt 1: 209-19.

Guissani, D.A., A.J. Forhead and A.L. Fowden, 2005. Development of cardiovascular function in the horse fetus. J. Physiol.(565): 1019-1030.

Kleiber, M., 1961. The Fire of Life – an introduction to animal energetics. Wiley, New York, NY, USA.

Meyer, H. and L. Ahlswede, 1976. Über das intrauterine Wachstum und die Körperzusammensetzung von Fohlen sowie den Nährstoffbedarf tragender Stuten. Übers. Tierernährung 4: 263-292.

Noblet, J., W.H. Close, R.P. Heavens and D. Brown, 1985. Studies on the energy metabolism of the pregnant sow. 1. Uterus and mammary tissue development. Br J Nutr 53(2): 251-65.

Platt, H., 1978. Growth and maturity in the equine fetus. J R Soc Med 71(9): 658-61.

Silver, M. and R.S. Comline, 1975. Transfer of gases and metabolites in the equine placenta: a comparison with other species. J Reprod Fertil Suppl(23): 589-94.

Digestibility of calcium and phosphorous in yearlings

A. Jansson and F. Löf
Swedish University of Agricultural Sciences, Department of Animal Nutrition and Management, P.O. Box 7024, 75007 Uppsala, Sweden

Introduction

The 2007 NRC recommendations for feeding calcium (Ca) and phosphorous (P) to growing horses were increased by 35 and 25%, respectively above the 1989 NRC recommended levels. The increases were mainly based on one study suggesting that endogenous faecal losses of Ca and P are greater than previously reported (Cymbaluk *et al.*, 1989). Applying the 2007 recommendations, many Swedish diets are highly insufficient in Ca and P and substantial supplementation is needed. The aim of this study was to measure Ca and P digestibility in yearlings fed Swedish forage and oats diets.

Materials and methods

Four Standardbred yearlings (330-340 kg) at Västerbo Stud farm were used in a 16 day cross-over design with two diets, grass roughage alone (R) and grass roughage and oats (RO, 75:25 DM basis). Fecal collections were made for 10 h/day on days 10-16. Feed allowances and leftovers were weighed and analysed for Ca and *P* content. Analysis of variance was used (*P* for significance <0.05, presented values=LS means ± SEM).

Results and discussion

Ca intakes were 39±1.5 (range 23-55) and 27±1.5 (range 17-32) g/day on diet R and RO, respectively. *P* intake was 13±0.7 (range 12-15) and 18±0.7 (range 16-22) g/day. Apparent Ca and *P* digestibility were respectively 63±4% and 23±7% (diet R) and 49±4% and 21±7% (diet RO) (ns). Using regression analyses, daily endogenous fecal Ca and *P* were estimated to be 23 and 9 mg/kg body weight, respectively. These results concur with the NRC 1989 recommendations. Although Ca and P availability may vary between different feed stuffs, using the 2007 NRC recommendations may result in overfeeding Ca and *P* with resultant increases in feed costs and potential for environmental pollution with excess excretory *P*.

References

Cymbaluk, N.F., G.I. Christinson and D.H. Leach, 1989. Nutrient Utilization by Limit- and *Ad Libitum*-fed Growing horses. Journal of Animal Science 67, 414-425.

National Research Council (NRC), 1989 and 2007. Nutrient requirement of horses, 5th and 6th edition, respectively. National Academy Press, Washington, DC, USA.

Part 2. Nutrition behaviour and welfare

Biological basis of behaviour in relation to nutrition and feed intake in horses

A.D. Ellis
Nottingham Trent University, School of Animal, Rural and Environmental Sciences, Brackenhurst, Southwell, Nottinghamshire, NG25 0QF, United Kingdom

Abstract

The term 'biological basis of behaviour' encompasses the physiological mechanisms which 'direct' and 'control' the body (the nervous and endocrine system) and which adapt and change due to both genetic programming but also due to adaptation (plasticity) allowing for individual interaction, perception and experience within the environment. The aim of this paper was to give an overview of these aspects and to assess behaviour in relation to nutrition and feed intake. Food intake is controlled by multiple energy-related homeostatic signals (such as the hormones ghrelin and leptin) as well as somato-sensory (taste, touch, smell) and motivational stimuli via the peripheral and central nervous system. Free-ranging horses tend to show 10-15 distinct feeding bouts within 24 hours and spend around 10-14 hours per day on feed intake behaviour. Voluntary intake in dry matter ranges between 1.3-1.7% of bodyweight (BW) for straws, 1.9% BW for hays and 2.6% BW for fresh cut grass. Physiological regulation of intake of sweet feed is unlikely to occur during the first 4-5 weeks of intake, as horses ingested up to 3.3-4.4% of bodyweight in dry matter when offered molassed chopped lucerne, chaff-pellet mix or pellets. Chews/kg fresh matter range from up to 3,400 for straw to around 2,700 for hays and 1,800 for chopped forages with corresponding ingestion times from 45-35-20 minutes/kg. Chewing rhythm (chews/min) remains fairly constant for each animal, while intake rates (g/min) are adjusted according to moisture content, fracture properties, dental condition and organoleptic perceptions. Concentrate feeds are ingested in 8 (pellets) – 18 minutes/kg (oats). Adding chopped forage to concentrate feed at a rate 20-38% increased feed intake time by 50-100%. Adding oil to feed also increased feed intake time. Foraging, geophagia (eating soil or sand), wood chewing and coprophagy may fulfil important functions in free-living horses. Stabled horses on low forage diets show increased ingestion of bedding and faeces as well as stereotypies, highlighting the inherent importance of accommodating the performance of species-specific feed intake behaviour in horse management systems for the welfare of horses therein. It is therefore suggested that structural fibre requirements should be specifically included in feeding systems for horses to allow expression of these intake behaviours, in addition to maintenance of gut function.

Keywords: time budgets, nervous system, ingestive behaviour, chewing, palatability

Introduction

Behaviour encompasses the actions and reactions, responses and general states of any living being. The term 'biological basis of behaviour' is often referred to but its meaning has been applied in multiple ways. In 1963, Tinbergen published his famous four areas (function, mechanism, development, evolution) to quantify the origin and progress of a behaviour (cited in Barnard, 2004). These areas have been adapted and re-named and defined by various scientific 'streams' and at times have been focused on in isolation, although in reality they are complimentary and all inclusive.

As an example, let us take the behaviour of foraging in the bedding of a stabled horse: firstly we need to ask 'what is the *purpose* of the behaviour?'; in this case the answer may be to find and select edible particles (the ultimate purpose of most behaviours being to 'survive and reproduce'). Secondly what is the *mechanism* behind the behaviour: the steps which lead to the behaviour (lowering head, sniffing, tasting, chewing, swallowing) and the physiological changes (hormonal and neural) triggering each step. The 'motivation' (an anthropomorphic term) to carry out the behaviour may be triggered by a signal of hunger or appetite stimulated following an increase in hormones, such as Neuropeptin Y (a powerful feeding stimulant) or Ghrelin (increases appetite) *via* the Central Nervous System (Gordon *et al.*, 2005). An 'innate' genetic basis of a behaviour can be defined as its *evolutionary* background and this may also be a genetically directed stimulus (neural/hormonal) which continues even if the need to 'find food' has been environmentally removed. Finally, there is also the *individual development*, unique to each animal which will have contributed to the 'formation' of a certain behaviour and this of course encompasses the environment (past and present) and a certain level of genetic *plasticity* but more importantly plasticity of nerve connections in the brain (acquiring of memory, learning) as well as *biochemical adaptations*. Learned 'successful' finding of food in the bedding but also life experiences of living outdoors and long term 'grazing/foraging' bouts in a social context of synchronised herd behaviour, could re-enforce foraging behaviour in a stable. Figure 1 sums up the various aspects of 'the biological basis of behaviour'.

The paper aims to give an overview of these aspects and to assess behaviour in relation to nutrition and feed intake (= purpose: essential to survival) in horses with some evaluation of how feed management may and does alter food intake and related behaviours in horses.

The nervous and endocrine system

Figure 1 highlights the important direct link between the nervous and endocrine systems and the final 'expression' of behaviour. The behaviour itself is the final manifestation of a physiological chain of events triggered by biochemical and electrical functions within the nervous system. Neural mechanisms and the hormonal system transmit information, often *via* various negative feedback loops, between the peripheral nervous system (PNS) and the central nervous system (CNS: brain and spinal cord). Several books describe these physiological mechanisms in great detail but this does not lie within the scope of this paper (McGreevy, 2004, Barnard, 2004). Figure 2 gives an overall, simplified outline of the major systems and functions involved, as a reference point for later discussion.

Knowledge of micro-physiological mechanisms determining equine behaviour and regulation is still extrapolated greatly from other species (mainly humans and rats). Yet, due to evolutionary species-specific variations, this may not always be appropriate. For example, McBride and Hemmings (2005) reported that the receptors in the equid brain for the neurotransmitter Dopamine-2, occurred around 10-fold more in horses than in rats; highlighting the genetic influence on physiological mechanisms. McBride and Hemmings (2005) also found changes in density and location of Dopamine 1 and 2 receptors in certain areas of the brain in stereotypic horses compared to a control group. They highlighted that this is likely to be a result of prolonged, repeated stereotypic behaviour patterns which have led to associative long-term potentiation (neural and synaptic plasticity). The stimulation of some dopaminergic neurons in the brain has been linked to 'reward' associations related to a certain behaviour/experience.

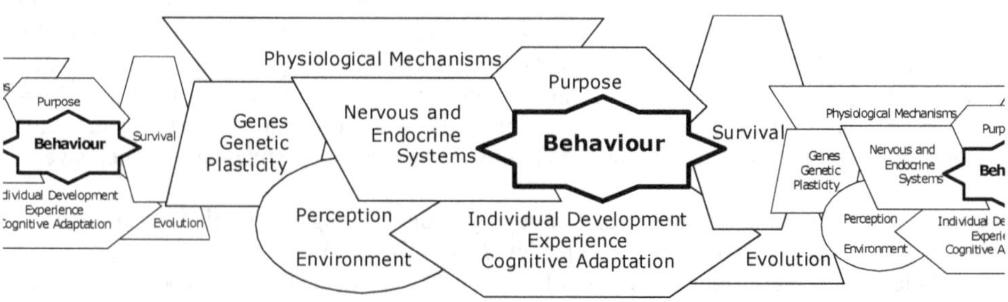

Figure 1. Continuous chain of multiple factors influencing the occurrence, characteristics, adaptation and expression of a behaviour.

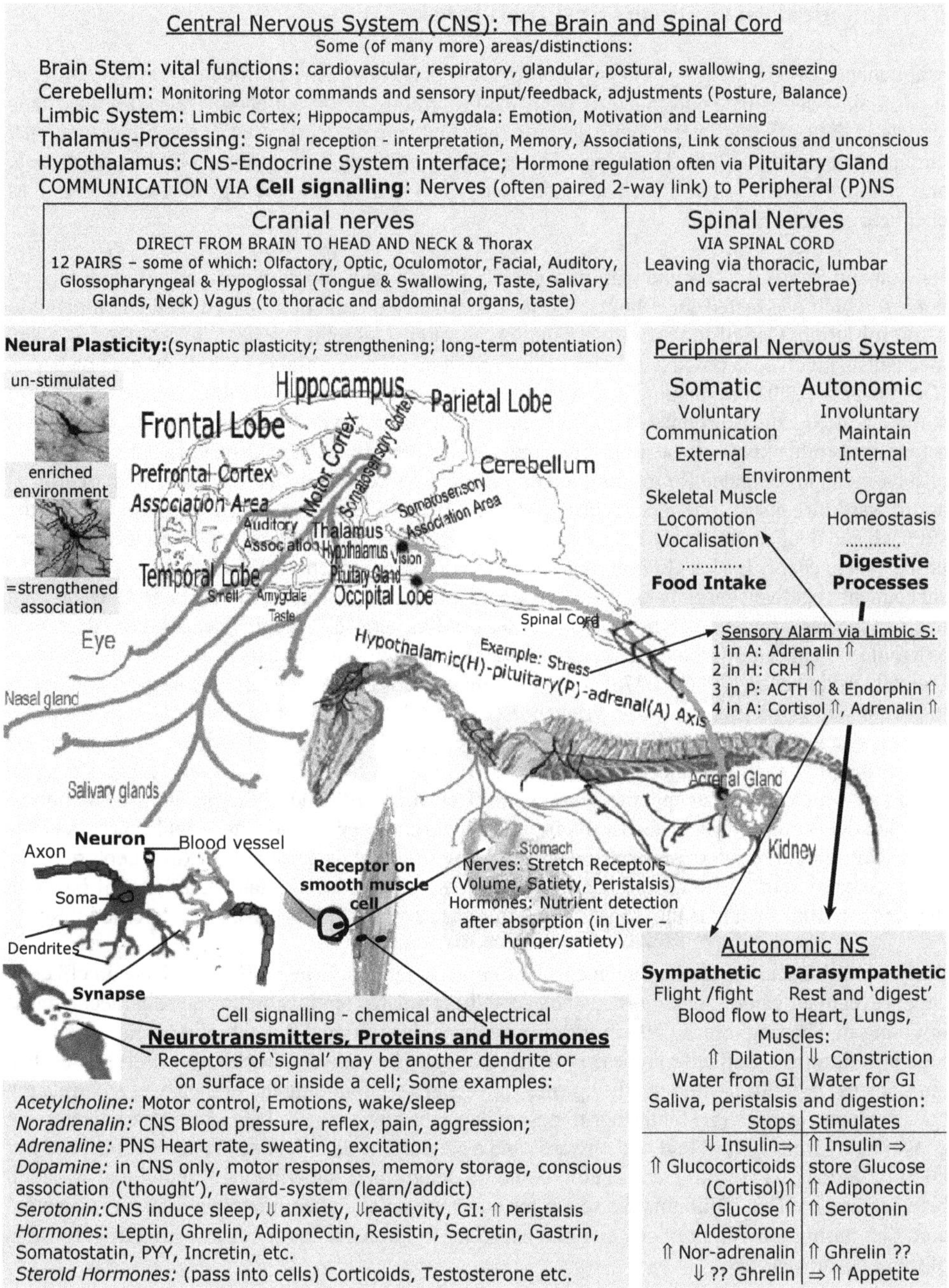

Figure 2. An abbreviated overview of the nervous system and examples of mechanisms and compounds which influence and facilitate the expression of behaviours: a nutritional 'slant' (CTH-corticotrophin releasing hormone, ACTH-adrenocorticotrophic Hormone; GI: Gastrointestinal Tract, PPY-Peptide Y36).

Physiological mechanisms and food intake

Maintenance of bodyweight via energy homeostasis plays an important role in controlling onset of food intake behaviour. Food intake is controlled by multiple afferent energy-related homeostatic signals in the CNS that interact with specific hypothalamic nuclei ordered into complex neuronal circuits (Lopez *et al.*, 2008). These include somatic signals from the sensory nerves like smell, taste and vision and neural signals such as those from stretch receptors in the digestive tract, as well as chemical signals (Table 1).

Ralston and Baile (1982) found that intra-gastric infusion with a high glucose meal before feeding ponies, which had fasted for 4 hours, led to a dose-related reduction in feed intake compared to a control group. One of the appetite regulating hormones, ghrelin has been investigated as a key mechanism involved in possibly leading to the behaviour of feed refusal in very fit competition horses (Gordon *et al.*, 2007a,b). Sakurai *et al.* (1998) first described ghrelin as the only known hormone to stimulate orexigenic neuronal cell bodies present in the lateral hypothalamus of the brain. Orexin is a neurotransmitter known to be involved in regulating sleep and feed intake patterns but interestingly also has been investigated for its involvement in addictive behaviours via its mechanism of arousal and reward circuits in the CNS (Winrow *et al.*, 2010). Ghrelin is released from the cells in the stomach and duodenum shortly before and at the start of feed intake bouts and its production falls in proportion to energy intake. Gordon *et al.* (2006) confirmed a reduction of voluntary food intake with subsequent negative energy balance after 8 weeks of intensive exercise in horses. But, as reported in other species, ghrelin levels actually increased during exercise and leptin levels overall decreased as would be expected to compensate for increased energy requirements (Gordon *et al.*, 2007a). Overall Gordon *et al.* (2006, 2007a) could not find obvious reasons for appetite suppression linked to ghrelin and leptin levels in very fit race horses. The effect of the upregulation of the sympathetic nervous system with increased production of glucocorticoids (e.g. cortisol) is likely to elucidate this area further. For example, injections with ACTH resulted in significant short-term release in leptin in horses, which acts as an appetite suppressant (Cartmill *et al.*, 2003). Studies in rats highlighted that chemical composition of feedstuffs influenced ghrelin production, with soluble carbohydrates decreasing ghrelin production, and lowering appetite as well as affecting orexin receptors in the brain (Sakurai *et al.*, 1998) and this may be the reason for reduced intake and delay of intake in the study by Ralston and Baile (1982). Therefore feed composition needs to be addressed further.

Body condition and metabolic rate due to fat to muscle ratio have been shown to have an effect on appetite regulating hormones. Gordon *et al.* (2007b) compared leptin, ghrelin and adiponectin levels in young fit Thoroughbreds (TB) in training versus older resting TB's. Ghrelin levels were much higher in fit horses than unfit horses ($P<0.01$) and leptin levels were much lower in fit horses than unfit horses ($P<0.001$). Although the age may have had a confounding factor there was a correlation between body fat index (%) for the hormones produced by adipose cells, with leptin clearly increasing considerably from 16% of body fat onwards and a concurrent clear decline in adiponectin. At lower body fat ratios (8-14%) little correlation could be determined, showing great individual variation between horses. This highlights the scope for developmental and environmental factors (maternal diet, diet during growth, previous diet and exercise) to influence the production and effectiveness of these hormones.

Treiber *et al.* (2005) demonstrated such effects of neural and hormonal plasticity when they found that a group of foals fed for a prolonged period on a high sugar and starch diet had lower insulin sensitivity than a corresponding group fed on a fat and fibre diet ($P=0.007$). In terms of physiological adaptation both groups may also have undergone some changes in neurotransmitter, chemical and hormonal metabolism. Insulin resistance and its related conditions (laminitis, obesity, growth disorders, pancreatic and pituitary disease) may not be the only systemic changes due to diet, although research is understandably focusing on this area. It highlights how much we have yet to learn in

Table 1. Some hormones (H) neurotransmitters (NT) and peptides (P) directly involved in regulating food intake (Sakurai et al.*, 1998; Barnard, 2004; Carlson, 2007; Winrow* et al.*, 2010).*

Produced in	Name		Via	Function
Stomach	Gastrin	H		Increases release of HCl and pepsin into stomach
Stomach and hypothalamus	Neuro peptide Y (NPY)	P	Hypothalamus	Major potent and fast *trigger of food intake*; also works as block to pain transmission, induces calming effect
Stomach and small intestine	Ghrelin	H	Hypothalamus	*Increases feed intake*, rises with anticipation of food, reduces relative to feed intake, counter acts Leptin and PYY
	Peptide Y 36 (PPY)	P	Hypothalamus Pancreas	Potent and fast feed intake inhibitor, counteracts NPY; increases relative to feed intake, greater release after high protein meal (humans); increases release of digestive juices
Hypothalamus	Orexin	NT		Transmitter of NPY and PPY signalling in brain;
Duodenum	Secretin	H	Pancreas	Increases as acidity from stomach reaches duodenum, Increases release of bicarbonate into pancreatic fluid
	Incretin	H	Pancreas	Reacts to sugars & starches in GI, increases Insulin production
	Somato-statin	H	Stomach Pancreas	Stops gastrin release; halts digestion; stops secretin and cholecystokinin release, *increases appetite*; reduces glucagon release
Duodenum and jejunum	Chole-cystokinin (CCK)	H	Liver, pancreas medulla oblongata	Increase bile flow and digestive enzyme secretion Send signal of satiety; halts food intake
Adipocytes (fat cells)	Leptin	H	GI and hypothalamus Pituitary liver, cell mitochondria blood vessels & immune cells	Long term regulation of energy homeostasis; reduces/blocks NPY and increases production of α-MSH – *reduces food intake*; Stimulates Fatty Acid (FA) oxidation in liver and increases FA receptors Promotes inflammation and production of TH1 cells (lymphocytes)
	α-MSH	P	Hypothalamus pituitary	Involved in *appetite suppression*
	Resistin	H	Liver	Reduces sensitivity to insulin in liver, therefore increases glycogenolysis, gluconeogenisis; increases insulin resistance (metabolic syndrome)
	Retinol binding protein	H	Skeletal muscle tissue	Supresses glucose uptake; enhances glucose release by liver
	Adipo-nectin	H	Plasma	Protective against arthrosclerosis; reduces gluconeogenises; increases glucose uptake in normal condition (in obese: reduced production)

relation to behaviour. By focusing on the hormones which we know influence satiety, appetite and hunger, but which often also influence insulin and glucagon metabolism (Table 1) we can draw on some limited experience from horses and other species. But there are hundreds of additional motivational pathways that lead to final expression of food intake and foraging behaviour and possibly 'unwanted' behaviour which may be related to food intake; whether it be excessive intake or lack thereof. For example, more understanding of physiological mechanisms is required in areas of somato-sensory perceptions, reward mechanisms related to food intake (orexin and dopamine receptors in the brain), social programming, circadian rhythm and oral stereotypies. Until these are fully understood, applied research focuses on the physical expression of these behaviours, cataloguing, associating and analysing behavioural responses under various conditions.

Natural feeding patterns

Studies of feed intake in horses can be divided into field observations, which assess the grazing patterns, sward selection and time budgets in semi-wild, managed and un-managed environments and into assessment of voluntary intake of feedstuffs with various physical and chemical compositions by domesticated/housed animals. Feed intake behaviour in the following paragraphs includes all ingestive behaviours both stationary and during locomotory movement (foraging). However, definitions of 'feed intake' are not always clearly specified in papers and this will be discussed at times when comparing results.

Free-ranging horses tend to show *10-15 distinct feeding bouts within 24 hours* and resting or non-feeding bouts are generally of no more than 3-4 hours duration (Ralston, 1984; Vulink, 2001). 'Natural' feed intake behaviour is often reported incorrectly to be around 16-18 hours per day, but this is an absolute and short term maximum and rarely occurs as an average of 24 hours (e.g. 70% of *day* time observed and falsely extrapolated to 70% of 24 hours). It may occur towards the end of the summer or in early spring: both times which have been identified as showing increased metabolism linked to feed intake related hormones. For example, Crowell-Davis *et al.* (1985) report up to 70% of time spent grazing in Welsh pony mares, however, this data was collected during 4 x 15 minute focal samples taken between 05:00 and 21:00 hours only. However, the maximum period has been focused on in order to highlight the great discrepancy between feeding time of stabled and free-living horses. The majority of studies reported have not recorded 24 hour time budgets (Table 2).

A variety of studies show a *range of 40-60% of 24 hours spent grazing* (9-16 hours, occasional 18 hours in spring) depending on time of year (Duncan, 1980; Arnold, 1985; Van Dierendonck, 1996; Berger *et al.*, 1999; Vulink, 2001; Souris *et al.*, 2007; Edouard *et al.*, 2009). There are only a few 24 hour studies and these have shown that horses spend around 60-70% of their day on feeding behaviour and only 40-50% of night time (darkness) on feeding (Vulink, 2001; Boyd, 1988; Berger *et al.*, 1999; Edouard *et al.*, 2009). Table 2 highlights that *average time per 24 h has most frequently been recorded at 10-12 hours* and in order to extrapolate day-sampling to 24 h time budgets *a ratio of 1.4:1 of day to night time spent on feeding activity* has been applied according to Berger *et al.* (1999) and Vulink (2001). Time recorded is affected by time of year, area (climate, space, plants), period of observation (daylight, day and night), type of recording (observation, activity recorders) extent of flies/mosquitoes (Duncan, 1980; Vulink, 2001), predators (Van Dierendonck, 1996) and by weather patterns. Nevertheless when comparing data from observations around the world similar patterns emerge (Figures 3, 4, 5, 6).

Berger *et al.* (1999) recorded behaviour of 4 semi-feral Przewalsky horses (from a 12 horse herd) in a nature park in Germany over a whole year, using an electronic detector (Ethosys) which records behaviour every second for 24 hours and thus can claim to have the most complete data set over 24 hour periods at various times of the year. Results confirm previous studies of a polyphasic daily feeding pattern throughout day and night with increased activity during dusk and dawn and the

Table 2. Average daily (24 h) feed intake times of free living horses reported and extrapolated daily feed or activity times.

Author	Horses	Observed	Average % of time spent grazing as reported	Average hours / 24 h ± s.d. as reported	Average hours/24 h[1] calculated
Duncan (1980)	Semi-feral Carmargue horses, France	Day-time, summer (12 h)	51 to 55		12-13
		Day-time, winter (8 h)	61		14
Crowell-Davis *et al.* (1985)	Welsh ponies	4 x 15 minutes, day time (05:00-21:00 hours)	59-70		
Van Dierendonck *et al.* (1996)	Feral Przewalsky (Takhi), Mongolia 45 ha adaptation enclosure	Day-time, summer (15 h)	40±8		9.3±2
		Day-time, winter (8 h)	41-49±15		12±3.6
Vulink (2001)	Feral Konik horses, nature reserves, NL	24 hours summer	56	14±3	
		24 hours winter	55	12±1	
Boyd *et al.* (1988)	Semi-feral Przewalsky horses, 12 ha pasture, Virginia, USA	24 hours, summer Note: Horses had access to concentrate feed pellets between 8:00-12:00	46±14	11±3.4	
Magnussen *et al.* (1994)	Semi-feral, Iceland horses, Iceland	Day light (18 h) August	54		12.5
		Day light (15 h) September	80		17
		Day time Average Summer	62		12.7
Souris *et al.* (2007)	Feral Przewalsky (Takhi) released, after 1 year in adaptation enclosure, Mongolia	Day time May-September	46		9.48
Berger *et al.* (1999)	Semi-feral Przewalsky horses, nature park, Germany	24 h continuous (chewing)			
		Summer		6	
		Spring		9	
		24 h continuous (activity)			
		Summer		16	
		Spring		13	
Edouard *et al.* (2009)	Domesticated horses; paddock	24 h continuous Summer		14	
		Mean ± s.d.	51±8	12.7±2	12.5±2.5

[1]Assuming day to night ratio 1.4:1 in summer and 1.2:1 in winter.

lowest activity in the period prior to dawn (Boyd, 1988a; Van Dierendonck, 1997; Vulink, 2001). Daytime activity and feed intake peak at around 1.4-1.5 times of night in late summer, reducing to 1.3 in autumn and 1.2 in winter and increasing back to 1.4 in spring (adapted from Berger *et al.*, 1999). However, the distinction between activity and food intake in this study leads to much lower average time spent on food intake of around 8 hours per day (only 55% of total activity)(Table 2). This may highlight that other observation studies included slow movement while grazing (foraging) and the head coming up while still chewing as feed intake while the recorder distinguished between this and chewing activity with head down. Neither may be completely accurate. Vulink (2001) shows a greater time spent on feeding behaviour (active and feeding; excluding playing, social behaviour, continuous walking) which comes much closer to the 'total activity' time recorded by Berger *et al.* (1999) (Figures 3 and 4).

Vulink (2001) observed a significant *increase in feeding times (mean +2 hours) for lactating mares during winter*, while pregnant and lactating mares showed an even greater feed intake time (+3

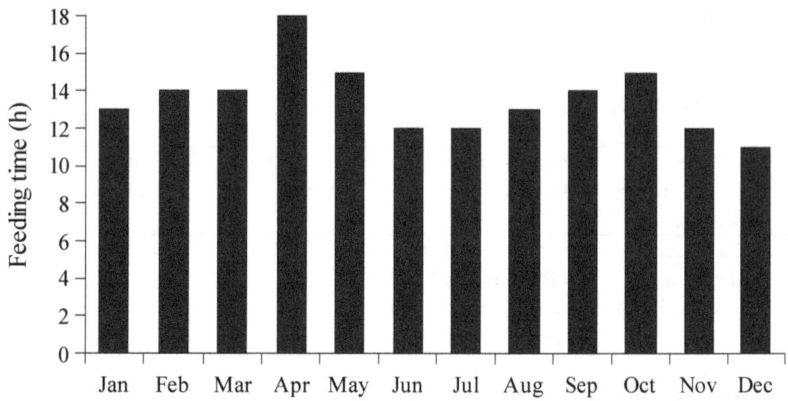

Figure 3. Average active time (walking and feeding, 24 h recording) for feral Konik horses at a 9,000 ha wet land nature reserve: Zoutkamperplaat, the Netherlands, according to time of year (adapted from Vulink, 2001).

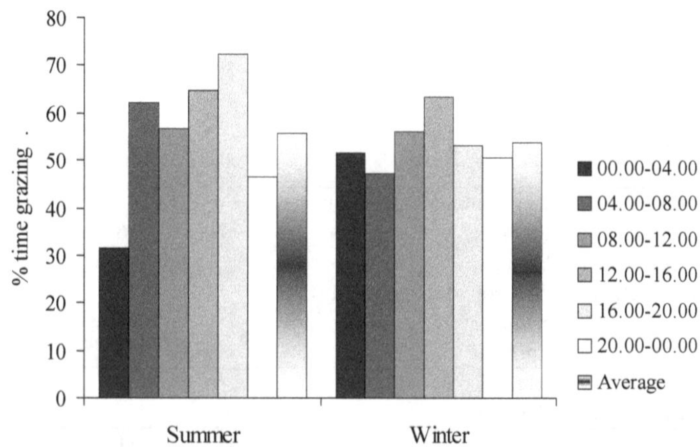

Figure 4. Distribution of active time (walking and feeding, 24 h recording) for feral Konik horses at a 5,500 ha nature reserve Oostvaadersplassen, the Netherlands, according to season (adapted from Vulink, 2001).

hours) with a sudden drop to 'normal' feeding time in the month prior to parturition (possibly due to the foal pushing against the digestive tract).

Souris *et al.*, (2007) highlighted a strong *effect of temperature* and time of day on the resting/feeding balance within the warmer months of the year (Figure 5). However, when looking at all year feed intake versus temperatures in Przewalsky horses in the same area the correlation with temperature is not strong and also highlights differences in herd behaviour which may be due to physiological status of horses (Figure 6).

A study by Boyd (1988) showed that *foals started nibbling grass from day 1 and that by day 3* they started to tear off grass copying their dam. By day 6, those which had some access started nibbling hay and they also showed coprophagy behaviour for the first time. These foals continued suckling until they were around 18 months old. Figure 7 highlights the *increase in feed intake behaviour over the first few months* and shows a large drop in suckling from months 1-2 with related increase in time spent grazing. Suckling time may be slightly inaccurate as 15 minute scan sampling was carried out, which could easily miss short suckling bouts.

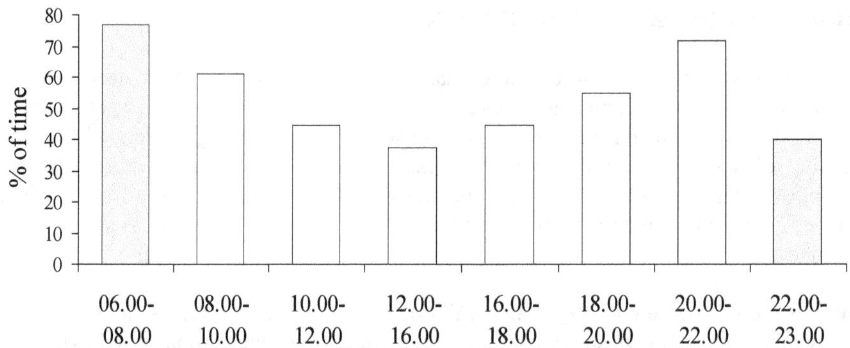

Figure 5. Percentage of time spent grazing at various times of day of newly released Thaki horses (Przewalsky) recorded during daylight hours, summer (White bars: Herd 1, adapted from Souris et al.*, 2007; Grey bars: Herd 2, adapted from Van Dierendonck, 1996).*

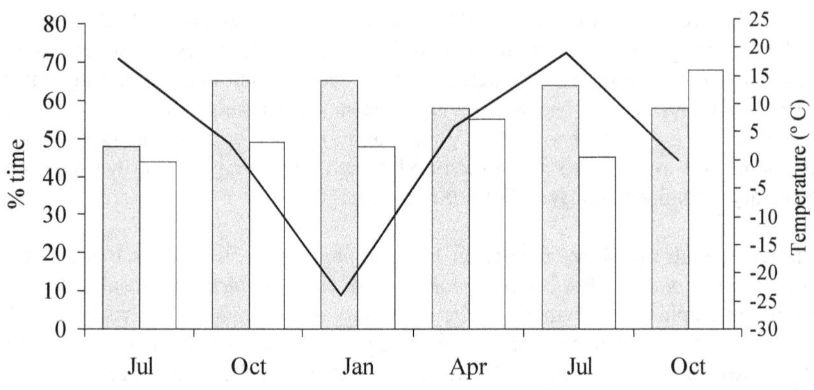

Figure 6. Variation in Feed Intake Behaviour (dawn to dusk) between 2 newly released Thaki herds (Herd 1: grey; Herd 2: white bars) and overlay of average temperatures (solid line) (adapted from Van Dierendonck, 1996).

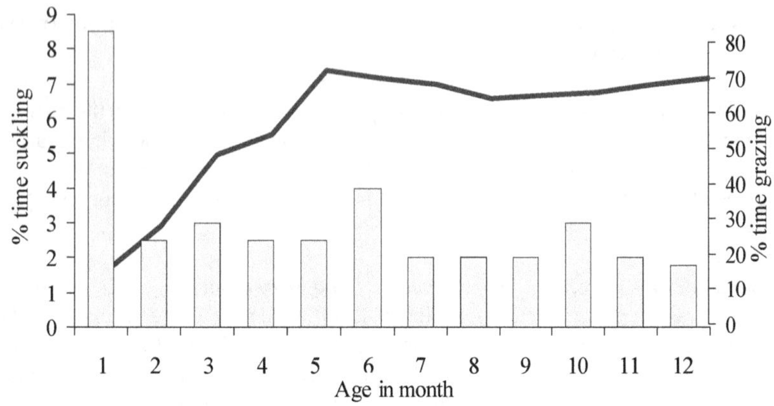

Figure 7. Feed intake behaviour of foals observed in 15 minute spot checks in the daytime (12 hours) (adapted from Boyd et al., *1988b) Suckling (bars) and grazing (solid line).*

Voluntary food intake in stabled horses

In domesticated horses provision of feed other than pasture is common and often necessary due to lack of pasture or due to management towards activities carried out with horses for sport. The scope of this paper does not extend to feed intake of horses used for industrial work. Voluntary food intake relates to the voluntary total consumption of *ad libitum* presented food. *Ad libitum* is defined as constant free availability of a food source (i.e. the horse never runs out before food is replenished). Table 3 gives an overview of a range of studies which offered horses forages to ascertain daily *ad libitum* intakes.

These studies record a voluntary dry matter (VDM) intake of 1.3 (for straws) to 3.1% of body weight. Dry matter intakes give some insight into the effect of moisture content and overall daily volume of intake but fresh matter intakes as fed (FM) are often wrongly omitted. After all, the horse 'experiences' and responds to the FM first of all.

The study by Edouard *et al.* (2009) reports data from pasture studies, but these confirm the results from studies using fresh cut forage. Edouard *et al.* (2009) and Naujek and Hill (2003) showed that horses prefer tall, long grass to short cropped swards. In contrast, horses on faeces contaminated pasture selected shorter sward areas, which may have been further away from dung patches and therefore less contaminated (Fleurence *et al.*, 2007). Recent studies using indirect methods to measure total grass intake per 24 hours (bodyweight gain, energy balance or alkane markers), highlighted that some ponies may ingest up to 5% of their bodyweight in DM/day, but further research in this area is necessary (Longland and Byrd, 2006; Smith *et al.*, 2007).

Table 3, together with the above results on free-ranging horses, highlights that horses *do not necessarily increase voluntary fresh matter intake to make up for reduced quality* although they have the option, unlike ruminants. The most detailed summary of previous studies has been collated by Dulphy *et al.* (1997) on 37 different grasses and forages which incorporated and confirmed earlier research (Willard *et al.*, 1977, Doreau, 1978; Martin-Rosset *et al.*, 1978, Doreau *et al.*, 1980; Vernet *et al.*, 1995). For dried forages there was no relationship between voluntary intake and physicochemical properties (e.g. structural carbohydrates: cell wall compositionor Acid Detergent Lignin content) of forages, although the intake for straw can be seen as considerably reduced. This may partially be attributed to the increased energy cost of consumption but also be linked to gut fill and motility (Vernet *et al.*, 1995). Furthermore no effect of season on voluntary intake, which is apparent in ruminants, could be seen in indoor housed horses (Dulphy *et al.*, 1997). The authors concluded

Table 3. Voluntary feed intake (I) of stalled or stabled horses fed forages ad libitum *(FM = fresh matter, DM = dry matter)(arranged in order of DM Content).*

References	Horses	Feed	FMI kg/ 500 kg horse	FMI g/kg BW	DMI g/kg $BW^{0.75}$	DMI %BW = g/100 kg
Houbiers and Smolders (1990)	TB Trotters	Fresh summer grass (DM 16%)	80	187	94	2.1
		Fresh spring grass (DM 14%)	84	195	99	2.3
	Warmblood	Fresh summer grass (DM 16%)	85	197	100	2.1
Chenost and Martin-Rosset (1985)		Fresh spring grass (DM 14%)	90	208	105	2.4
	Thoroughbred	Fresh hybrid ryegrass (DM 20%)	52	120	98	2.6
Dulphy *et al.* (1997)	Light horses	Fresh forages (n=16)	63	127	97	2.0
Edouard *et al.* (2009)	Light horses	Summer pasture: (DM 18%) tall, intermediate & short sward	57	115	94	2.1
	Mean fresh forages (DM 14-20%)		73±15	164±41	98±4	2.2±0.2
Bergero *et al.* (2002)	Ponies maintenance	Early harvest haylage (DM 56%)	24	48	85	2.7
	Ponies light work	Late harvest haylage (DM 63%)	22	44	87	2.8
	Ponies med. work	Late harvest haylage (DM 65%)	24	48	98	3.1
	Mean haylages, dry silage (DM 56-65%)		23±1	46±2	90±7	2.8±0.2
Martin-Rosset and Dulphy (1987)	Heavy horse yearlings	Hay medium quality	12	23	99	2.0
Vermorel *et al.* (1997)	Standardbreds	Hay late cut	10	19	82	1.7
		Hay medium	11	21	90	1.9
Dulphy *et al.* (1997) from various authors	Light horses	Lucerne hay (n=12)	13	25	107	2.2
		Grass hay (n=38)	13	25	93	2.2
		Straws (n=6)	7	14	61	1.3
Pearson *et al.* (2001)	Ponies	Molassed oat straw	12	22	68	2.2
	Mean hays (DM 80-88%)		11±2	22±4	86±17	1.9±0.3
		Mean straws (DM 90%+)	10±4	18±5	65±5	1.7±0.6
Pearson *et al.* (2001)	Ponies	Molassed dried lucerne chaff	18	35	155	3.3

that *physical appetite-regulating mechanisms are not apparent* in the short term in horses and point towards organoleptic (sensory) perceptions (Dulphy *et al.*, 1997).

Organoleptic perceptions include smell, taste, flavour, palatability and of course touch, which implicates particle length and fracture properties of the forages (Ellis, 2003). These somatosensory perceptions may explain the increased voluntary intake of Lucerne hays, haylages, molassed oat straw and molassed Lucerne chaff as well as the reduction in intake of untreated straws (Table 3, Pearson *et al.*, 2001). With long-stemmed conserved forages, water content and availability seem to have the greatest effect on voluntary intake, as can be seen from the reduced range (by 60%) between feed when converting from fresh matter to dry matter intake. However, the differences between fresh grass, hays, haylages and straws as well as the Lucerne chaff above show that merging all intake data into one range or mean does not serve any purpose in understanding processes. This applies even more for data including voluntary intake of treated forages and concentrate feed. For illustration the molassed lucerne chaff presented in Table 3 shows more affinity with intake rates and daily intake level with the diets presented in Table 4.

Table 4. Voluntary feed intake (I) of stalled or stabled horses fed complete diets including concentrates ad libitum *(FM = fresh matter, DM = dry matter).*

References	Horses	Feed	FMI kg/ 500 kg Horse	FMI g/kg BW	DMI g/kg $BW^{0.75}$	DMI %BW = g/100 kg
Argo *et al.* (2002)	8 Welsh Ponies, stabled 1 hour sand paddock/day	Complete diet: 30% straw chaff, 10% dried grass chaff, 50% pellets (straw & grass/P&min,vit) 10% molasses/oil	21	41	102	3.2
		As above: ground and pelleted	25	50	138	4.4
		Max intake ground and pelleted at day 26	28	56	156	4.9
Henneke and Callaham (2009)	Trial 1: 2 groups of 3 horses,	*Ad libitum* concentrate and hay	19	38	105	3.3
		Peak pellets intake at day 25				
		Hay intake at day 25	2	3	8	0.3
		Complete diet day 25	21	41	113	3.6
		Pellet intake at day 35	10	20	55	1.8
		Hay intake at day 35	7	13	38	1.2
		Complete diet day 35	17	33	93	3.0
	Trial 2: 2 groups of 3 horses, hay intake not measured	*Ad libitum* concentrate and hay	23	45	124	4.0
		Peak pellets Intake at day 8				
		Balanced pellet intake at day 60	17	33	91	2.9
		Ad libitum concentrate and hay while at grass from day 105 (week 15)	5	10	29	0.9
		Balanced pellet intake after day 20				
		Mean ± s.d. (complete diets only)	22±4	44±9	120±26	3.8±0.8

There are a limited amount of studies looking at total voluntary feed intake of horses fed concentrate feed. Some studies show that in the short term (1-5 weeks) ponies failed to regulate their intake according to energy requirements when fed *ad libitum* concentrate feed (Ralston, 1992; Cuddeford and Hyslop, 1996). Argo *et al.* (2002) showed recently that a complete chaff-concentrate diet will be eaten at a slightly higher rate than most forages but when pelleting the same diet, voluntary feed intake increased considerably and continuously until day 26, before horses were changed onto the chaff form of the diet. Interestingly the group which was offered the chaff diet first for 26 days did not reach the same extreme intake when switching onto the pellets after 26 days. This supports the hypothesis that after a period of 5-6 weeks on a diet, the long term regulation of feed intake according to chemical composition may have established itself somewhat over orgnoleptic properties. The amount of voluntary intake of *ad libitum* fed pellets reported by Henneke and Callaham (2009) in two trials supported this very high intake within the first 4 weeks, despite offering *ad libitum* hay simultaneously (Table 4). This points towards a somatosensory effect of sweet feed – a clear preference – which is coupled with a possible 'requirement' for a minimum daily time spent on ingestive behaviour (hypothesised by Ellis, 2003) leading to these high DM intakes. Both taste preference and requirement for chewing could be hypothesised to override any short term biochemical and hormonal regulation of feed intake. The bodyweight and condition of ponies/horses increased continuously in both studies. Leptin production as a suppressor of appetite is known to increase only after a considerable increase in bodyfat (adipocytes) (Gordon *et al.*, 2007b); i.e. possibly after 4-5 weeks in this case.

Henneke and Callaham (2009) point out that no laminitic episodes were recorded despite the high intakes. It is likely that the pellets (just as is the case in Argo *et al.*, 2002) contained low amounts of sugars and starches and in this context feeding bouts may be of interest. Argo *et al.*, (2002) observed an average of 20-25 feeding bouts with a duration of around 20 (pellets) to 30 (chaff) minutes over 24-hour periods. Research into glycaemic responses certainly shows a high increase after ingesting 2 kg of barley or oats (Jose-Cunilleras *et al.*, 2004), which would take around 25-30 minutes (Ellis *et al.*, 2001). It is unlikely that the voluntary regulation of length of bouts would guard against digestive (overspill of starches) or metabolic problems (hyper-glycemia followed by hyper-insulimea) when feeding a high starch or sugar concentrate. Nevertheless, the bout lengths for the pelleted diet were significantly shorter than for the chaff mix (Argo *et al.*, 2002). In comparison, free-living horses perform only around 10-15 distinct feeding bouts of durations of 30 -180 minutes in 24 hours (Ralston, 1984; Berger *et al.*, 1999).

Short-term ingestive behaviour

Factors affecting ingestive behaviour (chewing, swallowing, length of bouts, choice of feed, related) could be 'feed factors', such as physico-chemical and organoleptic properties of feed (particle length, elasticity, fracture properties, moisture, flavour and volume) or *animal factors*, such as age, temperament, previous experience, preferences, dental condition, physiological condition.

Short term intake rates (STIR) (chews/kg FM or DM, chews/min, intake rates in g/min, bites/min) are used to empirically assess intake behaviour and changes of such. Feed intake ethograms and time budgets are used to record and present behaviour during feed intake. This allows for individual bouts of behaviour to be expressed on bar plot charts or Markov-time chains. The bars represent the occurrence and the duration of acts and this gives a comparable impression (between different feed stuffs) of the 'rhythm and speed' at which ingestive behaviour takes place (Haccou and Meelis, 1995). Figure 8 shows an example of a Markov-time chain, highlighting initial *neophagia (aversion to a new food)* from a study which offered dried lucerne chaff and lucerne silage chaff (pH 4.5) to 9 horses for the first time (Ellis and Hill, 2000). Note that in the third phase, when a horse was offered both feed simultaneously the strong smell attracted it to the silage first before quickly rejecting this and then moving on to the dried lucerne chaff. In other studies, where smell or distinction between flavours may be less pronounced, offering feed in a 'cafeteria' style to assess palatability has resulted in the discovery that some horses have strong feeding side preferences and will always choose the favoured position first (Bottom *et al.*, 2004).

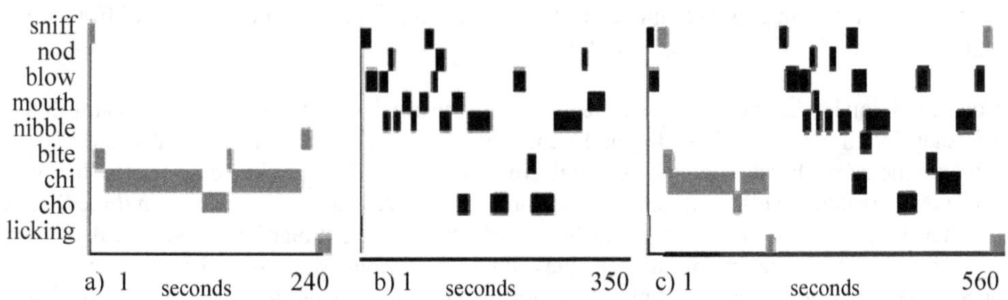

Figure 8. Markov-time chain for horse fed either a) dried lucerne (grey) b) lucerne silage (black bars) or c) offered a choice of both (Ellis and Hill, 2000).
Cho: chew outside trough; chi: chew inside trough [cannot distinguish bite and chews] mouth: mouthing, pushing food round/out with tongue; blow: snort/blowing air; nod: nodding head up and down).

Automatic recording equipment has been developed for measurement of ingestive behaviour (Meyer *et al.*, 1975; Chambers *et al.*, 1981; Rutter, 2000). Meyer *et al.* (1975) investigated chewing frequency in riding horses (450-550 kg) at two establishments. Measurements were taken by a telemetric heat sensor with electrodes attached to the upper jaw, below the medial eye corner at the lower edge of the mandibula. For hays a range of 3,000-3,500 chews/kg DM were recorded. The mean value for straw was 3,645 chews/kg DM and for concentrates a range of 832 chews kg/DM (oats) to 1,383 chews kg/DM (mixed concentrates) were reported. Grinding oats led to an increase in chews/kg and in a later report, pelleting of mixed feed decreased intake times from a mean of 18.6 min/kg (mixed concentrates) to around 8.31 min/kg (8 mm cubes) (Meyer, 1986). For all feedstuffs around 80-85 chews per minute were reported using the electronic recording system, indicating that the horse changes its intake rate according to feed properties while maintaining the chewing rhythm. This early study is supported by more recent results (Table 5). Meyer *et al.*, (1975) suggested that a reduction in dental chewing surface of the molars and smaller oral capacity in ponies is responsible for increases in chewing rate and this also influences intake rates. This relationship between bodyweight and bite rates as well as volume of bites has been confirmed recently by Fleurance *et al.*, (2009), showing significant differences between ponies and horses.

Within each horse chews/minute tend to stay within narrow parameters (overall range from 60-80 chews/minute). Meyer *et al.*, (1975) commented that it is the g/minute of intake which regulates chews/kg rather than changes in chewing rhythm. Figure 9 shows averages and standard deviations derived from Table 5. It highlights that intake time for long-stemmed forages is considerably higher than all chaffed and pelleted feed. Only when adding chaff at a minimum of 20% in weight of total feed could a significant increase in chews/kg be seen (Harris *et al.*, 2005; Ellis *et al.*, 2005). Ellis *et al.*, (2000) showed a significant decrease in chews/kg and intake time when adding either water or 250 g soaked sugar beet pulp to a cereal/chaff mix. This happened to a much lesser extent when soaking hay (Ellis, 2000).

Additional studies have shown that when adding fat to diets at a rate of 1 g/kg BW per day a very slight increase in total intake time is shown, but at a rate of 2 g/kg BW per day the intake time per kg feed more than doubled (Zeyner, 1999). Zeyner *et al.* (2006) confirmed this when intake times increased significantly on diets with added soya oil compared to high starch diets. This may be an effect of palatability but it may also due to a change in fracture properties of diets, with a change in 'chewing efficiency'. Chewing efficiency (particle breakdown) and with it the ability to form a bolus of required size and moisture content may have a considerable influence on feed intake (Ellis, 2003). Bonin *et al.* (2007) reported a great reduction in mandibular motion when ingesting pellets compared to hay, leading to incomplete occlusal surface contact of molars and this will further predispose horses to development of molar hooks.

Dental condition has been linked to a decrease in chewing efficiency and feed intake (Ralston, 2001). In a study looking at the effect of dental treatment, Ralston *et al.* (2001) found that only horses with extreme dental hooks showed a slight decrease in digestibility of a hay/concentrate diet. Ellis (2003) investigated the effect of dental condition on ingestive behaviour in horses in three 'before and after' studies involving a total of 23 horses which had large molar hooks without damage to gums or tongue at the beginning of each study. Intake behaviour was assessed for 7 forages and 7 concentrate feed after adaptation to the diets for 3 weeks. Following dental treatment (rasping of molar tables) horses intake rates were measured again 3 weeks later to allow for acclimatisation to new 'dental' efficiency. There was a significant decrease in chews/kg fresh matter and dry matter for whole grain cereals and cereal mixes ($P<0.05$) and a very significant reduction in chews/kg for hays and straws ($P<0.001$, Figure 10). Parallel to this a similarly significant increase in intake rates (g/minute) was recorded ($P<0.001$).

Table 5. Feed intake behaviour results adapted from various authors (original and extrapolated data given, when required information was present; ordered according to feed).

References	Horses	Feed	Chews/ min.	Chews /kg FM	Chews/ kg DM	Intake Rate (g DM/ min.)	Intake time for 1 kg FM (min.)
Meyer *et al.* (1975)	Horses (500 kg)	Straw	79	3,281	3,645	19	42
		Hays	85	2,720	3,200	28	32
Lengwenar *et al.* (1999)		Hay	60	2,383	2,871	21	40
		Dry haylage (81% DM)	60	2,208	2,725	21	37
Ellis *et al.* (2000)	Horses (550 kg)	Deseeded ryegrass hay-straw	74±3	3,457	3,842	20	45
		As above – chaffed (3 cm length)	73±2	3,640	4,000	20	46
Harris *et al.* (2005)	Horses (560 kg)	Clover hay				34	33
Ellis *et al.* (2000)	Horses (550 kg)	Medium cut mixture hay	68±3	1,932	2,147	33	26
		As above soaked 30 minutes, drained 20 minutes (50% DM)	74±4	1,128	2,255	34	15
		Lucerne chaff (2 cm length)	70±2	1,072	1,277	57	15
Argo *et al.* (2002)	Ponies (230 kg)	Full mixed diet (Table 5) pellets and chaff (50%)	83			17	
Ellis *et al.* (2005)	Mixed horses (550 kg)	Pellets, lucerne chop and long (4cm) straw chaff (30%)	68±9	1,074	1,343	53	16
		As above with straw chaff (20%)	68±10	896	1,120	60	14
		As above with straw chaff (10%)	68±9	735	919	75	11
		Pellets, lucerne chop (10%)	67±9	609	716	96	9
		Pellets (1cm, thin)	69±9	446	525	137	6
		Pellets (2 cm, thick)	65±11	472	555	119	7
Argo *et al.* (2002)	Ponies (230 kg)	Pellets (7 mm)	83			26	
Meyer *et al.* (1980)		Cereal mix – ground and pelleted	85	680	850	100	8
		As above, untreated cereal mix	75	1,106	1,383	60	15
Meyer *et al.* (1975)	Horses	Oats	84	1,227	1,394	54	17
Zeyner *et al.* (2006)	Horses (440 kg)	Barley-oat mix				52	17
Harris *et al.* (2005)	Horses (560 kg)	Sweet feed mix				83	14
		As above plus chaff (6%)				66	18
		As above plus chaff (35%)				40	28
Ellis *et al.* (2000)	Horses (550 kg)	High fibre pellets (1.5 cm)	60	1,074	1,220	61	18
		Whole barley	55	879	1,035	48	26
		Micronised whole barley	60	742	874	70	17
		Mix of pellets, barley, dried lucerne and 200 ml water	77	460	639	123	11
		Mix of pellets, barley, dried lucerne and 250 g wet sugarbeet	64	334	499	132	11

FM: fresh matter, DM: dry matter.

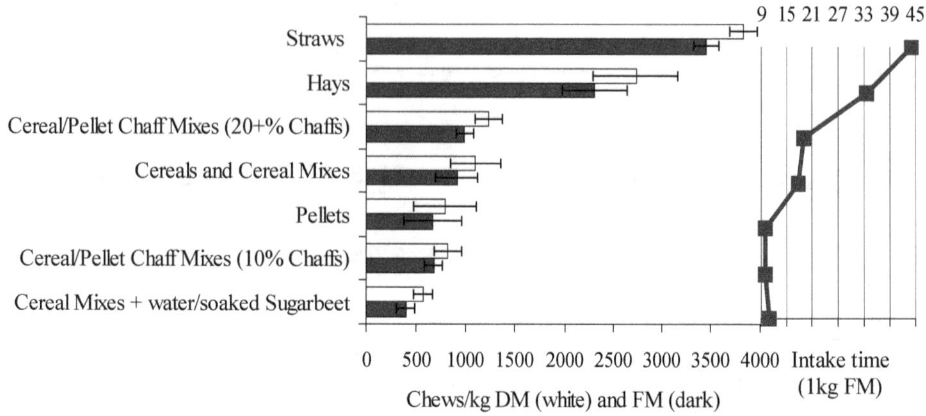

Figure 9. Chewing behaviour (± s.d.) according to dry matter and fresh matter intake per type of feed and mean intake time for FM using results presented in Table 5 (excluding soaked hay) (Ellis, 2000).

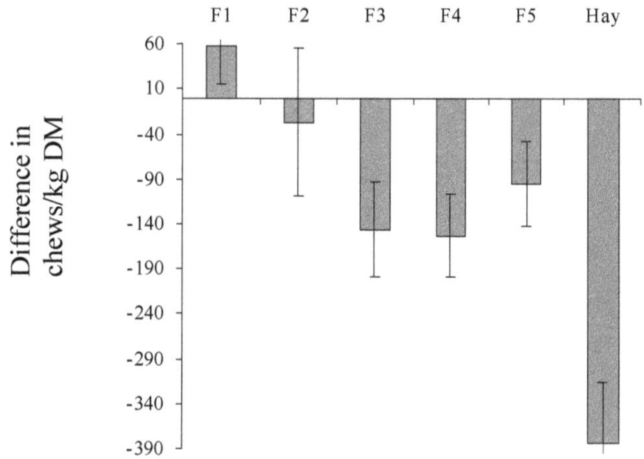

Figure 10. Mean difference (± s.e.) in chews/kg DM after dental treatment (where before dental treatment is set as 0) (Ellis et al., *2000).*
F1: short-chop heat dried lucerne, F2: pellets, F3: whole barley, F4: bruised micronised barley; F5: F1-F4 + wet sugar beet pulp).

Feacal particle size analysis showed no effect of dental treatment on mean particle size and distribution in 12 of the 13 horses where this was measured. In addition no change in digestibility or passage rates could be determined (Ellis *et al.*, 2001). All this points towards horses adjusting the speed at which they ingest feed (g/minute) according to dental efficiency as well as other somato-sensory perceptions, in line with findings by Ralston *et al.* (2001). Only when chewing causes pain or due to an 'inability' to adapt chewing behaviour will feed intake be impaired (Ralston *et al.*, 2001; Ellis *et al.*, 2003).

Palatability

Palatability can be summarised as the overall sensory perception of a feed by the animal (appearance, taste, smell, temperature, texture and consistency). As highlighted above, these pre-gastric factors, together with previous experience of a feed, are the first and most direct regulators of food intake and chewing behaviour in horses (Ralston, 1984; Ellis, 2003). In foals a strong cultural transmission of food preferences from the mother has been established and social facilitation also plays an important role in total voluntary intake (Houpt, 1990). Horses are also very able selectors of feedstuffs, isolating, picking out and consuming or rejecting very small particles (both in size and amount) from a total compound mix.

When offering horses single mixtures of wet sugar-beet pulp plus distillery by-products (wheat or maize or barley distiller's grains and malt residual culms) abnormal patterns of ingestive behaviour were recorded (Hill and Braithwaite, 1999), similar to those shown by Ellis *et al.,* (2000) (Figure 8). The rate of ingestion of molassed sugar-beet pulp alone was fastest while barley distiller's grains were mostly rejected. These were short term studies to assess neophagia and palatability without acclimatisation. In a more longitudinal design, Moore-Coyler and Longland (2000) observed a significant decrease in total voluntary dry matter intake (39 g/kg BW0.75) of wet clamp silage compared to dry hay-silage or hay (79 and 63 g DM/kg BW0.75 respectively). The clamp silage was offered over a continued period of time but adaptation of intake rates did not take place. However, Müller *et al.* (2007) found that 4 horses previously accustomed to wet haylage, but not silage, significantly and repeatedly preferred silage in a cafeteria style study. Horses were offered 4 different forms of conserved forage, made from the same crop of grass into either hay (88% DM), dry haylage (68% DM; pH 5.8), wet haylage (57% DM; pH 5.6) or silage (31% DM; pH 4.9). Particle length, fermentation characteristics but also previous experience and plant species may have affected preference (Müller *et al.*, 2007).

Preferences due to flavours have been studied to a limited extent in horses. Randall (1978) offered foals sweet (sucrose), sour (acetic acid), salty (NaCl) and bitter (quinine) solutions at various concentrations. Preference for a sweet solution, containing between 1.25 and 10 g sucrose per litre was clearly established, while concentrations above or below this were treated indifferently. Higher concentrations of the other solutions were rejected. Distillery by products were also offered to horses at varying levels of inclusion (0, 0.25, 0.50, 0.75 and 1.00) to assess palatability, showing a decline in intake rates proportional to increasing inclusion (Hill and Braithwaite, 1999).

Goodwin *et al.* (2005) looked at a large variety of flavours, by adding these to a cereal by-product mix. From an original twelve flavours, banana, carrot, cherry, cumin, fenugreek, oregano, peppermint and rosemary were designated as being well 'accepted' by horses. Individuality between the 8 test horses was highlighted when one rejected coriander completely and three rejected/partially rejected nutmeg and Echinacea. The well accepted flavours were then tested in 2-flavour choice preference tests. Results showed the following ranking for preference: (1) Fenugreek (2) Banana (3) Cherry (4) Rosemary (5) Cumin (6) Carrot (7) Peppermint (8) Oregano. Only two exceptions to this ranking occurred: when paired with peppermint, fenugreek 'lost' although overall peppermint was quite low in ranking (against human perceptions perhaps). In addition carrot was also preferred when paired with rosemary (Goodwin *et al.*, 2005). Incorporating the two most preferred flavours (fenugreek and banana) into mineral pellets, led to a significant increase in intake speed (decrease in consumption time) during a field test.

Foraging and food related behaviours

Foraging, geophagia (eating of soil or sand), wood chewing and coprophagy may be a normal part of the repertoire of feed intake behaviour in the horse, even if not always perceived as such. These behaviours are seen in wild and free-ranging horses and fulfil important functions, much speculated upon, although generally little understood. Overall there is a belief that these are motivated by discomfort in the digestive tract or through motivations triggered via the CNS due to a lack of certain nutrients and/or a lack of fibre. The physiological; metabolic pathways behind this are largely unknown. In domesticated horses crib-biting and windsucking as well as tongue flicking and other oral stereotypies are perceived as 'abnormal' and 'unwelcome' coping mechanisms and these are rarely seen in un-domesticated horses. Their link to food intake behaviour is well established (Cooper *et al.*, 2005).

Foraging, as we have seen in free-ranging horses forms an important part of feed intake activity. In stabled horses the supply of concentrate feed and conserved forages considerably reduces time spent on feed intake behaviour (e.g. 500 kg horse: 8 kg hay and 2 kg cereal mix = max 6 hours feed intake versus on pasture – free ranging = 10 -12 hours feed intake). Ellis *et al.* (2006) found that 18 Warmblood horses kept on a low forage diet spent a significant amount of time on foraging in bedding and many developed regular coprophagic behaviour. Both groups (high and low forage) spent additional time on foraging behaviour in the stable, which balanced out to a very similar overall daily time spent on food intake and related behaviour (Figure 11).

Goodwin *et al.* (2002) and Thorne *et al.* (2005) showed that horses spent more time on foraging behaviour when offered a multiple choice of forages. Goodwin *et al.* (2007) expanded this study by offering horses a choice between a stable with a single forage and a stable with multiple forages. Again horses after initial inspection of both areas preferred the multiple foraging area. Mills *et al.* (2000) reported that horses, which had a choice of 2 bedding areas (straw/woodshavings/paper) spent 11% and 15% (±1.8) of time on bedding related behaviour on straw while only spending 0.98 and 1.33% of time on bedding related behaviour on shavings and paper respectively (when straw was the second choice). Bedding related behaviour dropped to 1.65% of time on paper and 1.19% of time on shavings when these two beddings were the only choice. Mills *et al.* (2000) concluded that his study demonstrated that straw bedding allows the expression of a wider number of motivationally significant activities inherent to the horse.

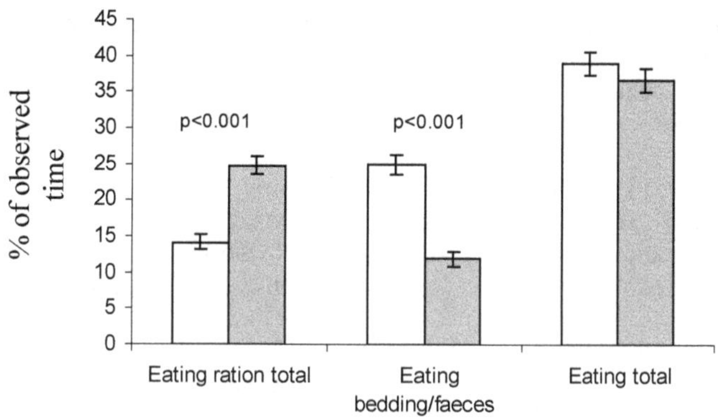

Figure 11. Time budgets (% of observed time (8:00-16:00, including morning and lunch feed only) for feed intake behaviour of horses on a low fibre (LF, white bars) and high fibre (HF, grey bars) diet (Ellis et al.*, 2006).*

Conclusion and recommendations: feed intake and welfare

Inherent feed intake and related motivationally significant activities, as highlighted in the first chapter, are underlined by a strong purpose, the physiological mechanisms which regulate feed intake and related behaviours and individual adaptations and experiences. Inability to carry out the behaviour of foraging and chewing at a required rate leads to excessive feed intake, bedding eating, faeces eating and has been strongly implicated in the development and increased performance of stereotypic behaviours. In addition the horses' feed intake behaviour does not seem to adjust in order to guard against possible negative impacts on digestive and metabolic processes from feed which are rich in sugars and starches, whether these be high quality grassland, chopped and molassed forages or concentrate feed. This highlights the importance in regulating feed intake of horses on lush grassland or with possible predisposition for laminitis. Our understanding of the regulation of equine feed intake behaviour in relation to biochemical and metabolic processes is still limited but strongly supported by knowledge gained from behavioural and observational studies. The inherent importance of performing species specific feed intake behaviour must be seen as a key point in safeguarding the welfare of horses. It is therefore suggested that structural fibre requirements should be included in all feeding systems for horses (instead or as well as average DM intake/kg BW).

References

Argo, C.M., J.E. Cox, C. Lockyer and Z. Fuller, 2002. Adaptive changes in the appetite, growth and feeding behaviour of pony mares offered ad libitum access to a complete diet in either a pelleted or chaff-based form. Animal Science 74, 517-528

Arnold, G.W., 1985. Comparison of the time budgets and circadian patterns of maintenance activities in sheep, cattle and horses grouped together, Applied Animal Behaviour Science 13,19-30.

Barnard, C., 2004. Animal Behaviour, Pearson Education Limited, Harlow, United Kingdom.

Berger, A., K. Scheibe, K. Eichhorn, A. Scheibe and J. Streich, 1999. Diurnal and ultradian rhythms of behaviour in a mare group of Przewalski horse (*Equus ferus przewalskii*), measured through one year under semi-reserve conditions, Applied Animal Behaviour Science, 64, 1-17.

Bergero, D., P.G. Peiretti and E. Cola, 2002. Intake and apparent digestibility of perennial ryegrass haylages fed to ponies either at maintenance or at work, Livestock Production Science, 77, 325-329.

Bonin, S.J., H.M. Clayton, J.L. Lanovaz and T. Johnston, 2007. Comparison of mandibular motion in horses chewing hay and pellets, Equine Veterinary Journal 39, 258-262

Bottom, S.H., H. Owen, R.E. Lawson, P.A. Harris and S. Hall, 2004. Equine feeding side-preference incidence and age effect. In: Alliston J., M. Moore-Couler, A. Hemmings and J. Hyslop (eds.) Emerging Equine Science, BSAS Publication No.32, Nottingham University Press, United Kingdom, p45.

Boyd, L., 1998. The 24-h time budget of a takh harem stallion (*Equus ferus przewalskii*) pre- and post-reintroduction, Applied Animal Behaviour Science, 60, 291-299.

Boyd, L. and N. Bandi, 2002. Reintroduction of takhi, Equus ferus przewalskii, to Hustai National Park, Mongolia: time budget and synchrony of activity pre- and post-release, Applied Animal Behaviour Science, 78, 87-102.

Boyd, L.E., 1988a. Time budgets of adult Przewalski horses: Effects of sex, reproductive status and enclosure, Applied Animal Behaviour Science, 21, 19-39.

Boyd, L.E., 1988b. Ontogeny of behaviour in Przewalski horses, Applied Animal Behaviour Science, 21, 41-69.

Boyd, L.E., D.A. Carbonaro and K.A. Houpt, 1988. The 24-hour time budget of Przewalski horses, Applied Animal Behaviour Science, 21, 5-17.

Carlson, N.R., 2007 Physiology of Behaviour, 7th ed., Pearson Allyn and Bacon, London, United Kingdom.

Carson, K. and D.G.M. Wood-Gush, 1983. Equine behaviour: II. A review of the literature on feeding, eliminative and resting behaviour. Applied Animal Ethology 10, 179-190.

Cartmill, J.A., D.L. Thompson, L.R. Gentry, H.E. Pruett and C.A. Johnson, 2003. Effects of dexamethasone, glucose infusion, adrenocorticotropin, and propylthiouracil on plasma leptin concentrations in horses, Domestic animal endocrinology, 24, 1-14.

Chambers, A.R.M., J. Hodgson and J.A. Milne, 1981. The development and use of equipment for the automatic recording of ingestive behaviour in sheep and cattle, Grass and Forage Science, 36, 97-105.

Chenost, M. and C. Demarquilly, 1985. Measurement of Herbage Intake by Housed Animals, In: Leaver, J.D. (1985) Herbage Intake Handbook, 2nd ed., British Grassland Society, Berkshire, United Kingdom.

Chenost, M. and W. Martin-Rosset, 1985. Comparaison entre especes (mouton, cheval, bovin) de la digestibilite et des quantites ingerees des fourrages verts, Annales de Zootechnique, 34 (3), 291-312.

Cooper, J.J., N. Mcall, S. Johnson and H.P.B. Davidson, 2005. The short-term effects of increasing meal frequency on stereotypic behaviour of stabled horses, Applied Animal Behaviour Science, 90, 351-364.

Crowell-Davis, S.L., K.A. Houpt and J. Carnevale, 1985. Feeding and Drinking Behavior of Mares and Foals with Free Access to Pasture and Water, Journal of animal science, 60, 883-889.

Cuddeford, D. and J.J. Hyslop, 1996. Intake and digestibility of a high fibre concentrate offered ad libitum to ponies and donkies, Proceedings of the 47th Annual Meeting of the EAAP, Lillehamer, Wageningen Pers, Wageningen, the Netherlands, p 296.

Doreau, M., 1978. Comportement alimentaire du cheval a l'ecurie. Ann. Zootech. 27, 291-302.

Doreau, M., W. Martin-Rosset and M. Petit, 1980. Activites alimentaires nocturnes du cheval au paturage. Ann. Zootech. 29, 299-304.

Dulphy, J.P., W. Martin-Rosset, H. Dubroeucq and M. Jailler, 1997a. Evaluation of voluntary intake of forage trough-fed to light horses. Comparison with sheep. Factors of variation and prediction, Livestock Production Science, 52, 97-104.

Dulphy. J.P., W. Martin-Rosset, H. Dubroeucq, J.M. Ballet, A. Detour and M. Jailler, 1997b. Compared feeding patterns in ad libitum intake of dry forages by horses and sheep, Livestock Production Science 52, 49-56.

Duncan P., 1980. Time-budgets of Camargue horses. Behaviour 72: 26-48.

Edouard, N., G. Fleurance, B. Dumont, R. Baumont and P. Duncan, 2009. Does sward height affect feeding patch choice and voluntary intake in horses?, Applied Animal Behaviour Science, 119, 219-228.

Ellis, A.D. and J. Hill, 2000. Intake Behaviour of Horses when offered short chopped high temperature dried lucerne and short chopped lucerne silage, Measuring Behaviour, Nijmengen, the Netherlands. p. 55.

Ellis, A.D., K. Fagence, E.M. Warr and J. Hill, 2000. Effect of dental condition on short term intake rates and faecal particle size in adult horses, Proceedings of the 51st European Association of Animal Science, Den Haag, the Netherlands, p. 375.

Ellis, A.D., E. Warr and J. Hill, 2001. Kinetics of forage feed particles in horses before and after dental treatment, Proceedings of the 52nd European Association of Animal Science, Budapest, Hungary, p. 339.

Ellis, A.D. 2003. Ingestive and Digestive Processes in the Horse, PhD Thesis, Writtle College, Essex University, United Kingdom.

Ellis, A.D., S. Thomas, K. Arkell and P. Harris, 2005. Adding chopped straw to concentrate feed: The effect of inclusion rate and particle length on intake behaviour of horses, Equine nutrition conference Hannover 2005, Pferdeheilkunde 21. 35-36.

Ellis, A.D., C.K. Visser and C.G. Van Reenen, 2006. Effect of a high concentrate versus high fibre diet on behaviour and welfare in horses, Proceedings of the 40th International Congress of the ISAE, Cranfield University Press, University of Bristol, United Kingdom, p. 42.

Fleurance, G., P. Duncan, H. Fritz, J. Cabaret, J. Cortet and I.J. Gordon, 2007. Selection of feeding sites by horses at pasture: testing the antiparasite theory. Appl. Anim. Behav. Sci. 108, 288-301.

Fleurance, G., H. Fritz, P. Duncan, I.J. Gordon, N. Edouard and C. Vial, 2009. Instantaneous intake rate in horses of different body sizes: influence of sward biomass and fibrousness. Appl. Anim. Behav. Sci. 117, 84-92.

Goodwin, D., H.P.B. Davidson and P. Harris, 2002. Foraging enrichment for stabled horses, Effects on behaviour and selection, Equine Veterinary Journal, 34, 686-691.

Goodwin, D., H.P.B. Davidson and P. Harris, 2005. Selection and acceptance of flavours inconcentrate diets for stabled horses. Applied Animal Behaviour Science 95, 223-232.

Goodwin, D., H.P.B. Davidson and P. Harris, 2007. Responses of horses offered a choice between stables containing single or multiple forages, Veterinary Record, 160, 548-551.

Gordon, M.E. and K.H. McKeever, 2005. Diurnal variation of ghrelin, leptin, and adiponectin in Standardbred mares, Journal of Animal Science, 83, 2365-2371.

Gordon, M.E., K.H. McKeever, C.L. Betros and H.C. Manso Filho, 2007a. Exercise-induced alterations in plasma concentrations of ghrelin, adiponectin, leptin, glucose, insulin, and cortisol in horses, The Veterinary Journal, 173, 532-540.

Gordon, M.E., K.H. McKeever, C.L. Betros and H.C. Manso Filho, 2007b. Plasma leptin, ghrelin and adiponectin concentrations in young fit racehorses versus mature unfit standardbreds, The Veterinary Journal, 173, 91-100.

Gordon, M.E., K.H. McKeever, S. Bokman, C.L. Betros, H. Manso-Filho, N. Liburt and J. Streltsova, 2006, Interval exercise alters feed intake as well as leptin and ghrelin concentrations in standardbred mares, Equine Vet J Suppl, 36, 596-605.

Haccou, P. and E. Meelis, 1995. Statistical Analysis of Behavioural Data, Oxford University Press, Oxford, United Kingdom.

Harris, P.A., M. Sillence, R. Inglis, C. Siever-Kelly, M. Friend, K. Munn and H. Davidson, 2005. Effect of short (<2 cm) lucerne chaff addition on the intake rate and glycaemic response of a sweet feed. Equine nutrition conference. Hannover 2005, Pferdeheilkunde 21, 87-88.

Henneke, D.R. and J.W. Callaham, 2009. Ad Libitum Concentrate Intake in Horses, Journal of Equine Veterinary Science, 29, 425-427.

Hill, J. and A.R. Braithwaite, 1999. Ingestive behaviour of horses offered distillery by-products, Proceedings of the British Society of Animal Science, BSAS, United Kingdom, p. 144.

Houbiers, H.J. and E.A. Smolders, 1990. Opname van vers gras van verschillende opbrengsten (Intake of fresh grass from various harvests). In: Praktijkonderzoek (1990) Paardenonderzoek 1990 (Equine Research 1990) Proefstation voor de Rundveehouderij, Schapenhouderij and Paardenhouderij (PR), Lelystad, The Netherlands.

Houpt, K.A., 1990. Ingestive Behaviour, Veterinary Clinics of North America: Equine Practice, 6, 319-337.

Jose-Cunilleras, E., L.E. Taylor and K.W. Hinchcliff, 2004, Glycemic index of cracked corn, oat groats and rolled barley in horses, Journal of Animal Science, 82, 2623-2629.

Kern, D.L. and J. Bond, 1972. Eating patterns of ponies fed diets ad libitum, Journal of Animal Science, 35, 285.

Lengwenar, O., U. Bartels, P. Fawick and A. Zeyner, 1999. Heu, Heulage, Briketts, Cobs: Das Verzehrverhalten des Pferdes im Vergleich (Hay, Haylage, Bricketts, Cobs: Intake behaviour of horses for comparison), Gottinger Pferdetage Zucht und Haltung von Sportpferden, Internationale Fachtagung, pp. 169-176.

López, M., R. Lage, A.K. Saha, D. Pérez-Tilve, M.J. Vázquez, L. Varela, S. Sangiao-Alvarellos, S. Tovar, K. Raghay, S. Rodríguez-Cuenca, R.M. Deoliveira, T. Castañeda, R. Datta, J.Z. Dong, M. Culler, M.W. Sleeman, C.V. Álvarez, R. Gallego, C.J. Lelliott, D. Carling, M.H. Tschöp, C. Diéguez and A. Vidal-Puig, 2008, Hypothalamic Fatty Acid Metabolism Mediates the Orexigenic Action of Ghrelin, Cell Metabolism, 7, 389-399.

Longland, A.C and B.M. Byrd, 2006. Pasture Nonstructural Carbohydrates and Equine Laminitis1-3, J. Nutr. 136: 2099S-2102S.

Magnusson J. and A.G. Thorhallsdottir, 1994. Horse grazing in northern Iceland – behaviour and habitat selection. Livestock Production Science 40, 77-86.

Martin-Rosset, W. and J.P. Dulphy, 1987. Digestibility interactions between forages and concentrates in horses: Influence of feeding level – Comparison with sheep, Livestock Production Science, 17, 263-276.

Martin-Rosset, W., M. Doreau and J. Cloix, 1978. Etude des activids d'un troupeau de poulinieres de trait et de leurs poulains au paturage. Ann. Zootech. 27, 33-45.

McBride, S.D. and A. Hemmings, 2005. Altered mesoaccumbens and nigro-striatal dopamine physiology is associated with stereotypy development in a non-rodent species, Behavioural brain research, 159, 113-118.

McGreevy, P. (ed.), 2004. Equine Behaviour, A guide for Veterinarians and Scientists, Saunders Limited, London, United Kingdom.

Meyer, H., L. Ahlswede and H.J. Reinhardt, 1975. Untersuchungen uber Fressdauer, Kaufrequenz und Futterzerkleinerung beim Pferd 'Studies on the duration of feeding, masticatory frequency and communition of feed in horses' Deutsche Tierartztliche Wochenschrift, 82 (2), 54-58.

Meyer, H., M. Coenen, H. Teleb and D. Probst, 1986. Beiträge zur Verdauungsphysiologie des Pferdes, 15.Untersuchungen über Futterzerkleinerung von Futterinhaltsstoffen im Kopfdarm des Pferdes (Studies on communition of feed contents in the mouth of the horse) Journal of Animal Physiology and Animal Nutrition, 56, 266-275.

Mills D.S., S. Eckley and J.J. Cooper, 2000. Thoroughbred bedding preferences, associated behaviour differences and their implications for equine welfare. Journal of Animal Science 70: 95-106.

Moore-Colyer, M.J.S. and A.C. Longland, 2000. Intakes and *in vivo* apparent digestibilities of four types of conserved grass forage by ponies, Animal Science, 71, 527-534.

Müller, C.E. and P. Udén, 2007. Preference of horses for grass conserved as hay, haylage or silage. Animal Feed Science and Technology 132, 66-78.

Naujeck, A. and J. Hill, 2003. Influence of sward height on bite dimensions of horses. Anim. Sci. 77, 95-100.

Pearson, R.A., R.F. Archibald and R.H. Muirhead, 2001. Comparison of the effect of forage type and level of feeding on the digestibility and gastrointestinal mean retention time of dry forages given to cattle, sheep, ponies and donkeys. British Journal of Nutrition 95: 88-98.

Pearson, R.A. and J.B. Merrit, 1991. Intake, digestion and gastrointestinal transit time in resting donkies and ponies and exercised donkeys given ad libitum hay and straw diets, Equine Veterinary Journal, 23, 339-343.

Ralston, S.L., 1992. Regulation of feed intake in the horse in relation to gastrointestinal disease, Pferdeheilkunde, 15-18.

Ralston, S.L., D.L. Foster, T. Divers and F. Hintz, 2001. Effect of dental correction on feed digestibility in horses, Equine Veterinary Journal, 33, 390-393.

Ralston, S.L. and C. A. Baile, 1982. Gastrointestinal stimuli in the control of feed intake in ponies. J. Animal Sci. 55: 243.

Ralston, S.L., 1984. Controls of Feeding in Horses, Journal of Animal Science, 59, 1354-1361.

Randall, R.P., D.C. Schurg and D.C. Church, 1978. Responses of horses to sweet, salty, sour and bitter solutions, Journal of Animal Science, 47, 51-55.

Rutter, S.M., 2000. Graze: A program to analyse recordings of the jaw movements of ruminants, Behaviour Research Methods, Instruments and Computers, 32, 86-92.

Sakurai, T., A. Amemiya, M. Ishii, I. Matsuzaki, R.M. Chemelli, H. Tanaka, S.C. Williams, J.A. Richardson, G.P. Kozlowski, S. Wilson, J.R. Arch, R.E. Buckingham, A.C. Haynes, S.A. Carr, R.S. Annan, D.E. McNulty, W.S. Liu, J.A. Terrett, N.A. Elshourbagy, D.J. Bergsma and M. Yanagisawa, 1998. Orexins and orexin receptors: a family of hypothalamic neuropeptides and G protein-coupled receptors that regulate feeding behavior, Cell 92 pp. 573-585

Smith, D.G., R.W. Mayes, T. Hollands, D. Cuddeford, H.H. Yule, C.M. Malo Ladreiro and E. Gillen, 2007. Validating the alkane pair technique to estimate dry matter intake in equids, The Journal of Agricultural Science, 145, 273-281.

Souris, A., P. Kaczensky, R. Julliard and C. Walzer, C. 2007. Time budget-, behavioral synchrony- and body score development of a newly released Przewalski's horse group Equus ferus przewalskii, in the Great Gobi B strictly protected area in SW Mongolia. Applied Animal Behaviour Science, 107, 307-21.

Sweeting, M.P., C.E. Houpt and K.A. Houpt, 1985. Social facilitation of feeding and time budgets in stabled ponies, Journal of Animal Science, 60, 369-374.

Tinbergen, N., 1963. On Aims and Methods in Ethology, Zeitschrift für Tierpsychologie, 20: 410-433. cited In: Barnard, C., 2004. Animal Behaviour, Pearson Education Limited, Harlow, United Kingdom.

Thorne, J.B., D. Goodwin, M.J. Kennedy, H.P.B. Davidson and P. Harris, 2005. Foraging enrichment for individually housed horses: Practicality and effects on behaviour. Applied Animal Behaviour Science, 94(1-2), 149-164.

Treiber, K.H., R.C. Boston, D.S. Kronfeld, W.B. Staniar and P.A. Harris, 2005. Insulin resistance and compensation in Thoroughbred weanlings adapted to high-glycemic meals, Journal of animal science, 83, 2357-2364.

Van Dierendonck M.C., N. Bandi, D. Batdorj, S. Dugerlham and B. Munkhtsog, 1996. Behavioural observations of reintroduced Takhi or Przewalski horses (Equus Ferus Przewalskii) in Mongolia. Applied Animal Behaviour Science 50 (2): 95-114.

Vermorel, M., J. Vernet and W. Martin-Rosset, 1997. Digestive and energy utilisation of two diets by ponies and horses, Livestock Production Science, 51, 13-19.

Vernet, J., M. Vermorel and W. Martin-Rosset, 1995. Energy cost of eating long hay, straw and pelleted food in sport horses. J. Anim. Sci. 61, 581-588.

Vulink J.T., P. Cornelissen, H.J. Drost and H.H.T. Prins, 2001. Hindgut fermentation is not an evolutionary dead end: comparative feeding ecology of cattle and horses In: Vulink, J.T. (ed.), Hungry Herds: management of temperate lowland wetlands by grazing. Proefschrift Rijksuniversiteit Groningen. PhD Thesis, Ministerie van Verkeer and Waterstaat, Lelystad, the Netherlands.

Winrow, C.J., K.Q. Tanis, D.R. Reiss, A.M. Rigby, J.M. Uslaner, V.N. Uebele, S.M. Doran, S.V. Fox, S.L. Garson, A.L. Gotter, D.M. Levine, A.J. Roecker, P.J. Coleman, K.S. Koblan and J.J. Renger, 2010. Orexin receptor antagonism prevents transcriptional and behavioral plasticity resulting from stimulant exposure, Neuropharmacology, 58, 185-194.

Zeyner, A., 1999. Moglichkeiten des Einsatzes von Fetten bei der Pferdefutterung (Possibilities for the use of fats in equine nutrition) Der praktische Tierarzt, Collegium Veterinarium XXIX, pp. 25-30.

Zeyner, A., C. Hoffmeister, A. Einpanier, J. Gottschalk, O. Lengwenat and M. Illies, 2006. Glycaemic and insulinaemic response of Quarter Horses to concentrates high in fat and low in soluble carbohydrates. Equine Exercise Physiology 7. Equine Vet. J., Suppl. 36, 643-647.

Electromyographic evaluation of masseter muscle activity in horses fed different types of roughage

I. Vervuert[1], N. Brüssow[2], M. Bochnia[1], T. Hollands[3], D. Cuddeford[4] and M. Coenen[1]

[1]Institute of Animal Nutrition, Nutrition Diseases and Dietetics, Faculty of Veterinary Medicine, University of Leipzig, An den Tierkliniken 19, 04103 Leipzig, Germany

[2]Institute of Animal Nutrition, University of Veterinary Medicine, Foundation, Bischofsholer Damm 15,30173 Hannover, Germany

[3]Dodson & Horrel Limited, Old Furnace Site Kettering Road, Islip, Kettering NN14 3JW, Ringstead, United Kingdom

4Department of Veterinary Studies, University of Edinburgh, Bush Farm, Roslin EH2S5 9RG, United Kingdom

Abstract

The aim of the study was to obtain information about masseter muscle activity, the main determinant of salivary flow during the chewing process. Four horses were offered the following diets: cracked corn (dry matter intake (DMI): 0.26% of BW), hay *ad libitum* (DMI: 2.8±0.5% of BW); haylage *ad libitum* (DMI: 2.4±0.4% of BW) and a straw/alfalfa chaff ([SAC] (DMI: 3.1±0.5% of BW). Feed intake time (min/kg) and chewing frequency (chews/kg) were recorded by direct observation or by using a modified halter. The activity of the masseter muscle was determined by EMG (IED®) and the following measurements made: Amplitude (muscle action potential = MAP, expressed in V) and duration of MAP (expressed in sec). Feed intake time and chewing frequency for roughages were different compared to cracked corn in the following order: haylage > hay > SAC > cracked corn. The intake of hay or haylage was associated with intense masseter muscle activity (MAP: hay 9.82±1.7 V, haylage 11.4±3.3 V, duration MAP: hay 0.31±0.04 sec, haylage 0.30±0.04 sec). Similar intense chewing was measured for the SAC (MAP 12.6±3.8 V), although duration of the chewing cycle was relatively short (0.22±0.03 sec) which is possibly related to limited fibre length. In contrast to roughages, concentrates are consumed rapidly, with less intense masseter muscle activity as reflected by the low amplitude of EMG (MAP 4.9±1.5 V). This is associated with low salivary flow rates that are likely to negatively affect gastric digestion.

Keywords: chewing behaviour, masseter muscle, saliva, feed intake

Introduction

The feed intake time and chewing frequency for different feedstuffs is well described in horses, suggesting a high speed intake rate for concentrates, and a more prolonged consumption time for forages (Meyer and Coenen, 2002). The speed of intake varies between ~10 min/kg for concentrates and ~45 min/kg for roughage. In addition, the intake of roughage is linked with the highest saliva production. A lack of roughage in the ration causes several health disturbances like gastric ulcers, hindgut acidosis, and behavioural problems. In general, ration formulation should always consider at least a minimum of roughage intake. The German recommendations which are actually under revision will state a daily minimum of 1.5 kg 'roughage' (based on DM)/100 kg BW. In order to offset the problems associated with the rapid consumption of concentrate it is common practice to recommend that chopped roughage is mixed with the concentrate or, that roughage be fed before providing the concentrate (Brüssow *et al.*, 2005, Bochnia *et al.*, 2008).

However, information about the masseter muscle activity as the main determinant of salivary flow rate during chewing process is lacking. In an ongoing experiment, electromyographic (EMG) evaluation of masseter muscle activity was investigated in healthy horses fed different types of roughages.

Material and methods

Four horses (mean age: 9.8±0.5 years, mean BW: 563±47 kg) were offered the following diets: hay ad libitum (hay: DM 891 g/kg, crude fiber: 345 g/kg DM); haylage *ad libitum* (haylage: DM 745 g/kg, crude fiber: 384 g/kg DM); straw alfalfa chaff (SAC: DM 878 g/kg, crude fiber: 257 g/kg DM, particle length: >3.15 cm: 42%, >2.0 cm: 21%, <2 cm: 37.1%). In a second trial, four horses (age 4 years, mean BW: 448±41 kg) were used to fed hay *ad libitum* (hay: DM 886 g/kg, crude fiber: 352 g/kg DM) and cracked corn (DMI: 0.26% of BW). Each diet was fed for 11-21 days. Feed intake time and chewing frequency were recorded daily by manual counting or by a modified halter. The activity of the masseter muscle was determined by EMG (IED®) once in each horse per feeding trial. The following parameters were obtained by EMG: Amplitude (muscle action potential = MAP, expressed in V) and duration of MAP (expressed in sec).

Results

Dry matter intake varied between 2.4% of BW for feeding haylage ad libitum and 3.1% of BW for feeding SAC *ad libitum* (Table 1).

Consumption time and chewing activity for roughages are different in comparison to the intake of cracked corn in the following order: haylage > hay > SAC > cracked corn (Table 2). The intake of SAC revealed high MAP amplitudes accompanied by a short duration cycle of MAPs, whereas the consumption of hay and haylage was linked with high MAP amplitudes and a prolongation of chewing cycle (Table 2).

Discussion

The intake of hay or haylage is associated with an intensive masseter muscle activity which is supposed to stimulate salivary flow rate. A similar intensification of the chewing process is supposed for the SAC although the chewing frequency for SAC is relatively low which is possibly related to the short fibre length (particle length: >3.15 cm: 42%, >2.0 cm: 21%, <2 cm: 37.1%) or due to the lower crude fiber intake (crude fiber: 257 g/kg DM). However, similar results were obtained by

Table 1. Mean (±SD) maximum dry matter intake (DMI, % of BW) for feeding hay, haylage or SAC ad libitum *in horses under maintenance conditions.*

	Hay *ad libitum*	Haylage *ad libitum*	SAC *ad libitum*
DMI (% of BW)	2.8±0.5	2.4±0.4	3.1±0.5

Table 2. Feed intake time (min/kg DM), chewing frequency (chews/kg DM), amplitude MAP (V) and duration MAP (sec) for the different feedstuffs (means ± SD).

Diet	DMI (min/kg DM)	Chews/ kg DM	Amplitude MAP (V)	Duration MAP (sec)
Cracked corn	18.2±6.21^{a}	1,184±278^{a}	4.9±1.5^{a}	0.23±0.02^{a}
Hay *ad libitum*	39.3±15.3^{b}	2,661±600^{b}	9.82±1.74^{b}	0.31±0.04^{b}
Haylage *ad libitum*	52.0±10.4^{b}	3,378±572^{b}	11.4±3.26^{b}	0.30±0.04^{b}
SAC *ad libitum*	23.5±4.18^{c}	2,038±325^{c}	12.6±3.82^{b}	0.22±0.03^{a}

Means in the same column with unlike superscripts are different with $P<0.05$.

feeding alfalfa chaff with a fiber length of ~40% below 2 cm, but a fiber content of 335 g/kg DM (Brüssow *et al.*, 2005) suggesting the importance of the particle size of roughages. In contrast to roughage, concentrates like cracked corn are consumed rapidly, with a less intensive masseter muscle activity as reflected by the low amplitude of EMG. This condition is associated to a low saliva flow rate, likely to impair gastric digestion.

In conclusion, a minimum of roughage intake as well as the particle size of the different roughages have to be considered in adequate diet formulation for horses.

References

Bochnia, M., I. Vervuert and M. Coenen, 2008. The impact of different feeding orders of hay chaff combined with a concentrate meal on feed intake pattern in horses. Proc. Soc. Nutr. Physiol. Band 17, 91.

Brüssow, N., K. Voigt, I. Vervuert, T. Hollands, D. Cuddeford and M. Coenen, 2005. The effect of the order of feeding oats and chopped alfalfa to horses on the rate of feed intake and chewing activity. Pferdeheilkunde, 21, 37-38.

Meyer, H. and M. Coenen, 2002. Pferdefütterung. Parey Verlag, Berlin, Germany.

Increasing foraging opportunities improves welfare and reproduction efficiency in Arab breeding mares

H. Benhajali[1], M.-A. Richard-Yris[1], M. Ezzaouia[2], F. Charfi[3] and M. Hausberger[1]
[1]Université de Rennes, UMR CNRS 6552, Ethologie animale et humaine, 35042 Rennes, France
[2]FNARC, Haras National de Sidi Thabet, 2020 Sid Thabet,Tunisia
[3]Département de Biologie animale, 1068 Tunis, Tunisia

Abstract

One hundred Arab mares were randomly divided into 2 groups: the experimental and the control groups and separated during the day in 2 bare paddocks at the same density. Hay was provided in haynets only to the experimental mares. Focal and scan samplings were used to determine the time budget of the mares during the period from 9 am to 3 pm and study their social behaviour. Mares were tested for oestrous detection by teasing with a stallion and were then examined clinically by rectal palpations and ultrasound before being mated naturally or inseminated. ANOVA, GLM and non parametric tests were used to analyse data. Experimental mares' time budgets included more time spent feeding ($P<0.01$), less time spent in locomotion ($P<0.01$), stand resting ($P<0.01$) and alert standing ($P<0.01$). Experimental mares showed more positive social interactions ($P<0.01$) and less aggressions ($P<0.01$). The treatment affects significantly the conception rate of the mares (81% vs. 55%, $P<0.01$).

Keywords: mare, foraging opportunities, behaviour, welfare, reproduction

Introduction

Many animals such as horses, primates and pigs spend a large portion of their daily activity budget in the search for and consumption of food in their natural habitats (Herbers, 1981). In captivity, food is generally offered to these animals and foraging opportunities are often restricted. As a consequence, captive animals spend less time feeding than their free ranging counterparts. The reduction of the time spent foraging was associated with the emergence of stereotypies and abnormal behaviour in many species. This aspect is even more important in species such as horses which have evolved specific digestive physiology and anatomy around their natural diet and feeding behaviour of trickle feeding for long periods (Harris, 1999).

The present study aimed to study the effect of increasing foraging opportunities on behaviour, welfare and reproduction efficiency of breeding mares housed in individual boxes but allowed outside for 6 hours a day in a bare paddock.

Material and methods

The experiment was conducted between the 1st of April and 11th of June 2006 at the national breeding facility of Sidi Thabet, located 20 km from Tunis, Tunisia. Mares were housed in individual boxes at night where they received barley grain (4 kg per day) and hay (10 kg per day) every morning and evening. They were released every day from 9 am to 3 pm in a paddock where free access to water and limited shelter were provided. No food was available then, but some freshly cut grass was left on the ground around 2 pm every day. It was generally entirely consumed by the mares in the following hour.

We used 100 purebred Arab mares, aged 4 to 21 years ($\mu = 8.49 \pm 4.96$). They had been present at the stud for 1 week at least and 3 weeks at most. They were housed all the time in individual boxes before the experiment.

Mares were randomly divided into 2 groups: the experimental and the control groups and were separated during the day in 2 bare paddocks at the same density. In the experimental group (n=50), hay was provided in nylon haynets. These were filled with hay and hung up in the experimental paddock every morning before the arrival of the mares. In the control group (n=50), no hay was provided in the paddock. All mares were kept in individual boxes for the night, where hay was provided in the evening for the control mares, in order to maintain similar intakes in both groups. Twenty minute animal focal samplings and scan samplings were used to determine the time budget of the mares during the period from 9 am to 3 pm and study their social behaviour.

Mares were tested for oestrous detection by teasing with a stallion and were then examined clinically by rectal palpations and ultrasound before being mated naturally or inseminated.

Threehundred focal sampling (6000 min), 3,300 individual scan sampling (6000 min) and 62 group observations (1240 min) corresponding to 100 mares and equally balanced over the groups were recorded. ANOVA, GLM and non parametric tests were used to analyse data.

Results

Experimental mares' time budgets included more time spent feeding (65.12%±2.40% vs. 29.75%±2.45%, $P<0.01$), less time spent in locomotion (11.70±1.31% vs. 23.56±1.34%, $P<0.01$), stand resting (11.76±2.57% vs. 27.52±2.62%, $P<0.01$) and alert standing (5.23±1.2% vs. 14.71±1.23%, $P<0.01$). Experimental mares showed more positive social interactions ($P<0.01$) and less aggression ($P<0.01$). The treatment significantly affected the conception rate of the mares (81% vs. 55%, $P<0.01$).

Discussion

The increase of the time spent foraging in the experimental group can of course be explained by the availability of hay in the experimental paddock and may explain the observed decrease of the time spent stand resting as also observed by Duncan (1980) and Boyd and Bandi (2002). The lower time spent in locomotion and in alert standing in the experimental group could be a sign of a lowered level of stress in this group. Indeed, locomotion increases in stressful conditions (Houpt and Houpt, 1989) and vigilance behaviour such as alert standing may be an indicator of acute (Morgan and Tromborg, 2007) or chronic stress (Carlstead *et al.,* 1993) and are used to assess emotionality in horses (Wolff *et al.*, 1997).

If the basic assumption of a chronic stress response in the control group is valid, the effect of the hay-treatment on experimental mares' fertility is not unexpected. There is evidence that stress can affect both oestrous expression and the maintenance of pregnancy in several species (Dobson and Smith, 2000).

Conclusion

This study shows that the increase of foraging opportunities during the day affects the time budget, social behaviour and reproduction efficiency of mares. Behavioural changes generated by the increase of foraging opportunities seem to indicate a better welfare in the treated mares.

References

Barnett, J.L. and P.H. Hemsworth, 1991. The effects of individual and group housing on sexual behaviour and pregnancy in pigs. Anim. Reprod. Sci. 25, 265-273.

Boyd, L. and N. Bandi, 2002. Reintroduction of takhi, *Equus ferus przewalskii*, to Hustai National Park, Mongolia: time budget and synchrony of activity pre- and post-release. Appl. Anim. Behav. Sci. 78, 87-102.

Carlstead, K., J.L. Brown and W. Strawn, 1993. Behavioural and physiological correlates of stress in laboratory cats. Appl. Anim. Behav. Sci. 38, 143.158.

Duncan, P., 1980. Time-budgets of Camargue horses. II. Time-budgets of adult horses and weaned sub-adults. Behaviour 72, 26-49.

Dobson, H. and R.F. Smith, 2000. What is stress, and how does it affect reproduction? Anim. Reprod. Sci. 60, 743.52.

Harris, P.A., 1999. How understanding the digestive process can help minimise digestive disturbances due to diet and feeding practices. In: Harris, P.A., G. Gomarsall, H.P.B. Davidson and R. Green (eds.), Proceedings BEVA Foraging opportunity and horse welfare specialist meeting on Nutrition and Behaviour, pp. 45-50. Equine Veterinary Journal, Newmarket, United Kingdom.

Herbers, J.M., 1981. Time resources and laziness in animals. Oecologia 49, 252-262.

Houpt K.A. and T.R. Houpt, 1989. Social and illumination preferences of mares. Journal of Animal Science 66, 2159-2164.

Macfarlane, M.S., K.M. Breen, H. Sakurai, B.M. Adams and T.E. Adams, 2000. Effect of duration of infusion of stress-like concentrations of cortisol on follicular development and the preovulatory surge of LH in sheep. Anim. Reprod. Sci. 63, 167.175.

Mann, G.E., 2001. Pregnancy rates during experimentation in dairy cows. Vet. J. 161, 301.305.

Morgan, K.N. and C.T. Tromborg, 2007. Sources of stress in captivity. Applied Animal Behaviour Science 102, 262-302.

Pluske, J.R. and I.H. Williams, 1996. Split weaning increases the growth of light piglets during lactation. Austr. J. Agr. Res. 47, 513-523.

Wolff, A., M. Hausberger and N. Le Scolan, 1997. Experimental tests to assess emotionality in horses. Behavioural Processes 40, 209-221.

Influence of physical structure of haylage on equine eating behaviour

C.E. Müller
Department of Animal Nutrition and Management, Swedish University of Agricultural Sciences, 753 23 Uppsala, Sweden

Abstract

Forage related factors such as physical structure may influence eating time and eating behaviour of horses, which is important from an equine welfare perspective. The physical structure of the forage is influenced by a variety of factors. The effect of increasing plant maturity at harvest of haylage, and the effect of mechanical treatment of the grass at harvest (cut *vs.* long-stemmed forage) of haylage were investigated in two separate studies, where the influence on equine eating time and eating behaviour were studied. Feeding haylage harvested at an early plant maturity resulted in a shorter eating time compared to haylages harvested at later plant maturities. Feeding cut or long-stemmed haylage resulted in very small differences in equine eating time and behaviour. In conclusion, harvest at a later plant maturity may produce a forage that increases the eating time for the horse, but the hygienic (and nutritive) quality of the forage also needs to be considered.

Keywords: ingestion, behaviour, haylage, plant maturity

Introduction

As the horse has evolved as a specialized grass eater over millions of years (Janis, 1976), forage is the most important feed in the equine diet both from a behavioural and digestive point of view. Free-living horses spend 0.57 (Duncan, 1980) to 0.75 (Salter and Hudson, 1979) of their time feeding, and eating time is an important equine welfare issue. Feeding low amounts of forage results in short eating times, which may cause both development of stereotypic behaviour (McGreevy *et al.*, 1995) and digestive upsets (e.g. Archer and Proudman, 2006). Apart from the amount of forage, other forage-related factors that may influence eating behaviour and eating time in horses has not been studied to any large extent and needs to be investigated further (Cuddeford, 2005; Müller and Udén, 2007). A forage related factor that may affect eating time and eating behaviour of equines is the physical structure of the forage, which is influenced by a variety of factors including plant maturity and mechanical treatment of the grass at harvest. The influence of plant maturity at harvest on equine eating behaviour has not been studied. However, increased plant maturity at harvest has been reported to have a negative influence on conservation and hygienic quality of wrapped forages (Behrendt *et al.*, 1997; Müller, 2009a). In an attempt to improve the conservation of wrapped forages, the crop is sometimes chopped in conjunction with baling as chopping has been reported to improve lactic acid fermentation and bale density (Borreani and Tabacco, 2006). Information about the effect of this mechanical treatment of the crop on the ingestive behaviour of horses is scarce. With this background, two experiments were conducted; the first experiment compared the influence of cut and long-stemmed haylage on equine eating time and eating behaviour; and the second experiment compared the influence of increasing plant maturity at harvest of haylage on equine eating time and behaviour.

Material and methods

Experiment I: cut vs. long-stemmed haylage

Round bales of long-stemmed (0.5 of particles >30 cm) and cut (knife distance 7 cm, 0.85 of particles <30 cm) haylage was produced simultaneously from the primary growth of a grass sward consisting mainly of Timothy (*Phleum pratense*). The dry matter (DM) content of both haylages was 567 g/

kg, and the content of neutral detergent fibre (NDF) averaged 555 g/kg DM. The haylages were fed to ten adult horses in a cross-over study with two periods. Horses were observed during two of four daily meals, where eating time (min/kg DM), chewing rate (chews/min) and swallowing rate (swallows/min) were recorded. From this data, the number of chews/kg DM and the number of chews/swallow were calculated. Differences in eating behaviour between cut and long-stemmed haylage were analysed using SAS Mixed Models Procedure, where the repeated measurements on the same horse were accounted for in the statistical model (Müller, 2009b).

Experiment II: plant maturity at harvest

Round bales of haylage were produced from the primary growth of a grass-clover sward on three occasions; 8th June, 2nd July and 5th August in 2009. The proportion of red clover increased from 0.003 in June over 0.04 in July to 0.26 in August. The remaining proportions consisted of timothy and meadow fescue (*Festuca pratensis*). Haylage DM content was 549, 573 and 583 g/kg and NDF content was 522, 610 and 637 for June, July and August haylages respectively. The haylages were fed to twelve adult horses in a cross-over study with three periods. Horses were observed during one of four daily meals and behaviour was recorded as described in Experiment I. Differences in eating behaviour between June, July and August haylages were analysed using SAS Mixed Models Procedure, where the repeated measurements on the same horse were accounted for in the statistical model.

Results and discussion

Experiment I: cut vs. long-stemmed haylage

Eating time and eating behaviour for cut and long-stemmed haylage are presented in Table 1. Eating time did not differ between cut and long-stemmed haylage (Table 1). Horses were chewing (but not eating) cut haylage slightly faster than long-stemmed haylage. However, differences in chewing rate were also present among horses (P<0.001), displaying individual variation in this variable. The differences in chewing rate and number of chews/kg DM between the haylage types were small (Table 1), but may have a biological importance if multiplied by the amount of haylage fed daily. This study was reported in full by Müller (2009b).

Experiment II: plant maturity at harvest

Eating time and eating behaviour for primary growth haylages harvested in June, July and August are presented in Table 2. Horses were eating, chewing and swallowing the June haylage faster than

Table 1. Effect of forage type on eating behaviour in horses fed cut or long-stemmed haylage.

Variable	Cut haylage		Long-stemmed haylage		Level of significance (P)
	Mean	Std. dev.	Mean	Std. dev.	
Eating time, min/kg DM	28	3.9	30	6.2	NC
Chewing rate, chews/min	84	3.8	82	3.9	0.01
Swallowing rate, swallowings/min	1.9	0.35	1.8	0.45	NS
No. of chews/kg DM	2368	346.9	2441	454.9	<0.0001
No. of chews/swallowing	51	8.5	52	11.8	NS

NC: programme did not converge due to too small variation, NS: not significant.

Table 2. Effect of forage type on eating behaviour in horses fed haylage harvested in June, July and August.

Variable	June harvest		July harvest		August harvest		Level of significance (P)
	Mean	Std dev	Mean	Std dev	Mean	Std dev	
Eating time min,/kg DM	29^a	4.4	37^b	7.4	36^b	9.8	<0.0001
Chewing rate, chews/min	84^a	5.9	78^b	6.3	77^b	7.4	<0.0001
Swallowing rate, swallowings/min	1.9^a	0.46	1.3^b	0.48	1.2^b	0.43	<0.0001
No. of chews/kg DM	2472^a	388.0	2947^b	546.5	2969^b	660.1	<0.0001
No. of chews/swallowing	51^a	11.7	65^b	22.7	81^c	34.4	<0.0001

[a,b,c] Values with different superscript letters diverge at the P-level presented.

the July and August haylages. Also, the number of chews before the horse swallowed increased with increasing plant maturity, which was probably an effect of the increased proportion of stems and therefore fibre in the haylage. A later cut of the primary growth may produce a forage that increases the eating time for the horse, but the hygienic (and nutritive) quality of the forage also needs to be considered. The influence of plant maturity at harvest on the hygienic quality of wrapped forages is currently being studied.

Conclusions

Feeding cut or long-stemmed haylage resulted in very small differences in equine eating time and behaviour. Feeding primary growth haylage harvested in June resulted in a shorter eating time compared to feeding primary growth haylages harvested in July and August. Thus, eating time can be increased by increasing plant maturity at harvest of haylage, but the effect of this practice on hygienic quality also needs to be addressed.

References

Archer, D.C. and C.J. Proudman, 2006. Epidemiological clues to preventing colic. The Vet J172, 29-39.

Behrendt, U., T. Müller and W. Seyfarth, 1997. The influence of extensification in grassland management on the populations of microorganisms in the phyllosphere of grasses. Microbiol. Res. 152, 75-85.

Borreani, G. and E. Tabacco, 2006. The effect of a baler chopping system on fermentation losses of wrapped big bales of alfalfa. Agr. J 98, 1-7.

Cuddeford, D., 2005. Voluntary food intake by equids. Pferdeheilkunde 21, 7-8.

Duncan, P., 1980. Time-budgets of Camargue horses. II. Time-budgets of adult horses and weaned sub-adults. Behaviour 72, 26-49.

Janis, C., 1976. The evolutionary strategy of the equidae and the origins of rumen and cecal digestion. Evol. 30, 757-774.

McGreevy, P.D., P.J. Cripps, N.P. French, L.E. Green and C.J. Nicol, 1995. Management factors associated with stereotypic and redirected behaviour in the Thoroughbred horse. Equine Vet. J. 27 (2), 86-91.

Müller, C.E. and P. Udén, 2007. Preference of horses for grass conserved as hay, haylage or silage. Anim. Feed Sci. Technol. 132, 66-78.

Müller, C.E., 2009a. Influence of harvest date of primary growth on microbial flora of grass herbages and haylage, and on fermentation and aerobic stability of haylage conserved in laboratory silos. Grass Forage Sci. 64, 328-338.

Müller, C.E., 2009b. Long-stemmed vs. cut haylage in bales – Effects on fermentation, aerobic storage stability, equine eating behaviour and characteristics of equine faeces. Anim. Feed Sci. Technol. 152, 307-321.

Salter, R.E. and R.J. Hudson, 1979. Feeding ecology of feral horses in western Alberta. J Range Management 32, 221-225.

The effects of soaked and un-soaked high fibre pelleted feeds on chewing time to swallow in the horse and feed volume changes *in vitro*: a pilot study

C. Lawrence[1], N. Blackie[2], A.D. Ellis[3] and S. Tracey[2]
[1]Allen and Page, Norfolk Mill, Thetford, Norfolk, IP25 7SD, United Kingdom
[2]Writtle College, Lordship Road, Chelmsford, CM1 3RR, United Kingdom
[3]School of Animal Rural and Environmental Sciences, Nottingham Trent University, NG25 0QF, United Kingdom

Abstract

This study aimed to investigate the relationship of volume changes of high fibre pelleted feed *in vitro* and chewing rates *in vivo* with regard to potential oesophageal obstruction in the horse. Four horses were fed five high fibre pelleted feed daily for a one week adaptation period according to manufacturers recommendations (feed 1-4 dry and feed 5 soaked for 60 seconds). At the end of each week, the chewing time (seconds) for each horse was measured for 6 consecutive feed boluses from start of intake. *In vitro* the pellets were put into volumetric flasks and 300 ml of water was added to 600 ml of feed volume. The *in vivo* experiment showed a significant difference ($P<0.01$) in the mean time taken to chew and swallow one bolus for soaked (28.67±15.3 seconds; mean ± SD) and un-soaked (37.84±14.6 seconds ± SD) high fibre feeds. *In vitro* feed 5 showed the highest expansion after 60 seconds. Feeds 3 and 4 had the highest chewing times to swallow *in vivo* and lowest expansion rates *in vitro*; therefore these feeds may be the least likely to lead to oesophageal obstruction due to swelling of pellets if fed dry.

Keywords: pellets, feed expansion, ingestion, choke

Introduction

Oesophageal obstruction occurs when the oesophagus becomes blocked, usually by feed impactions causing distress and discomfort to the horse and this could lead to secondary complications. This condition could be attributed to dry and fibrous feed stuffs (Hance *et al.*, 1997; Frape, 1998). This study aimed to investigate the relationship of volume changes of high fibre pelleted feed *in vitro* and chewing rates *in vivo* with regard to potential oesophageal obstruction in the horse.

Material and methods

The *in vivo* trial took place over 5 weeks. Four horses (Thoroughbreds; 6-16 years old) were acclimatised to each feed for 5 consecutive days and measurements were taken in the morning of day 6. All horses were offered each of the feeds at 1000 g daily (feed 1-5, see Table 1). Feed 1 to 4 were fed un-soaked and feed 5 soaked according to manufacturers instructions (1000 g = 4,200 g soaked, for 60 seconds) Horses were fed from the ground in a rubber feeding bucket. The chewing time (seconds) for each horse was measured from its first mouthful until the first bolus was swallowed and 6 consecutive feed boluses were recorded.

In vitro the same 5 high fibre feeds were put into volumetric flasks and 300 ml of water was added to 600 ml of feed. All data were normally distributed and differences between groups were assessed by Analysis of Variance and Tukey *post hoc* test using SPSS.

Table 1. Feed composition.

Feed	Type	Treatment	Protein (%)	Oil (%)	Fibre (%)	Dry matter g/kg	Cube diameter (mm)
1	Cubes	Un-soaked	10.0	2.75	20	87	6
2	Cubes	Un-soaked	9.0	3.50	20	87	6
3	Nuts	Un-soaked	9.0	3.20	20	86	6
4	Pencils	Un-soaked	9.5	3.20	21	86	6
5	Pencils	Soaked	8.5	3.00	27	85	4

Results

The *in vivo* experiment showed a significant difference ($P<0.01$) in the mean time taken to chew and swallow one bolus for soaked (28.67±15.3 seconds; mean ± SD) and un-soaked (37.84±14.6 seconds ± SD) high fibre pelleted feeds (Table 2.)

There was a significant difference in expansion ($P<0.001$) between feed 5 and feeds 1-4 (Figure 1).

Table 2. Mean chewing times and standard deviation for feeds 1-5.

Feed	Unsoaked (secs)	Soaked (secs)	Standard deviation
1	38.96		13.58
2	35.29		11.05
3	38.00		14.45
4	38.79		18.45
5		28.67	15.30
Mean unsoaked	37.84		14.60

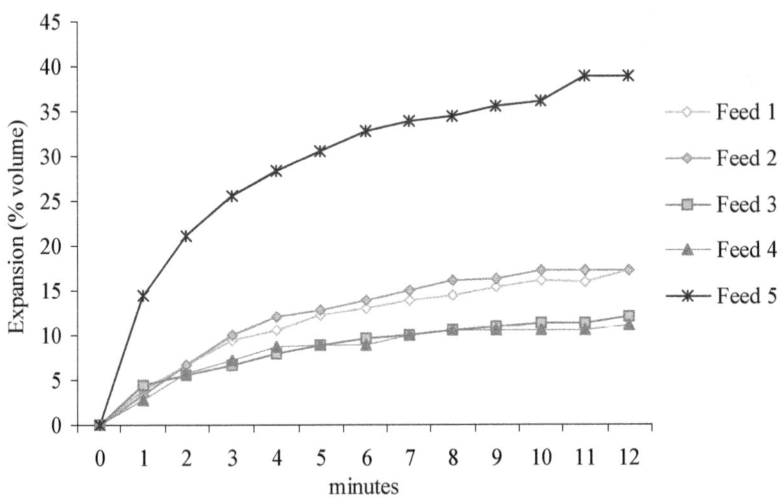

Figure 1. Increase in feed volume per feed following soaking 600 ml of each feed with 300 ml of water for 12 minutes in vitro.

Discussion

Feed 2 had the lowest chewing time (35.29±11.05) of the unsoaked feeds and it was noted that the pellet appeared softer and crumbled more easily than the other feeds which may have reduced the time needed to chew and form a bolus. Hill (2002) notes that soaked feed causes a reduction in chewing times and feed 5 (soaked) showed a reduction in chewing time to swallow by 23%. This confirms results by Ellis (2003) who found that adding wet sugarbeet pulp to concentrate feed led to a considerably faster intake rates. Ellis (2003) suggests that increasing moisture content of short particle feed leads to a reduction in chewing due to a decreased need for saliva production. Shumacher and Shumacher (1995) also report that the amount of saliva produced is dependent on the moisture content of the feed. Feeds 1,3 and 4 (dry) were all 6 mm in diameter and demonstrated similar chewing times to swallowing per bolus (Table 2).

Feed 5 showed the highest expansion after 60 seconds whereas feed 1-4 expanded much slower and to a lesser extent. Manufacturers instructions recommend a 60 second soak time for feed 5. In conclusion feeds 3 and 4 had the highest chewing times *in vivo* and lowest expansion rates *in vitro*; therefore these feeds may be the least likely to lead to oesophageal obstruction due to swelling of pellets if fed dry. Whereas feed 5 had the highest expansion rate after 60 seconds and further research may be required to identify optimal soaking times.

References

Ellis, A.D., 2003. Ingestive and digestive processes in horses. PhD thesis. Chelmsford; Writtle College, Essex University, United Kingdom.

Frape, D., 1998. Equine nutrition and feeding (3rd Edition). Blackwell, Oxford, United Kingdom.

Hance, S.R., J. Noble, S. Holcomb, B. Rush-Moore and W. Beard, 1997. Treating choke with oxytocin. Proceedings of the annual convention of the AAEP. 338-339.

Hill, J., 2002. Effect of level of inclusion and method off of presentation of a single distillery by-product on the processes of ingestion of concentrate feeds by horses. Livestock Production Science, 75, 209-218.

Schumacher, J. and S. Schumacher, 1995. Diseases of salivery glands and ducts of the horse. Equine Veterinary Education. 7, 313-319.

Preference of forage feeding position in stabled horses: a pilot study

V. Webster and A.D. Ellis
Nottingham Trent University, School of Animal, Rural and Environmental Sciences, Brackenhurst, Southwell, Nottinghamshire, NG25 0QF, United Kingdom

Introduction

The majority of stabled horses are fed forage in hay nets above the ground. This is contrary to their natural feed intake position. Physiologically this leads to a use of different muscles in the neck and back and it may impair the ability to clear particles via the mucociliary escalator. The aim of this study was to investigate preference of forage eating position in stabled horses.

Material and methods

Ten horses (Bodyweight 659±59 kg, age 7-17 years), previously accustomed to hay nets, were acclimatized for 2 weeks to receiving their forage from the floor below the hay net. During the study 2 kg of forage were presented in the hay net (1.5 m high, holes 16 cm^2) and 2 kg on the floor directly below the hay net in the morning and evening (no concentrate feed). Intake behaviour was recorded by direct observation for four days for one hour from point of feeding. A paired t-test (minutes) and Chi-square test (first choice) were used to analyse data (SPSS, 2007).

Results and discussion

First choice and during the first 20 minutes of observations all horses preferred to eat from the floor ($P<0.001$). Overall a weak trend favouring the hay on the floor was recorded (floor: 16 min ± 1.3 s.e.; hay net: 13±1.8 min; $P=0.08$) with a significant individual preference for eating from the floor in 5 horses (floor: 16±2.3 min; hay net: 9±1.4 min; $P<0.05$). Results may confirm a preference for multiple choices of sites (Goodwin *et al.*, 2002). However, ethogram analysis indicated that most horses looked up and around at 8 to 15 minutes into observations, then chose the hay net for several minutes, before returning to the hay on the floor for short periods. It is, therefore, possible that horses' peripheral vision was compromised when eating with the head down in an enclosed stable leading to changes in eating position. Further studies, including physiological measures, are needed to establish if preferences do exist and to show relevance to health.

References

Goodwin, D. Davidson, H.P.B. and Harris, P. (2002). Foraging enrichment for Stabled Horses: Effects on Behaviour and Selection. Equine Veterinary Journal. 37, 7, 686-691.

Effects of diet on behaviour of Standardbred horses in training

A. Jansson
Swedish University of Agricultural Sciences, Department of Animal Nutrition and Management, 750 07 Uppsala, Sweden

Introduction

Some trainers claim that concentrates, i.e. starch, might have a 'heating' effect. A recent study also shows that horses appear calmer on diets including more fibre and fat (Nicol *et al.*, 2005). The aim of this study was to investigate the effect of diet on behaviour of Standardbred horses in training.

Materials and methods

Six Standardbred geldings in training were fed a grass haylage-only diet (H) and a haylage concentrate diet (HC, 50:50, concentrate 82% oats) in a change-over design for 29.5 days. Feed allowances were estimated to be iso-caloric and nitrogenous. On days 17 and 29 the temperament during standardised exercise was assessed by the drivers using a scale (0-11.5 cm, corresponding to dull-very hot). On day 29, the behaviour (eating haylage or standing resting) was registered 30, 60 and 120 min post exercise and a novel object test (NOT) was performed. In the NOT, the horses were kept individually in a paddock where a folded tarpaulin (0.75 m^2, green in period 1 and white in period 2) was placed. The time required for the horses to get as close as 1 m with the head from the tarpaulin was measured (maximum 6 min allowed). At the end of each period, all horses were fitted with a step frequency meter and voluntary motion was estimated for 24 h. Chi2 analysis and analysis of variance was used ($P<0.05$ for significance).

Results and discussion

The present study could not detect any differences between diets in the exercise temperament (LSmeans 6.3 vs. 6.9 on diet H and HC, respectively, SEM 0.3), NOT (140 vs. 170 s, SEM 21) and voluntary motion (14.9 vs. 15.9 motions/min, SEM 1.3) but eating was a more common post exercise on diet H.

References

Nicol, C.J., A.J. Badnell-Waters, B. Bice, A. Kelland, A.D. Wilson and P.A. Harris. 2005. The effects of diet and weaning method on the behaviour of young horses. Appl. Anim. Beh. Sci. 95 (3/4), 205-221.

Effect of diet on plasma tryptophan and serotonin in horses: a link to behaviour?

A.D. Ellis[1], G. Piccione[2], C. Giannetto[2], E.K. Visser[3] and D. Alberghina[2]
[1]Nottingham Trent University, School of Animal, Rural and Environmental Sciences, Brackenhurst, Southwell, Nottinghamshire, NG25 0QF, United Kingdom
[2]Department of Experimental Sciences and Applied Biotechnology, Faculty of Veterinary Medicine, University of Messina, 98168 Messina, Italy
[3]Animal Sciences Group, Wageningen University and Research Centres, P.O. Box 65, 8200 AB Lelystad, the Netherlands

Introduction

In humans the neurotransmitter serotonin (5-HT) is known to be involved in physiological and behavioural processes, including mood, appetite, depression, anxiety and obsessive compulsive disorders (Lucki, 1998). Serotonin is synthesised via tryptophan hydroxylase and therefore tryptophan (TRP) levels affect serotonin metabolism. That pathway may explain possible dietary effects of tryptophan on equine behaviour (Grimmet and Sillence, 2005). This study measured the effect of two isoenergetic diets with different forage to concentrate ratios on plasma TRP and 5-HT levels in horses.

Material and methods

Thirty-six Dutch Warmblood horses (equal mares and geldings), were kept on a high forage (HF, n=18, concentrate to haylage 1: 3) or high starch (HS, n=18, concentrate to haylage 3:1) diet at light (week 1-4) and medium (weeks 5-8) exercise (45 minutes [8 minutes canter, 15 minutes trot] on a 'trainings mill' 40 m diameter). In Week 8, 3 hours after feeding 30% of daily diet, blood samples were taken. Concentrations of TRP and 5-HT in plasma were measured by high-performance liquid chromatography (Alberghina *et al.*, 2010). Behaviour was scored by independent observers (Scores 1: very easy - 4: very difficult) for leading and during blood sampling (for details see Ellis *et al.*, 2006). Results were analysed according to diet and gender (Chi-square, ANOVA, Statistica, 5.5).

Results and discussion

Plasma 5-HT concentrations and pH were higher in HF than HS horses (P<0.05 and P<0.01 respectively). Mares had higher plasma TRP levels than geldings (P<0.05) and this increase was attenuated by diet. Dietary tryptophan levels were not measured but a crude estimate (after feed manufacturer and McDonald *et al.*, 2007) returns a ratio of 1.4:1 for HF:HS diets. Behavioural observations showed an effect from diet but not gender (Ellis *et al.*, 2006). Two HS horses developed crib-biting behaviour following a further 4 weeks of the trial. When walking off after exercise HS horses were more excited and less manageable (score 1.6) than HF horses (score 1.0) (P<0.05). Results confirmed previous work showing that plasma 5-HT is influenced by gender and diet but the pathways to show whether this therefore may be a good marker of serotoninergic activity and its relationship to behaviour in horses need to be researched further.

References

Alberghina, D., C. Giannetto, E.K. Visser and A.D. Ellis, 2010. Effect of diet on plasma tryptophan and serotonin in trained mares and geldings. Veterinary Record 166, 133-136.

Ellis, A.D., C.K. Visser and C.G. Van Reenen, 2006. Effect of a high concentrate versus high fibre diet on behaviour and welfare in horses, Proceedings of the 40th International Congress of the ISAE, University of Bristol, p. 42, Cranfield University Press, Abstract.

Grimmett, A. and M.N. Sillence, 2005. Calmative for the excitable horse: a review of L-tryptophan. Vet J. 170 (1):24-32.

Lucki, I., 1998. The spectrum of behaviors influenced by serotonin. Biological Psychiatry 44, 151-62.

McDonald, P., R.A. Edwards, J.F.D. Greenhalgh and C.A. Morgan, 2002. Animal Nutrition, Pearson Prentice Hall, Essex, United Kingdom.

Walsh, A.E., A.D. Oldman, M. Franklin, C.G. Fairburn and P.J. Cowen, 1995. Dieting decreases plasma tryptophan and increases the prolactin response to d-fenfluramine in women but not men. Journal of Affect Disorders 21, 89-97.

Part 3. Promoting health and preventing disease

On the interaction between nutrition, health and disease

M. Coenen
Institute of Animal Nutrition, Nutrition Disease and Dietetic, Faculty of Veterinary Medicine, University Leipzig, Germany, Gustav Kühn Str. 8, 04159 Leipzig, Germany

Abstract

The profile of a safe equine ration is characterised by (a) covering requirements, (b) ingestible in terms of volume and palatability, (c) adequate for the species with due regard for the specific conditions in digestion and metabolism, (d) providing feeds of good hygienic quality, (e) compatible for the individual, (f) safe regarding the quality of products (meat, milk) and (g) minimizing impact of feeding on the environment. A conflict with those basic characteristics results in an animal response to compensate for deficits in nutrition. Visible consequences are absent or negligible if compensation is sufficient but a health issue is evident if a tissue function or cellular processes cannot be adapted any longer. Requirement data are sufficient to avoid improper energy and nutrient intakes. However obesity in equines show that there is a lack of precision in some requirement data and the transfer of those data from science to practice. Diets low in structured fibre and high in starch are a major problem particularly in the 'urbanised' horses. Undesirable substances of biotic (e.g. microorganisms) or non-biotic origin (xenobiotics) can create impacts on health. Poor feed hygiene can be of great importance; feeds with deficits in hygienic quality are frequently fed to horses resulting in both respiratory and GI tract compromise. Furthermore, unhygienic feeds affect the liver, kidney and muscle. The liver detoxifies undesirable substances, and may become damaged so-doing, whereas the kidney's excretory capabilities may be overwhelmed or damaged by nutrients in great excees such as nitrogen, selenium or by toxins (Ochratoxin). The orchestra of communication of muscle cells with the extracellular environment and intracellular metabolism can be disturbed by unbalanced substrate and electrolyte supply. Oxidative stress can cause tissue damage. Further systems to respond to nutritional factors are the endocrine and immune systems.

Keywords: equine digestion, liver, kidney, toxins, immune system, feed management

Introduction

The domestication of equids during the 4th- 3rd millennium BC was initiated by the need for food. During that time, horses still relied on natural feed sources, as it is assumed that strategies for feed and food conservation had yet to be developed both in human and animal nutrition. The ability to cope with the seasonality of feed availability made equines extremely advantageous for human use. The intensification of horses for meat and milk production and as a means of transport and provision of draught power, were accompanied by changes in horse nutrition. First reports on the use of grain for horses date from the 14th century BC, when the Roman cavalry used concentrates comparable to mixtures fed to horses nowadays. As the ability to maintain the fitness of horses regardless of season grew in importance, so the need for knowledge of feed conservation methods and their application to a wide range of available resources, such as the production of leaf hay from sources including trees like beech or willow, also developed.

Various physiological similarities between horses and man resulted in some commonality in the diseases experienced by both species. Aristotle (384-322 BC) was one of the first to systematically describe animals and list their diseases, i.e. colics. The words colic and ileus both have Greek roots (κωλικο, ειλος, (Anders, 2005)). During the following centuries our knowledge of several diseases increased but was far removed from a systematic discipline for a long period. Severino (Naples, 1645, *Zootomia democritea*) described the equine pancreas. Spallanzani (Pavia, 1780: *Dissertazioni di fisica animale e vegetale*) was a pioneer regarding gastric function. However, it was not until the 19th century, partly driven by the fast development of chemistry, that digestive functions and their modifying factors were elucidated. Colin (1854) first presented details describing protozoa in the caecum of the horse. Many digestion trials for example, those by Patterson (Patterson, 1897) produced quantitative data on gut function and founded work on equine energy metabolism as well as on performance limiting disease (Carlström, 1931; Zuntz and Hagemann, 1894).

Shortages in nutrients like calcium were detected as reasons for failures in the function of specific tissues (Montgomerie *et al.*, 1929). Work on selenium deficiency (Stowe, 1967) marked the shift from studying an individual nutrient towards a focus on the interactions between nutrients and their roles in the pathogenesis of various dysfunctions. Knowledge of foetal maturation markedly increased (Fowden and Silver, 1995; Meyer and Ahlswede, 1976) and expanded existing knowledge of growth. Increasing insights into partial digestibility and the microbial community of the intestine (Julliand *et al.*, 1999; Meyer *et al.*, 1982) on the one hand and on the complexity of biochemical responses to exercise (Lindholm and Saltin, 1974) on the other, were driving forces to link digestion and performance. The role of nutrition in horse behaviour, as well as nutrition being responsible for various 'diseases of civilisation' such as gastric ulcers and obesity (Ralston and Baile, 1982) (Hoffman *et al.*, 2003; Nadeau *et al.*, 2000a) highlighted the complexity of nutrition and its potential impact on well being. The disclosure of events inside the cells is an upcoming challenge which will provide further insights into the nutrition-health network (McGivney *et al.*, 2009; Moore *et al.*, 2008)

Basic nature of the diet

In principle it is a simple profile that characterizes a safe ration:

1. Covering requirements.
2. Ingestible in terms of volume and palatability.
3. Adequate for the species with due regard for the specific conditions in digestion and metabolism.
4. Providing feeds of good hygienic quality.
5. Compatible for the individual.
6. Safe regarding the quality of products (meat, milk).
7. Minimizing impact of feeding on the environment.

Covering requirements

Some diseases are associated with a mismatch in feed or nutrient allowances with respect to the animals' requirements. It should be noted that the quality of data on defining requirements is sufficient to avoid nutrient deficiencies. However, primary deficiencies still may occur, e.g. big head disease (Luthersson *et al.*, 2005). Feeding practice and management may interact with the nutritional status and the tabled nutrient uptake is not realized as was impressively shown by the depressed iron status of stabled foals (Brommer and Sloet van Oldruitenborgh-Oosterbaan, 2001). Genetic complement can pre-dispose a horse to an increased risk of suffering from a shortage of a specific nutrient (McGorum *et al.*, 2009; Mohammed *et al.*, 2007).

High performance demands can result in huge changes in requirements; this is the case for water and electrolytes. Again, possible deficiencies are not necessarily a consequence of a lack in knowledge of sodium or chloride requirements, but may reflect inappropriate management of suitable compensatory nutrient supplies. In fact, feeding strategies regarding electrolyte, protein and antioxidative agents for high performance horses requires further research.

Further improvements in the precision of data defining nutrient requirements for mares is needed. The factorial approach to defining energy and protein requirement for breeding mares is probably incomplete. Protein enriched feeding of mares improves their reproductive efficiency (Van Niekerk and Van Niekerk, 1997a,b,c,d, 1998a,b,c)

The adjustment of nutrient supply to requirements includes avoiding nutrient excess. The most prominent disease in that direction is obesity. (Wyse *et al.*, 2008). Over consumption of nutrients may be performance limiting, e.g. very high protein, or the cause of acute fatalities e.g. selenium intoxication. Energy and trace elements are the most relevant regarding over consumption.

Ingestible in terms of volume and palatability

Fortunately, dry matter intake (DMI) is rarely a limiting factor in horse feeding. So far, in contrast to dairy cow nutrition, the conflict between energy requirement and DMI has not been a problem in rationing horses. Cases of unacceptably low body condition scores are caused by man made limitations on intake, rather than an inability of the horse to ingest sufficient DM for his needs. Low palatability can reduce DMI; it is a common experience that a specific batch of hay is not accepted. However, it remains unclear whether this lack of acceptance is most frequently due to botanical composition, physical-chemical properties, or some harmful anti-nutritional factor that has remained undetected.

Adequate for the species with due regard for the specific conditions in digestion and metabolism

The caloric needs of horses can easily be balanced by supplementation of forage with grain. Certain guidelines advocate the intensive use of cereals in equine diets (Richards *et al.*, 2006; Southwood *et al.*, 1993). However, energy rationing based on this tenet neglects the blueprint of the equine digestive tract and the role of the hind-gut in equine nutrition. Energy derived from fibre is the first choice of the hindgut fermenter. Consequently, the proportion of roughage in the ration defines the adequacy or inadequacy of the diet for equines regardless of the type of production or performance expected. Providing no less than 15 g of structured roughage (88% dry matter) per kg body weight daily is a useful benchmark which considers aspects of digestive physiology and behaviour (Coenen and Vervuert, 2010).

Providing feeds of good hygienic quality

Feed is vector for microbes and some of them produce toxins in the feed or even in the intra-intestinal environment. Inorganic dust and microbial spores together represent the respirable particle load that challenges the respiratory tract (Ainsworth *et al.*, 2009; Deaton *et al.*, 2006; Elfman *et al.*, 2009). The hygienic standard of feeds, including bedding, is a key factor in safe nutrition (Keller *et al.*, 2007; Wichert *et al.*, 2008). Systems to evaluate feed quality based on microbial data are well documented (Kamphues *et al.*, 2009). For example, it is commonly accepted that a high count for yeast like *Candida* limits the suitability of feeds. The application of high standards to the hygienic quality of horse feeds is an essential element in ameliorating risks from contaminated feed. Toxin contamination of feeds has three different origins:

1. Bacterial toxins: the most important, and potentially fatal, toxin is produced by *Clostridium botulinum* (Gerber *et al.*, 2006; Kinde *et al.*, 1991; Venner, 1999; Whitlock and Buckley, 1997). Manure management in animal production and soil contamination of feed are risk modulating factors. The mildly acidic but strictly anaerobic environment in moist, wrapped roughages can elicit toxin production and cases of contamination of fresh feed by *Clostridia*. The role of *Clostridia* in the pathogenesis of grass sickness is object of controversial debate (McCarthy *et al.*, 2004). Other *Clostridium* related toxins are discussed as a trigger for atypical myoglobinuria in horses. Recently, corynetoxins gained attention following more than 4,000 fatalities in animals in Wales (Davis *et al.*, 1995).
2. Mycotoxins: numerous mycotoxins have been identified and are a serious health-related issue in human and animal nutrition. Particularly relevant to equines are Aflatoxins, Fumonisins and ergot alkaloids (Johnson *et al.*, 1997; Osweiler, 2001; Ross *et al.*, 1991; Uhlig *et al.*, 2007; Wilson *et al.*, 1990). Although reports on negative effects in horses by mycotoxin ingestion by

deoxanivalenol, zearalenone,are suspected, published data are insufficient to categorically state the role of these mycotoxins to equine feeding practice (Johnson *et al.*, 1997; Raymond *et al.*, 2005)

3. Toxic plants: pyrrolizidine alkaloids rank first among plant associated toxicoses in Europe. The main plant species responsible for such toxicosis are the ragworts (*Senneccio* sp), of which different varieties are widely distributed. Ragworts are particularly aggressive colonisers of bare ground, over-grazed pastures and roadside verges. Seed as well as whole plants may be presented (Cope *et al.*, 2004; Naude *et al.*, 2005) to the animals in part, as a constituent of hay, or accidentally available in the environment (Barr and Reagor, 2001)
4. Inorganic toxins and others: ionophores represent a group of feed additives, which are proven to be beneficial in poultry and ruminant feeding systems but are toxic in equines. Toxicity may happen through misuse of feed or mixing errors at the feed mill. (Aleman *et al.*, 2007). In principle, the horse is exposed to any deleterious substance which is emitted by industry or traffic or other human activities (Dey and Dwivedi, 2004; Spoo, 2001). However, a direct systemic risk seems to be absent. In this context, it needs to be considered that lifespan plays an enormous role in evaluating xenobiotics. Epidemiological studies in aged horses are needed to evaluate a chronic load of the animals by xenobiotics like Phtalates which are suspected to interact with the hormonal communication. It has been shown in humans that individuals are genetically different regarding their detoxifying capabilities (Neafsey *et al.*, 2009). If only for economic considerations, (due to the cost of breeding a horse and the expected time for its use) the application of genomics to equine nutrition research could be of benefit.

Compatible for the individual

Feeding management: The prevalence of aberrations in the respiratory tract of horses, incidences of colic and gastric mucosa damage (Durham, 2009; Egenvall *et al.*, 2008; Feige *et al.*, 2002; Knubben *et al.*, 2008; Luthersson *et al.*, 2009; Ramzan *et al.*, 2008; Tinker *et al.*, 1997) show an impact of nutrition on equine health beside the countable Joules and nutrients.

The well known studies on the link between hindgut dys-fermentation and laminitis show that even safe horse feeds can induce dysfunction or provoke risks for them (Garner *et al.*, 1975; Pollitt and Davies, 1998). It is evident that in particular, stabled horses are at risk from the man made structure of the single meal. The horse cannot – in contrast to a horse on pasture – compensate for false dimensions in the meal size for a very simple reason: The horse is 'constructed' for continuous feed intake and any corresponding regulatory network is calibrated to this principle of survival. There is a quite voluminous literature on glucose and insulin responses in horses showing the dynamics of postprandial changes which are less pronounced under natural conditions. The impact of meal size is linked to the starch content of the meal, Consequently, the starch intake per meal is an issue for gastric health (Luthersson *et al.*, 2009) and control of endocrine responses (Vervuert *et al.*, 2009). The recommendation to limit starch intake per meal to 1 g per kg body weight, recognizes the risks of high-starch meals and this has been substantiated by recently published studies (Luthersson *et al.*, 2009);(Vervuert *et al.*, 2009) on gastric ulcer and starch-dependent insulin responses.

Safe regarding the quality of products (meat, milk)

Horses, as farm animals are included in the human food chain, in cultures that embrace consumption of horse-meat. Horse milk is accepted for both human consumption and for cosmetics. This type of use implies that any kind of xenobiotic carry-over is a matter of issue regardless the influence of the xenobiotic on the animals itself (Penumarthy *et al.*, 1980); (Beldomenico *et al.*, 2001).

Minimizing impact of feeding on the environment

Nitrogen and minerals are excreted in a dose-dependent manner. The necessity of monitoring the flow of these substances in animal production is increasing due to environmental pollution concerns and legal restrictions (e.g. copper in complete feeds) are now in force. From a legal standpoint manure is not only a kind of fertilizer but is also a sink for non-used elements which load the soil where manure is spread. Considering the economy of nutrients and ecological implications, over consumption of potentially polluting nutrients should be avoided.

Mechanisms behind the feed/feeding: health axis

Although there is some overlap in the kind of communication between a dietary factor and the organism, it is reasonable to discuss some major structures.

Selected tissues

Gastro-intestinal tract

A local response, i.e. by *secretions* or *inflammatory reactions* are typical for immediate damage of superficial cells such as on the tongue or gut mucosa. Essential oils in some plants like *Thuja occidentalis* cause lesions, and more dangerously, lectins which are contained within a number of plants cause heamaglutination of red blood cells, and may be found in. the wood or leaves from robinia (*Robinia pseudoacacia*).

Feeds and nutrients may influence the *intra-intestinal environment* by interfering with secretory function. For example, in the stomach, concentrates may maintain a prolonged acid secretion due to non-complete/retarded acidification of gastric contents or by interaction with the microbial community.

The degree of starch fermentation determines the profiles of microbes and of volatile fatty acids in gastric and intestinal contents (De Fombelle *et al.*, 2003). Butyric acid can change discrete properties of the non-glandular mucosa in the stomach (Nadeau *et al.*, 2000b). Deranged fermentation is associated with marked changes in the microbial population (Milinovich *et al.*, 2006) and products of fermentation i.e. biogenic amines (Bailey *et al.*, 2003) which may penetrate the gut wall. It is likely that gut function controlling structures like neurones or gut hormone secretory cells (Bishop *et al.*, 1984; Freytag *et al.*, 2008; Hall *et al.*, 1982) are affected by these events or possibly by the absorption of products from digestion. The gut wall contains a substantial complement of active compounds. Serotonin containing cells are widely distributed throughout the GIT, storing high levels of a vasoactive agent (Ceccarelli *et al.*, 1995; Fink *et al.*, 2006).

Liver function

Urea synthesis, *gluconeogenesis, fatty acid processing, presentation of transport proteins, trace element and vitamin storage and xenobiotic management* are major functions of the liver. The increase in blood ammonia during hepatoencephalopathy, cell degeneration due to triglyceride accumulation, hyperbilirubinaemia or other events that are frequently detected by routine laboratory diagnostics show the essentiality of safe liver function, as well as the potentially adverse effects of inappropriate nutrition. Major dietary impacts are:

- energy intake: ⇨fatty acid/triglyceride storage
- protein intake: ⇨urea synthesis
- trace element intake: ⇨cellular uptake
- ⇨oxidative stress
- xenobiotics: ⇨glucuronide formation

Again, the first three points can be summarized as being a part of the requirement orchestra.

The glucuronidation of undesirable substances is an indispensable function of the liver and will define at least in part the sensitivity of the organism against xenobiotics. The initial phase of the processing of these substances is located in the intestinal wall and is represented by the Cytochrome P450. It is remarkable that in the equine, this protein with Cyp 3A as a major component is most highly expressed in the upper part of the duodenum and yet, is nearly absent in the ileum and hindgut. The mono-oxygenases in the gut mucosa and liver ensure the phase I processing of xenobiotics. The effects of these enzymes are an enforced accessibility of the substrate to the second phase, which is the formation of glucuronides. There are remarkable species differences regarding the formation of conjugates (Pandey *et al.*, 1990); the related transferases are inducible to a certain degree. However, the situation in the equine is not well defined. The horse has a liver Cytochrome 450 capacity comparable with the pig, but at a substantially lower level than is found in ruminants. (Nebbia *et al.*, 2003). The capacity of the equine liver to activate xenobiotic biotransformation via phase I and phase II processing is shown for different steroids (Dumasia and Houghton, 1981; Dumasia and Houghton, 1984; Houghton, 1977; Houghton and Dumasia, 1979; Houghton and Dumasia, 1980).

Hepatic modification of xenobiotics does not necessarily imply the detoxification of those substances. The formation of metabolites after intake of toxic substances which escape the liver or biliary export can lead to skin disease, often leading to photodynamic substances deposited in the skin resulting in photosensitivity (Berry and Merriam, 1970; Booth and McDowell, 1975; Cook, 1969; Cornelius *et al.*, 1975; Fadok, 1995; Ford and Gopinath, 1974; Moore *et al.*, 2008; Nation, 1989; Pascoe, 1973; Singh, 1970; Yeruham *et al.*, 1999).

Kidney function

A feature of the kidney is its flexibility. The kidney is responsible for the safe elimination of a wide range of non-utilised nutrients and metabolites (Schott, 2007). Therefore, it is unlikely to induce tissue damage by unbalanced nutrient intake. However, renal excretion profiles may indicate imbalances in nutrient intakes, and the ratio of a given substance to creatinine (which acts as a kind of renal excretion marker) is used for diagnosis. As the regulation of urine volume is at least partly independent of the amount of unused substances or nutrients which need to be safely exported, there can exist conflicting kidney challenge, (a) limiting or even reducing urine volume but (b) maintaining surplus excretion. The concentration by the kidney of urea, magnesium, phosphorus and particularly calcium may exceed their solubility in urine with the potential for crystalisation and formation of urinary calculi (Duesterdieck-Zellmer, 2007).

Muscle function

Muscle tissues tend to be unresponsive to short term fluctuations in nutrient intake. Energy and oxygen availability are organized in order to keep the animal ready for flight at any time. The requirements data recognise the quantitative aspect of energy and nutrient provision. The environment in which the muscle cell is embedded is influenced by nutritive status, yet this aspect of nutrition is often neglected. The interstitial fluid of muscle tissue influences the intracellular compartment, e.g. pH, ion movements through the cell membrane and functions for transport of substrates or synthesis which are mostly linked to the availability of energy via the sufficient presentation of ATPase. Exercise induces fast and intensive changes in ion homeostasis and distribution (Bayly *et al.*, 2006; Hess *et al.*, 2006; Hyyppa and Poso, 1998; Kronfeld *et al.*, 1999; Lloyd and Rose, 1995; Schuback *et al.*, 2002). On the other hand, protein and electrolyte intakes are effective modifiers of acid base-balance, in the relaxing and regenerating muscle. (Foreman, 1998; Reeta Poso and Hyyppa, 1999; Waller *et al.*, 2009). Consequently, water and amino acid and electrolyte supply to horses with high muscle activity is a decisive issue for maintaining muscle performance.

Muscle activity is associated with oxidative stress, a physiological event, and as a consequence, there is a need for substantial antioxidative capacitiy during and after exercise. This is mostly compensated by intakes of various anti-oxidants such as Vitamin E, selenium, carotene and by endogenous adaptations. Considering the weakness of defence mechanisms in both the respiratory and gastro-intestinal tracts as a by-product of strenuous exercise, it seems likely that even moderate deficits in feed quality (elevated counts of bacteria and mould) or unbalanced nutrient intake (high iron, iodine intake) may increase the load on the anti-oxidative mechanisms.

Superordinate systems

Hormone secretion

Insulin responses to glucose absorption are one of the best known feed-hormone interactions, but are only one among many. Amino acids seem to be neglected with regard to their potential to interact with hormone secretion (DePew *et al.*, 1994a; DePew *et al.*, 1994b; Sticker *et al.*, 2001). Water and electrolyte management or even mismanagement induces responses of ACTH, ADH and natriuretic peptide. The consequences of diet for gastrin secretion (high concentrate = high gastrin) (Cargile *et al.*, 2004; Sandin *et al.*, 1998; Smyth *et al.*, 1989) dictate the conclusion that *dietary factors provoke a complex response by hormones* which are linked to the GIT (Burcelin, 2005) Gastro-splanchnic-portal sensing is a general tool that responds effectively to nutrient intake in animals (Ertl *et al.*, 2008) but the response may differ depending on the nutrition strategy employed. A special role is the recognition of nutrients which arrive in portal blood and the liver (Burcelin *et al.*, 2000; Donovan, 2002; Fukaya *et al.*, 2007; Mithieux *et al.*, 2005; Thorens, 2008). Substrate peaks interact with liver perfusion and feed intake. It should be expected that in hindgut fermenters, adapted to a more continuous substrate appearance in the portal blood, splanchnic-portal sensing is different from that in the pig or rat. It is hypothesised that, glucose sensing after starch–rich meals is poorly developed in equines, but this still requires experimental confirmation. However, energy and nutrient intake induces an endocrine response as a kind of fingerprint, involving endocrine networks far away from digestion. Zearalenone and ergotalkaloides are prominent examples of compounds which disconnect physiological hormonal communication in the uterus, ovary and mammary tissue.

Immune system

Immunodeficiencies are well known in horses (Perryman, 2000; Perryman and McGuire, 1980). However little is known about the impact of nutrition specifically on the equine immune system. Shortages in certain nutrients like selenium, vitamin E and zinc, are evaluated as lowering the capability of the immune system in animals. Positive effects on immunity in horses are described for selenium and vitamin E but are mainly the result of compensated depletion (Baalsrud and Overnes, 1986; Knight and Tyznik, 1990). Nutrients believed to be involved with developing immunity include amino acids (Calder, 2006; Calder and Yaqoob, 1999), fatty acids and even prebiotics (Calder, 2002, 2007, 2008; Lomax and Calder, 2009). The latter are believed to restore the health of a compromised gut microflora and compromised microflora may result from reduced immune function. Nutrition of the mare during the last trimester of gestation may have a great influence on prepartum changes in foetal immune maturation which is rapidly completed post partum (Calder *et al.*, 2006). In general it can be concluded that a diet which is adjusted to the specific requirements of the horse will cover the needs of immune system function. Nevertheless, where necessary, immune function may be enhanced by provision of n3 fatty acids for horses at high risk of infection (high performance horses, intensive medical care, and chronic hypocaloric states) and those which experience tissue damage with accompanying losses of fluids, elevated amino acid availability may be beneficial. Hypogammaglobulinemia can serve as indicator for dietary countermeasures by dietary addition of high quality protein.

Acknowledgements

The author thanks A.C. Longland, C.M. Argo and J. E. Lindberg for their assistance with the language in this paper

References

Ainsworth, D.M., M. Matychak, C.L. Reyner, H.N. Erband and J.C. Young, 2009. Effects of *in vitro* exposure to hay dust on the gene expression of chemokines and cell-surface receptors in primary bronchial epithelial cell cultures established from horses with chronic recurrent airway obstruction. Am. J. Vet. Res. 70, 365-372.

Aleman, M., K.G. Magdesian, T.S. Petersonand and F.D. Galey, 2007. Salinomycin toxicosis in horses. J. Am. Vet. Med. Assoc. 230, 1822-1826.

Anders, K., 2005. Die Kolik des Pferdes – Ein Beitrag zur Geschichte der Haustierkrankheiten [Diss. vet. med.]. Berlin: Freie Universität.

Baalsrud, K.J. and G. Overnes, 1986. Influence of vitamin E and selenium supplement on antibody production in horses. Equine Vet. J. 18, 472-474.

Bailey, S.R., M.L. Baillon, A.N. Rycroft, P.A. Harris and J. Elliott, 2003. Identification of equine cecal bacteria producing amines in an *in vitro* model of carbohydrate overload. Appl. Environ. Microbiol. 69, 2087-2093.

Barr, A.C. and J.C. Reagor, 2001. Toxic plants. What the horse practitioner needs to know. Vet Clin North Am Equine Pract. 17, 529-546.

Bayly, W.M., J.K. Kingston, J.A. Brown, R.D. Keegan, S.A. Greene and R.H. Sides, 2006. Changes in arterial, mixed venous and intraerythrocytic concentrations of ions in supramaximally exercising horses. Equine Vet. J. Suppl. 294-297.

Beldomenico, H.R., E. Baroni, D.U. Campagnoli, M.E. Sigrist, M. Rubio and J.C. Boggio, 2001. Cadmium accumulation and distribution in slaughtered horse kidneys from the Argentine central region. Arch Environ. Contam. Toxicol. 41, 100-103.

Berry, J.M. and J.G. Merriam, Jr., 1970. Phototoxic dermatitis in a horse. (A case report). Vet. Med. Small Anim. Clin. 65, 251-257.

Bishop, A.E., N.P. Hodson, J.H. Major, L. Probert, J. Yeats, G.B. Edwards, J.A. Wright, S.R. Bloom and J.M. Polak, 1984. The regulatory peptide system of the large bowel in equine grass sickness. Experientia. 40, 801-806.

Booth, N.H. and J.R. McDowell, 1975. Toxicity of hexachlorobenzene and associated residues in edible animal tissues. J. Am. Vet. Med. Assoc. 166, 591-595.

Brommer, H. and M.M. Sloet van Oldruitenborgh-Oosterbaan, 2001. Iron deficiency in stabled Dutch warmblood foals. J. Vet. Intern. Med. 15, 482-485.

Burcelin, R., 2005. The incretins: a link between nutrients and well-being. Br. J. Nutr. 93 Suppl. 1, S147-156.

Burcelin, R., W. Dolci and B. Thorens, 2000. Glucose sensing by the hepatoportal sensor is GLUT2-dependent: *in vivo* analysis in GLUT2-null mice. Diabetes. 49, 1643-1648.

Calder, P.C., 2002. Dietary modification of inflammation with lipids. Proc. Nutr. Soc. 61, 345-358.

Calder, P.C., 2006. Branched-chain amino acids and immunity. J. Nutr. 136, 288S-293S.

Calder, P.C., 2007. Immunomodulation by omega-3 fatty acids. Prostaglandins Leukot. Essent. Fatty Acids. 77, 327-335.

Calder, P.C., 2008. The relationship between the fatty acid composition of immune cells and their function. Prostaglandins Leukot Essent. Fatty Acids. 79, 101-108.

Calder, P.C., S. Krauss-Etschmann, E.C. de Jong, C. Dupont, J.S. Frick, H. Frokiaer, J. Heinrich, H. Garn, S. Koletzko, G. Lack, G. Mattelio, H. Renz, P.T. Sangild, J. Schrezenmeir, T.M. Stulnig, T. Thymann, A.E. Wold and B. Koletzko, 2006. Early nutrition and immunity – progress and perspectives. Br. J. Nutr. 96, 774-790.

Calder, P.C. and P. Yaqoob, 1999. Glutamine and the immune system. Amino Acids. 17, 227-241.

Cargile, J.L., J.A. Burrow, I. Kim, N.D. Cohen and A.M. Merritt, 2004. Effect of dietary corn oil supplementation on equine gastric fluid acid, sodium, and prostaglandin E2 content before and during pentagastrin infusion. J. Vet. Intern. Med. 18, 545-549.

Carlström, M.R., 1931. Über die Ätiologie und Pathogenese der Kreuzlähme des Pferdes. Skand. Arch. Phys. 61, 161-224.

Ceccarelli, P., V. Pedini and A.M. Gargiulo, 1995. Serotonin-containing cells in the horse gastrointestinal tract. Anat. Histol. Embryol. 24, 97-99.

Coenen, M. and I. Vervuert, 2010. A minimum of roughage and a maximum of starch – necessary benchmarks for equine diets. Pferdeheilkunde. in Press,

Cook, W.E., 1969. Acute hepatic failure in the horse. Vet. Rec. 85, 421-422.

Cope, R.B., C. Camp and C.V. Lohr, 2004. Fatal yew (*Taxus* sp) poisoning in Willamette Valley, Oregon, horses. Vet. Hum. Toxicol. 46, 279-281.

Cornelius, C.E., J.A. Himes and R.R. Gronwall, 1975. Experimental liver diseases. Methods Achiev Exp. Pathol. 7, 22-55.

Davis, E.O., G.E. Curran, W.T. Hetherington, D.A. Norris, G.A. Wise, I.J. Roth, A.A. SeaWright and W.L. Bryden, 1995. Clinical, pathological and epidemiological aspects of flood plain staggers, a corynetoxicosis of livestock grazing Agrostis avenacea. Aust. Vet. J. 72, 187-190.

De Fombelle, A., M. Varloud, A.-G. Goachet, E. Jacotot, C. Philippeau, C. Drogoul and V. Julliand, 2003. Characterization of the microbial and biochemical profile of the different segments of the digestive tract in horses given two distinct diets. Animal Science. 77, 293-304.

Deaton, C.M., D.J. Marlin, L. Deaton, N.C. Smith, P.A. Harris, R.C. Schroter and F.J. Kelly, 2006. Comparison of the antioxidant status in tracheal and bronchoalveolar epithelial lining fluids in recurrent airway obstruction. Equine Vet. J. 38, 417-422.

DePew, C.L., D.L. Thompson, Jr., J.M. Fernandez, L.L. Southern, L.S. Sticker and T.L. Ward, 1994a. Plasma concentrations of prolactin, glucose, insulin, urea nitrogen, and total amino acids in stallions after ingestion of feed or gastric administration of feed components. J. Anim. Sci. 72, 2345-2353.

DePew, C.L., D.L. Thompson, Jr., J.M. Fernandez, L.S. Sticker and D.W. Burleigh, 1994b. Changes in concentrations of hormones, metabolites, and amino acids in plasma of adult horses relative to overnight feed deprivation followed by a pellet-hay meal fed at noon. J. Anim. Sci. 72, 1530-1539.

Dey, S. and S.K. Dwivedi, 2004. Lead in blood of urban Indian horses. Vet. Hum. Toxicol. 46, 194-195.

Donovan, C.M., 2002. Portal vein glucose sensing. Diabetes Nutr. Metab. 15, 308-312; discussion 313-304.

Duesterdieck-Zellmer, K.F., 2007. Equine urolithiasis. Vet. Clin. North Am. Equine Pract. 23, 613-629, vi.

Dumasia, M.C. and E. Houghton, 1981. Studies related to the metabolism of anabolic steroids in the horse: the identification of some 16-oxygenated metabolites of testosterone and a study of the phase II metabolism. Xenobiotica. 11, 323-331.

Dumasia, M.C. and E. Houghton, 1984. Studies related to the metabolism of anabolic steroids in the horse: the phase I and phase II biotransformation of 19-nortestosterone in the equine castrate. Xenobiotica. 14, 647-655.

Durham, A.E., 2009. The role of nutrition in colic. Vet. Clin. North Am. Equine Pract. 25, 67-78, vi.

Egenvall, A., J. Penell, B.N. Bonnett, J. Blix and J. Pringle, 2008. Demographics and costs of colic in Swedish horses. J. Vet. Intern. Med. 22, 1029-1037.

Elfman, L., M. Riihimaki, J. Pringle and R. Walinder, 2009. Influence of horse stable environment on human airways. J. Occup. Med. Toxicol. 4, 10.

Ertl, A.C., S. Mann, A. Richardson, V.J. Briscoe, H.B. Blair, D.B. Tate and S.N. Davis, 2008. Effects of oral carbohydrate on autonomic nervous system counterregulatory responses during hyperinsulinemic hypoglycemia and euglycemia. Am. J. Physiol. Endocrinol. Metab. 295, E618-625.

Fadok, V.A., 1995. An overview of equine dermatoses characterized by scaling and crusting. Vet. Clin. North Am. Equine. Pract. 11, 43-51.

Feige, K., A. Furst and M.W. Eser, 2002. [Effects of housing, feeding and use on equine health with emphasis on respiratory and gastrointestinal diseases]. Schweiz Arch. Tierheilkd. 144, 348-355.

Fink, C., M. Tatar, K. Failing, R. Hospes, M. Kressin and K. Klisch, 2006. Serotonin-containing cells in the gastrointestinal tract of newborn foals and adult horses. Anat. Histol. Embryol. 35, 23-27.

Ford, E.J. and C. Gopinath, 1974. The excretion of phylloerythrin and bilirubin by the horse. Res. Vet. Sci. 16, 186-198.

Foreman, J.H., 1998. The exhausted horse syndrome. Vet. Clin. North Am. Equine Pract. 14, 205-219.

Fowden, A.L. and M. Silver, 1995. Glucose and oxygen metabolism in the fetal foal during late gestation. Am. J. Physiol. 269, R1455-1461.

Freytag, C., J. Seeger, T. Siegemund, J. Grosche, A. Grosche, D.E. Freeman, G.F. Schusser and W. Hartig, 2008. Immunohistochemical characterization and quantitative analysis of neurons in the myenteric plexus of the equine intestine. Brain Res. 1244, 53-64.

Fukaya, M., A. Mizuno, H. Arai, K. Muto, T. Uebanso, K. Matsuo, H. Yamamoto, Y. Taketani, T. Doi and E. Takeda, 2007. Mechanism of rapid-phase insulin response to elevation of portal glucose concentration. Am. J. Physiol. Endocrinol. Metab. 293, E515-522.

Garner, H.E., J.R. Coffman, A.W. Hahn, D.P. Hutcheson and M.E. Tumbleson, 1975. Equine laminitis of alimentary origin: an experimental model. Am. J. Vet. Res. 36, 441-444.

Gerber, V., R. Straub and J. Frey, 2006. Equine botulism and acute pasture myodystrophy: new soil-borne emerging diseases in Switzerland? Schweiz Arch. Tierheilkd. 148, 553-559.

Hall, G.M., T.E. Adrian, S.R. Bloom and J.N. Lucke, 1982. Changes in circulating gut hormones in the horse during long distance exercise. Equine Vet J. 14, 209-212.

Hess, T.M., D.S. Kronfeld, R.A. Carter, K.H. Treiber, B.M. Byrd, W.B. Staniar, L.T. Smith, L.A. Gay and P.A. Harris, 2006. Does usefulness of potassium supplementation depend on speed? Equine Vet. J. Suppl. 74-79.

Hoffman, R.M., R.C. Boston, D. Stefanovski, D.S. Kronfeld and P.A. Harris, 2003. Obesity and diet affect glucose dynamics and insulin sensitivity in Thoroughbred geldings. J. Anim. Sci. 81, 2333-2342.

Houghton, E., 1977. Studies related to the metabolism of anabolic steroids in the horse: 19-nortestosterone. Xenobiotica. 7, 683-693.

Houghton, E. and M.C. Dumasia, 1979. Studies related to the metabolism of anabolic steroids in the horse: testosterone. Xenobiotica. 9, 269-279.

Houghton, E. and M.C. Dumasia, 1980. Studies related to the metabolism of anabolic steroids in the horse: the identification of some 16-oxygenated metabolites of 19-nortestosterone. Xenobiotica. 10, 381-390.

Hyyppa, S. and A.R. Poso, 1998. Fluid, electrolyte, and acid-base responses to exercise in racehorses. Vet. Clin. North Am. Equine Pract. 14, 121-136.

Johnson, P.J., S.W. Casteel and N.T. Messer, 1997. Effect of feeding deoxynivalenol vomitoxin)-contaminated barley to horses. J. Vet. Diagn. Invest. 9, 219-221.

Julliand, V., A. de Vaux, L. Millet and G. Fonty, 1999. Identification of Ruminococcus flavefaciens as the predominant cellulolytic bacterial species of the equine cecum. Appl. Environ. Microbiol. 65, 3738-3741.

Kamphues, J., M. Coenen, E. Kienzle, J. Pallauf, O. Simon and J. Zentek. Supplemente zu Vorlesungen und Übungen in der Tierernährung. 11. ed. Hannover: Schaper Verlag; 2009.

Keller, K.M., B.D. Queiroz, L.A. Keller, J.M. Ribeiro, L.R. Cavaglieri, M.L. Pereyra, A.M. Dalcero and C.A. Rosa, 2007. The mycobiota and toxicity of equine feeds. Vet Res Commun. 31, 1037-1045.

Kinde, H., R.L. Bettey, A. Ardans, F.D. Galey, B.M. Daft, R.L. Walker, M.W. Eklund and J.W. Byrd, 1991. Clostridium botulinum type-C intoxication associated with consumption of processed alfalfa hay cubes in horses. J. Am. Vet. Med. Assoc. 199, 742-746.

Knight, D.A. and W.J. Tyznik, 1990. The effect of dietary selenium on humoral immunocompetence of ponies. J. Anim. Sci. 68, 1311-1317.

Knubben, J.M., L. Gygax, J. Auer, A. Furst and M. Stauffacher, 2008. [Frequency of diseases and injuries in the Swiss horse population]. Schweiz Arch Tierheilkd. 150, 399-408.

Kronfeld, D.S., P.L. Ferrante, L.E. Taylor and W. Tiegs, 1999. Partition of plasma hydrogen ion concentration changes during repeated sprints. Equine Vet. J. Suppl. 30, 380-383.

Lindholm, A. and B. Saltin, 1974. The physiological and biochemical response of standardbred horses to exercise of varying speed and duration. Acta Vet. Scand. 15, 310-324.

Lloyd, D.R. and R.J. Rose, 1995. Effects of sodium bicarbonate on fluid, electrolyte and acid-base balance in racehorses. Br. Vet. J. 151, 523-545.

Lomax, A.R. and P.C. Calder, 2009. Prebiotics, immune function, infection and inflammation: a review of the evidence. Br. J. Nutr. 101, 633-658.

Luthersson, N., S. Chumekamrai, J.C. Estepa and E. Aguilera-Tejero, 2005. Secondary hyperparathoidism in ponies in Northern Thailand. Pferdeheilkunde. 21, 97-98.

Luthersson, N., K.H. Nielsen, P. Harris and T.D. Parkin, 2009. Risk factors associated with equine gastric ulceration syndrome EGUS) in 201 horses in Denmark. Equine Vet. J. 41, 625-630.

McCarthy, H.E., N.P. French, G.B. Edwards, I.R. Poxton, D.F. Kelly, C.E. Payne-Johnson, K. Miller and C.J. Proudman, 2004. Equine grass sickness is associated with low antibody levels to Clostridium botulinum: a matched case-control study. Equine Vet. J. 36, 123-129.

McGivney, B.A., S.S. Eivers, D.E. MacHugh, J.N. MacLeod, G.M. O'Gorman, S.D. Park, L.M. Katz and E.W. Hill, 2009. Transcriptional adaptations following exercise in thoroughbred horse skeletal muscle highlights molecular mechanisms that lead to muscle hypertrophy. BMC Genomics. 10, 638.

McGorum, B.C., I.S. Henderson, D. Stirling, R. Wallace, C. Haggart and A.E. Thomas, 2009. Vitamin K deficiency bleeding in a Standardbred colt. J. Vet. Intern. Med. 23, 1307-1310.

Meyer, H. and L. Ahlswede, 1976. Über das intrauterine Wachstum und die Körperzusammensetzung von Fohlen sowie den Nährstoffbedarf tragender Stuten. Übers. Tierernährung. 4, 263-292.

Meyer, H., M. Schmidt, G. Lindemann and H. Muuss, 1982. [Prececal and postileal digestibility of elements (Ca, P, Mg) and trace elements (Cu, Zn, Mn) in the horse]. Fortschr. Tierphysiol. Tierernährg. 13, 61-69.

Milinovich, G.J., D.J. Trott, P.C. Burrell, A.W. van Eps, M.B. Thoefner, L.L. Blackall, R.A. Al Jassim, J.M. Morton and C.C. Pollitt, 2006. Changes in equine hindgut bacterial populations during oligofructose-induced laminitis. Environ. Microbiol. 8, 885-898.

Mithieux, G., P. Misery, C. Magnan, B. Pillot, A. Gautier-Stein, C. Bernard, F. Rajas and C. Zitoun, 2005. Portal sensing of intestinal gluconeogenesis is a mechanistic link in the diminution of food intake induced by diet protein. Cell Metab. 2, 321-329.

Mohammed, H.O., T.J. Divers, B.A. Summers and A. de Lahunta, 2007. Vitamin E deficiency and risk of equine motor neuron disease. Acta Vet. Scand. 49, 17.

Montgomerie, R.F., W.H. Savage and E.C. Dodds, 1929. Tetany in Welsh mountain ponies. Vet. Rec. 9, 319-334.

Moore, R.E., D. Knottenbelt, J.B. Matthews, R.J. Beynon and P.D. Whitfield, 2008. Biomarkers for ragwort poisoning in horses: identification of protein targets. BMC Vet. Res. 4, 30.

Nadeau, J.A., F.M. Andrews, A.G. Mathew, R.A. Argenzio, J.T. Blackford, M. Sohtell and A.M. Saxton, 2000a. Evaluation of diet as a cause of gastric ulcers in horses. Am. J. Vet.Res. 61, 784-790.

Nadeau, J.A., F.M. Andrews, A.G. Mathew, R.A. Argenzio, J.T. Blackford, M. Sohtell and A.M. Saxton, 2000b. Evaluation of diet as a cause of gastric ulcers in horses. Am J Vet Res. 61, 784-790.

Nation, P.N., 1989. Alsike clover poisoning: A review. Can. Vet. J. 30, 410-415.

Naude, T.W., R. Gerber, R.J. Smith and C.J. Botha, 2005. Datura contamination of hay as the suspected cause of an extensive outbreak of impaction colic in horses. J. S. Afr. Vet. Assoc. 76, 107-112.

Neafsey, P., G. Ginsberg, D. Hattis and B. Sonawane, 2009. Genetic polymorphism in cytochrome P450 2D6 (CYP2D6): Population distribution of CYP2D6 activity. J. Toxicol. Environ. Health B Crit. Rev. 12, 334-361.

Nebbia, C., M. Dacasto, A. Rossetto Giaccherino, A. Giuliano Albo and M. Carletti, 2003. Comparative expression of liver cytochrome P450-dependent monooxygenases in the horse and in other agricultural and laboratory species. Vet. J. 165, 53-64.

Osweiler, G.D., 2001. Mycotoxins. Vet. Clin. North Am. Equine Pract. 17, 547-566, viii.

Pandey, A., A.M. Hassen, D.R. Benedict and D.W. Fitzpatrick, 1990. Effect of UDP-glucuronyltransferase induction on zearalenone metabolism. Toxicol. Lett. 51, 203-211.

Pascoe, R.R., 1973. The nature and treatment of skin conditions observed in horses in Queensland. Aust. Vet. J. 49, 35-40.

Patterson, H.J., 1897. Horse feeding: Tests of the digestibility of oats, cornhay and the 'newcorn product'. Md. Agr. Expt. Sta. Bul. 51, 11-44.

Penumarthy, L., F.W. Oehme and R.H. Hayes, 1980. Lead, cadmium, and mercury tissue residues in healthy swine, cattle, dogs, and horses from the midwestern United States. Arch Environ. Contam. Toxicol. 9, 193-206.

Perryman, L.E., 2000. Primary immunodeficiencies of horses. Vet. Clin. North Am. Equine Pract. 16, 105-116, vii.

Perryman, L.E. and T.C. McGuire, 1980. Evaluation for immune system failures in horses and ponies. J. Am. Vet. Med. Assoc. 176, 1374-1377.

Pollitt, C.C. and C.T. Davies, 1998. Equine laminitis: its development coincides with increased sublamellar blood flow. Equine Vet. J. Suppl. 125-132.

Ralston, S.L. and C.A. Baile, 1982. Gastrointestinal stimuli in the control of feed intake in ponies. J. Anim. Sci. 55, 243-253.

Ramzan, P.H., T.D. Parkin and M.C. Shepherd, 2008. Lower respiratory tract disease in Thoroughbred racehorses: analysis of endoscopic data from a UK training yard. Equine Vet J. 40, 7-13.

Raymond, S.L., T.K. Smith and H.V. Swamy, 2005. Effects of feeding a blend of grains naturally contaminated with Fusarium mycotoxins on feed intake, metabolism, and indices of athletic performance of exercised horses. J. Anim. Sci. 83, 1267-1273.

Reeta Poso, A. and S. Hyyppa, 1999. Metabolic and hormonal changes after exercise in relation to muscle glycogen concentrations. Equine Vet J Suppl. 30, 332-336.

Richards, N., G. Hinch and J. Rowe, 2006. The effect of current grain feeding practices on hindgut starch fermentation and acidosis in the Australian racing Thoroughbred. Aust. Vet. J. 84, 402-407.

Ross, P.F., L.G. Rice, J.C. Reagor, G.D. Osweiler, T.M. Wilson, H.A. Nelson, D.L. Owens, R.D. Plattner, K.A. Harlin, J.L. Richard *et al.*, 1991. Fumonisin B1 concentrations in feeds from 45 confirmed equine leukoencephalomalacia cases. J. Vet. Diagn. Invest. 3, 238-241.

Sandin, A., K. Girma, B. Sjoholm, A. Lindholm and G. Nilsson, 1998. Effects of differently composed feeds and physical stress on plasma gastrin concentration in horses. Acta Vet. Scand. 39, 265-272.

Schott, H.C. Urinary Tract Disorders 2007.

Schuback, K., B. Essen-Gustavsson and S.G. Persson, 2002. Effect of sodium bicarbonate administration on metabolic responses to maximal exercise. Equine Vet. J. Suppl. 539-544.

Singh, R.P., 1970. A case of photosensitisation in horse. Indian Vet. J. 47, 450-451.

Smyth, G.B., D.W. Young and L.S. Hammond, 1989. Effects of diet and feeding on postprandial serum gastrin and insulin concentration in adult horses. Equine Vet. J. Suppl. 56-59.

Southwood, L.L., D.L. Evans, W.L. Bryden and R.J. Rose, 1993. Feeding practices in thoroughbred and standardbred racehorse stables. Aust. Vet. J. 70, 184-185.

Spoo, W., 2001. Industrial chemicals and the horse. Vet. Clin. North Am. Equine Pract. 17, 501-515.

Sticker, L.S., D.L. Thompson, Jr. and L.R. Gentry, 2001. Pituitary hormone and insulin responses to infusion of amino acids and N-methyl-D,L-aspartate in horses. J. Anim. Sci. 79, 735-744.

Stowe, H.D., 1967. Serum selenium and related parameters of naturally and experimentally fed horses. J. Nutr. 93, 60-64.

Thorens, B., 2008. Glucose sensing and the pathogenesis of obesity and type 2 diabetes. Int. J. Obes. Lond. 32 Suppl 6, S62-71.

Tinker, M.K., N.A. White, P. Lessard, C.D. Thatcher, K.D. Pelzer, B. Davis and D.K. Carmel, 1997. Prospective study of equine colic incidence and mortality. Equine Vet. J. 29, 448-453.

Uhlig, S., T. Vikoren, L. Ivanova and K. Handeland, 2007. Ergot alkaloids in Norwegian wild grasses: a mass spectrometric approach. Rapid Commun. Mass Spectrom. 21, 1651-1660.

Van Niekerk, F.E. and C.H. van Niekerk, 1997a. The effect of dietary protein on reproduction in the mare. I. The composition and evaluation of the digestibility of dietary protein from different sources. J. S. Afr. Vet. Assoc. 68, 78-80.

Van Niekerk, F.E. and C.H. van Niekerk, 1997b. The effect of dietary protein on reproduction in the mare. II. Growth of foals, body mass of mares and serum protein concentration of mares during the anovulatory, transitional and pregnant periods. J. S. Afr. Vet. Assoc. 68, 81-85.

Van Niekerk, F.E. and C.H. van Niekerk, 1997c. The effect of dietary protein on reproduction in the mare. III. Ovarian and uterine changes during the anovulatory, transitional and ovulatory periods in the non-pregnant mare. J. S. Afr. Vet. Assoc. 68, 86-92.

Van Niekerk, F.E. and C.H. van Niekerk, 1997d. The effect of dietary protein on reproduction in the mare. IV. Serum progestagen, FSH, LH and melatonin concentrations during the anovulatory, transitional and ovulatory periods in the non-pregnant mare. J. S. Afr. Vet. Assoc. 68, 114-120.

Van Niekerk, F.E. and C.H. van Niekerk, 1998a. The effect of dietary protein on reproduction in the mare. V. Endocrine changes and conception during the early post partum period. J. S. Afr. Vet. Assoc. 69, 81-88.

Van Niekerk, F.E. and C.H. van Niekerk, 1998b. The effect of dietary protein on reproduction in the mare. VI. Serum progestagen concentrations during pregnancy. J. S. Afr. Vet. Assoc. 69, 143-149.

Van Niekerk, F.E. and C.H. van Niekerk, 1998c. The effect of dietary protein on reproduction in the mare. VII. Embryonic development, early embryonic death, foetal losses and their relationship with serum progestagen. J. S. Afr. Vet. Assoc. 69, 150-155.

Venner, M., 1999. [Botulism in horses]. Dtsch. Tierarztl. Wochenschr. 106, 73.

Vervuert, I., K. Voigt, T. Hollands, D. Cuddeford and M. Coenen, 2009. Effect of feeding increasing quantities of starch on glycaemic and insulinaemic responses in healthy horses. Vet. J. 182, 67-72.

Waller, A.P., G.J. Heigenhauser, R.J. Geor, L.L. Spriet and M.I. Lindinger, 2009. Fluid and electrolyte supplementation after prolonged moderate-intensity exercise enhances muscle glycogen resynthesis in Standardbred horses. J. Appl. Physiol. 106, 91-100.

Whitlock, R.H. and C. Buckley, 1997. Botulism. Vet. Clin. North Am. Equine Pract. 13, 107-128.

Wichert, B., S. Nater, M.M. Wittenbrink, P. Wolf, K. Meyer and M. Wanner, 2008. Judgement of hygienic quality of roughage in horse stables in Switzerland. J Anim Physiol Anim Nutr (Berl. 92, 432-437.

Wilson, T.M., P.F. Ross, L.G. Rice, G.D. Osweiler, H.A. Nelson, D.L. Owens, R.D. Plattner, C. Reggiardo, T.H. Noon and J.W. Pickrell, 1990. Fumonisin B1 levels associated with an epizootic of equine leukoencephalomalacia. J. Vet. Diagn. Invest. 2, 213-216.

Wyse, C.A., K.A. McNie, V.J. Tannahill, J.K. Murray and S. Love, 2008. Prevalence of obesity in riding horses in Scotland. Vet. Rec. 162, 590-591.

Yeruham, I., Y. Avidar and S. Perl, 1999. An apparently gluten-induced photosensitivity in horses. Vet. Hum. Toxicol. 41, 386-388.

Zuntz, N. and O. Hagemann, 1894. Untersuchungen über den Stoffwechsel des Pferdes bei Ruhe und Arbeit. Landw. Jb. 27, 284-301.

Ration formulation in horses: the scope of interpretation

I. Vervuert[1] and D. Bergero[2]
[1]Institute of Animal Nutrition, Nutrition Diseases and Dietetics, Veterinary Faculty, University of Leipzig, Gustav-Kühn-Strasse 8, 04159 Leipzig, Germany
[2]Department of Animal Production, Epidemiology and Ecology, University of Torino, Via L. Da Vinci, 44 10095 Grugliasco (TO), Italy

Abstract

Recently, ration formulation in the horse has attracted much attention in order to maintain health, improve performance or to manage various health problems such as laminitis, colic, gastric ulcers or muscular disorders. To simplify the procedure, numerous ration calculation programs are available; however, the information provided regarding the sum of daily energy and nutrient intake needs to be carefully interpreted. In most cases, an over- or an undersupply of certain nutrients can be detected by performing a ration formulation, but to what extent such an imbalanced diet may cause health problems or not, needs to be evaluated by a nutritionist. In that context, information regarding nutrient allowance is of particular interest. On the other hand a well balanced diet with regards to energy and nutrient supply may also result in severe health problems, possibly related to inappropriate feedstuffs, poor quality ingredients, or an imbalanced forage/concentrate ratio. The present review focuses on the essential requirements and considers the adequacy as well as limitations of ration formulation in the horse.

Keywords: ration formulation

Information on feedstuffs

Amounts of forages and concentrates

For ration evaluation, information regarding the amount and type of feed given to the horse is necessary. Unfortunately, feedstuffs are very often offered on a volume basis, e.g. scoop of concentrate or 'slice' of hay and many owners are not truly aware of the weight of the feed given. In Germany and the UK, about 60% of horse owners are responsible for actually feeding their horses and only around 15% of those owners with horses in riding clubs or livery stables have this responsibility (Harris, 2000). Consequently, in many cases detailed information regarding the amount or nature of feed given, especially information on roughage intake is not available. This is a serious limitation to the evaluation of the actual diet and to improving feeding strategies in horses.

Feed composition

For dietary evaluation and to meet nutrient requirements, knowledge of the composition of the feed is essential. For commercial feeds, analysis of crude nutrients such as protein, oil and fibre and of macro and micro minerals is a routine procedure and information is freely available on the respective feed labels. The nutrient information required on the label is dependent on the regulations applicable in the country where the commercial feed is to be sold. In most cases, however, an extensive list of nutrients is usually available from other promotional information for commercial feed products. For cereals or grains, using tabular values or averages provides sufficient information on nutrients in most cases. However, feed composition of forages is challenging as their nutrient content can vary greatly (Table 1), depending on several factors such as climatic conditions, soil type, grass species, fertilization management, cutting time, harvesting technique, and other factors (Kienzle *et al.*, 2008).

Table 1. Composition of grass hay for horses from Bavaria (n=104) and Switzerland (n=52), adapted from Kienzle et al. *(2008).*

Nutrient	Bavaria				Switzerland			
	Mean ± SD	Median	Min	Max	Mean ± SD	Median	Min	Max
Protein, %	7.5±2.4	7.0	4.4	17.0	7.9±2.1	7.5	3.9	12.9
Fiber, %	31.4±3.4	31.8	18.8	36.6	30.0±2.8	30.5	22.4	36.2
Ca, g/kg	4.7±1.8	4.2	1.8	10.2	6.0±2.9	5.3	2.1	13.8
Mg, g/kg	1.8±1.0	1.5	0.7	5.3	1.5±0.6	1.4	0.2	2.8
P, g/kg	3.8±0.6	3.9	2.2	5.4	1.8±0.7	2.0	0.6	4.0
K, g/kg	20.0±6.0	20.2	5.3	34.0	15.4±5.1	14.6	7.1	28.9

Under practical conditions, ration formulation for a single horse is routinely performed using tabular values for forages despite the above mentioned limitations. For large horse units like stud farms, routine nutrient analysis of the respective forages is strongly recommended.

Demands of a ration for horses

For a balanced diet, the following points have to be considered:

- ration needs to be tailored to the individual's requirement;
- ration should fall within a horse's capacity for dry matter intake;
- ration should accommodate specific needs (e.g. adequate forage intake);
- ration should be tolerated well by the equine (e.g. limitation of starch and fat, nutrient tolerances).

Tailoring to the individuals requirement

There are several sources from which to obtain information on energy and nutrient requirements in horses, e.g. like NRC (2007), INRA (1990) or GEH (1994). For ration formulation and adequate assessment of the ration, further information on actual body weight, optimal body weight, body condition and the use or lifestage of the horse (maintenance, different stages of growth, performance or different stages of breeding) is essential. In most cases, obtaining information on the actual body weight is uncomplicated (e.g. mobile scaling service); however information about the optimal body weight and body condition is sometimes a barrier, especially in obese horses. Wyse *et al.*, (2008) reported a lack of awareness amongst horse owners that prevented them from accurately classifying their horse's body condition of their horses and this was most common in owners of obese horses. As optimal body weight of the horse should be the basis for the calculation of daily energy and nutrient intake the lack of awareness of obesity and optimal bodyweight and condition by horse owners, leads to misjudgement of particularly energy requirement. Energy requirement is also misjudged in relation to the classification of the work intensity in exercising horses (Coenen and Vervuert, 2001). Horse owners most often overestimate the amount of exercise undertaken by their horses e.g. a horse owner may classify exercise such as a 20 min walk, 30 min trot and 10 min gallop as medium to high working intensity, whilst in terms of energy expenditure the work intensity is actually low. Consequently, energy requirement is often overestimated by horse owners due to the misinterpretation of work load. On the other hand, energy and nutrient requirements should be only be used as an indicator or true requirements as external (e.g. ambient temperature) and internal factors (e.g. easy keepers, aged horses, dental condition stress) might increase or decrease the actual respective requirements. Dietary evaluation should also consider factors which may impair intestinal absorption of minerals and other nutrients, which can certainly increase the requirement for specific elements. For example phytic acid, which is known as an antinutritional factor hinders intestinal absorption of P, Ca, and other minerals including Fe, Mn, and Zn (McDowell, 2003). The proportion of P contained as phytate ranges from 59% to 70% in cereal seeds and 20% to 46% in legume seeds (McDowell, 2003). In horses, P appears to be mainly absorbed in the lower large intestine; therefore, phytate is more easily absorbed because it is degraded by bacteria in the hindgut (Schryver *et al.*, 1972; Matsui *et al.*, 1999). The addition of phytases did not improve P digestibility in horses (Morris-Stoker *et al.*, 2001; Van Doorn *et al.*, 2004a), but they did increase Ca digestibility in horses receiving a phytate-rich diet (Van Doorn *et al.*, 2004b).

Capacity for dry matter intake

In diary cows, the capacity for dry matter intake (DMI) is a common issue due to the high energy requirements of lactation. In horses, DMI should be adequate to satisfy energy requirements even during situations with a high energy demand such as lactation or strenuous exercise (Table 2).

In horses under maintenance conditions, mean voluntary DMI is about 2.6% of BW when hay is provided (Bochnia *et al.*, 2008),whilst the DMI of grazing horses varies between 3.2 up to 5.2% of BW (Smith *et al.*, 2007). Although the data is equivocal, DMI in lactating mares is reported to be in the region of 3-3.5% (Meyer and Coenen, 2002). In exercising horses, DMI varied between 2.5-3% of BW, however feeding schedule before and after exercise is of greater concern than limitations by DMI capacity (Vervuert, 2008).

However, many factors influence voluntary DMI (Figure 1) which have to be considered to ensure that the estimation of DMI, whilst within the expected DMI capacity is likely to be an accurate reflection of the true DMI for that individual.

For sport horses, the maximal daily DMI is not the specific target in most practical cases, because in reality high concentrate rations are provided, in spite of the capacity of the horse to consume a more fibrous ration. This practice increases the risk of the occurrence of stereotypies as it underestimates the intangible necessity, for horses to spend considerable time chewing and in other feeding related activities. The total estimated time required for horses to consume a daily ration in a stable ranges from 6-5 to 1.5 hours (completely pelleted diets), which is very far from the 10 hours or more spent by free ranging horses in the same activity. The 'non nutritional' aspects of feeding horses are pivotal and cannot be ignored without detriment to the health and welfare of horses.

Roughage intake

Horses are classified as non-ruminant 'roughage grazers' and the starch and oil fraction of plants plays only a minor role in their nutrition under normal circumstances (adapted from Ellis and Hill, 2005). A lack of roughage in the ration contributes to several health disturbances including gastric ulcers, hindgut acidosis, and behavioural problems (Meyer and Coenen, 2005). In general,

Table 2. Typical ration formulation in lactating mares (3rd month of lactation).

Feedstuff	Kg as fed	DM, kg	Digestible energy (MJ DE)
Hay	12	11.6	88
Compound feed	4	3.6	50
Soybean meal	0.3	0.27	4
Vegetable oil	0.3	0.3	11
Total	16.6	15.8	153

Mean DMI capacity: 18-21 kg (3-3.5% of BW), requirement lactating mare (3rd month): 153 MJ DE.

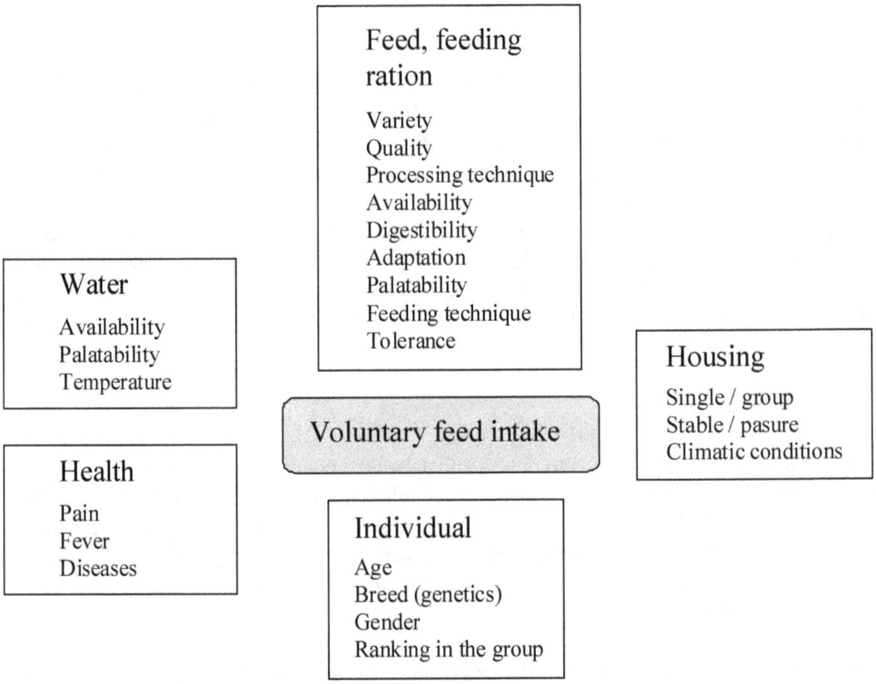

Figure 1. Significant factors affecting voluntary DMI in horses (adapted from Kamphues et al., *2009).*

ration calculations should always incorporate a minimum quantity of long roughage, such as fresh grass (DM content: <20%), hay (DM content >80%), haylage (DM content >45%) or silage (DM content <45%). In contrast to silages, haylages show a restricted fermentation profile and higher pH values which may result in overgrowth of yeasts after opening the respective bales thereby limiting consumption time (Kamphues *et al.,* 2009). When compared to hay, well preserved haylages or silages have less aeroallergens which is clearly favourable for horses with respiratory diseases. The German recommendations, which are currently under revision, suggest a daily minimum of 1.5 kg roughage (based on DM)/100 kg BW although the term 'roughage' is quite imprecise.

Limitation of starch intake

Cereal carbohydrates comprise an important part of the diet of horses with a high daily energy requirement. Results from different surveys indicate that horses competing to a high level of performance are fed an average of 5.5 and up to 7.3 kg of grain-containing concentrate per day (Richards *et al.,* 2006, Williamson *et al.,* 2007). Several health problems such as gastric ulcers, colic or hindgut acidosis are associated with the intake of high amounts of grains. To reduce the risk of a significant ileal outflow of starch into the hindgut, it is recommended that highly processed grains and compound feeds be fed (Vervuert, 2009). As in humans, the glycaemic response of the horse to starchy foods is of interest for its impact on exercise performance, behaviour, obesity, insulin resistance, laminitis and osteochondrosis. It has been shown that a starch intake of less than 1.1 g/kg BW per meal produced only moderate glucose and insulin responses even though highly processed cereals were used (with an assumed precaecal starch digestibility of >90%) in order to provide the maximum protection against an undesirable bacterial starch break-down in the gastrointestinal tract (Vervuert, 2009). For this reason a meal size of 0.3 kg/100 kg BW is recommended for compound feeds or grains with a starch content of between 30% and 40%. Even further limitation to meal size is necessary in feedstuffs with starch levels exceeding 40%.

Limitation of fat intake

For horses, especially performance horses, there is increasing interest in increasing the lipid content of the diet usually in the form of vegetable based oil (Harris *et al.*, 1999). There are numerous potential advantages to fat supplementation of the equine diet including higher energy density (Meyer and Coenen, 2002), lowered heat production during exercise due to improved efficiency in the utilisation of metabolisable energy (Kronfeld *et al.*, 1994), a reduction in breathing effort due to lower carbon dioxide production (Kronfeld *et al.*, 1994), higher resting muscle glycogen stores (Hambleton *et al.*, 1980), metabolic adaptation to fat supplementation in muscle (Orme *et al.*, 1997) that facilitates a glucose sparing effect during exercise (Treiber *et al.*, 2008). However, fat intake should be limited to a daily maximum <1 g/kg BW as a higher fat intake is associated with an impaired fiber digestibility in the hindgut (Zeyner *et al.*, 2002).

Energy and protein tolerances

A prior aim of ration formulation is to detect imbalances in energy and nutrient intake. Knowledge of specific individual requirements helps to detect deficient, marginal, adequate, high or excessive nutrient intake provided by the actual ration. In the case of a detected deficiency, supplementation of the respective nutrient is recommended. However, different energy evaluation systems (e.g. Germany: digestible energy, France: net energy) exist in Europe which have to be considered (e.g. Martin-Rosset, 2008a).

In recent years, obesity linked to inappropriately high energy intake has become of some concern. The lack of awareness of obesity by horse owners has already been discussed. In healthy horses, an excess of energy intake (often in combination with a lack of exercise) will result in obesity and potentially an increased risk of developing metabolic syndrome, therefore ration formulation can help to balance energy intake.

Protein requirement is basically estimated either by a factorial method from metabolic data and then appropriate nutrient conversion rates for calculation, or/and by whole animal trials (e.g. feeding experiments). An extensive review of protein requirements in the equine has been recently published by Martin-Rosset (2008b).

Whilst an excess of protein is often of concern to horse owners, in practice excessive protein intake in healthy horses has little impact on health or exercise performance. Excess nitrogen is simply removed by the kidneys in the form of ammonia, which is then converted by the liver to urea and which is finally excreted by the kidneys. Practically, protein intake should not exceed 2 g digestible crude protein/kg BW (Meyer and Coenen, 2002) as conversion from nitrogen into urea is an energy demanding process. Furthermore, it should be emphasized that a significant protein reduction in the diet is recommended in the management of liver (Bergero and Nery, 2008) or kidney diseases (Jarvis, 2009). However, requirements for essential amino acids are not well defined except for lysine mainly in the growing horse (Martin-Rosset, 2008b).

Mineral and vitamin tolerances

However, all nutrients, whether essential or nonessential can adversely affect animal health when amounts in the diet become excessive. The prevention of excessive intake of nutrients e.g. minerals (Figure 2) is therefore a fundamental aim of ration evaluation. However, this requires the levels at which a nutrient becomes toxic in the equine to be established to aid diet evaluation and ration formulation. The maximum tolerable level of a mineral or vitamin is defined as the dietary level that, when fed for a defined period of time, will not impair animal health and performance (NRC, 2005). The concern of excessive intake differs between minerals and vitamins as shown in Table 3.

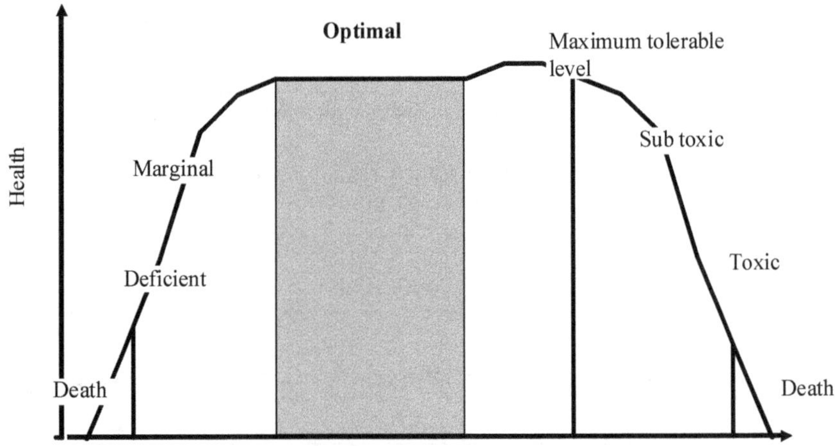

Figure 2. Mineral intake and health (adapted from McDowell, 2003).

Table 3. Maximum tolerable levels of minerals and vitamins in the feed for equines (expressed as mg/kg or % of DM, McDowell, 2003).

Element	Concern for health	mg/kg DM, % of DM or IU/kg DM
Calcium, %	Medium	2
Phosphorus, %	Medium	1
Magnesium, %	Low	0.8
Potassium, %	Medium	3
Sodium Chloride, %	High	3-6
Copper, mg/kg	Low	800
Zinc, mg/kg	Low	500
Selenium, mg/kg	High	2
Iodine, mg/kg	Medium	5
Iron, mg/kg	Medium	500
Vitamin A, IU/kg	Medium	16,000
Vitamin D, IU/kg	High	(2,200), possibly lower
Vitamin E, mg/kg	Low	Not assessed

Besides the total intake of a nutrient in a ration, differences in the chemical and physical forms can affect the availability and so maximum tolerances of elements and have to be considered in this context. For example, Fe is far more available as ferrous sulphate than as ferric oxide (McDowell, 2003). The interrelationships between minerals or other factors should also be considered (Table 4).

For example, Se interferes with sulphur due to similarity in chemical structure. In horses, high Se intake (Table 5), but still below the maximum tolerable levels, has been shown to increase Se content in the hoof horn (Figure 3) due to the substitution of sulphur with Se as part of the S-S disulphide bridges in keratin (the principle protein present in hoof and hair) which may affect the functionality of keratin (Coenen *et al.*, 1996).

Interference of absorption as a result of mineral interaction may be overestimated. For example, Zn may reduce the absorption of Cu, as it competes directly for the same transport mechanisms in the

Table 4. Supposed interrelationships of trace elements with other elements (adapted from Kamphues et al., *2009).*

Trace element	Interrelationship with minerals or other factors
Cu	Mo, S, Ca, Fe, Zn, Cd, Ag, phytic acid
Fe	Ca, P, Cu, Zn, phytic acid
Mn	P, phytic acid
Se	S
Zn	Ca, Cu, phytic acid

Table 5. Ration evaluation a dressage horse (BW 550 kg) performing for the Olympics 2008 (Vervuert and Coenen, unpublished).

Components/day	DM intake (kg)	Cu (mg)	Zn (mg)	Se (mg)
Hay	5.5	36	154	0.16
Compound feed	4.5	120	175	2.00
Magnolythe S100	0.12	96	408	1.80
Elytan	0.04	8	58	
Horsal	0.1	50	250	1.00
Total ration	11.46	310	1,045	4.96[a]
Requirement GEH		93	620	1.55

[a]0.43 mg Se/kg DM.

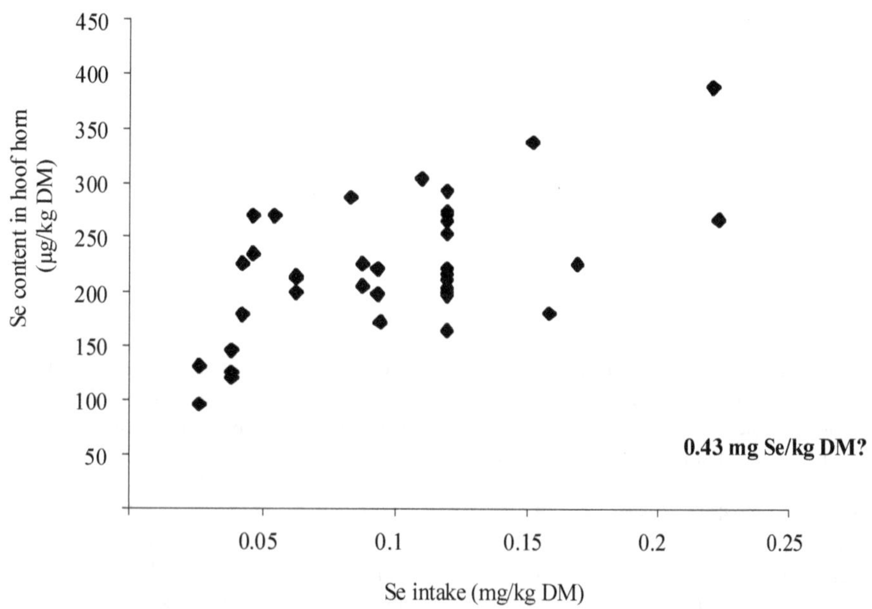

Figure 3. Relationship between Se intake (mg/kg DM) and Se content (µg/kg DM) in hoof horn (Coenen and Spitzlei, 1996).

small intestine. Furthermore, Zn can also impair Cu absorption indirectly by enriching the level of metallothionein, a protein which binds Cu, thereby limiting its rate of transport through mucosal cells (McDowell, 2003). Although studies in horses are rare, it seems that a very high level of dietary Zn is necessary to impair Cu absorption. Foals fed a diet containing either 29 (control) or 250 mg/kg of zinc maintained normal serum Cu and Zn concentrations for 14 to 15 weeks, whereas those fed 1000 or 2,000 mg Zn/kg became hypocupremic (Bridges and Moffitt, 1990). Coger *et al.* (1987) found no differences in apparent Cu absorption or tissue concentrations (liver, kidney or spleen) feeding growing horses either a diet with 38 mg Zn/kg DM (control) or 1,170 mg Zn/kg DM. In this study, excessive Zn intake resulted in a drop in apparent Zn absorption from 89% (control diet) to 6% (high Zn diet) which is a supposed mechanism to avoid Zn toxicosis.

Table 6. Control points to monitor an adequate ration concept.

Item	Method	Explanatory notes
DMI	Weighing of feeds	Clarifies amounts of actual feed fed
	Sensory exam of feeds	Hygienic quality: Depression in feed intake by spoilage
	Botanical exam of feeds	Detects inappropriate feeds
	Scaling BW or BCS	Provides retrospective information on energy intake
Well-tolerance	Sensory and microbial exam of feeds	Analysis of moulds/mycotoxins as spoilage can impact health
	Botanical exam of feeds	Toxic plants
	Examination of feeding management	Forage to concentrate ratios Appropriate energy sources Attention: feeding schedule before and after exercise
	Weighing feeds	Control of amounts of concentrates and forages,
	Faeces quality	Deviations from normal conditions e.g. starch flow in the hindgut
Energy intake	Measurement of BW or BCS	Provides information on energy intake or diseases
Protein intake	Blood urea or blood nitrogen	
Ca, P, Mg, Na, Cl	Plasma, serum + urine Calculation of fractional excretion (FE) by the following formula: FE: Electrolyte urine/electrolyte plasma or serum x creatinine plasma or serum/creatinine urine	Limited information for Ca and P, relative sensitive indicator for Mg, Na and Cl
Trace elements	Plasma, serum	Reliable indicators for Cu and Se
Vitamin D and E	Plasma, serum	Reliable
Carotenoids	Sensory exam of roughage	The degree of green colour in a roughage is a reliable indicator of its carotene content (carotene content in the roughage may adequate to maintain adequate Vitamin A supply)

Control of ration concept

An optimal ration formulation should provide adequate energy and nutrient intake from well-tolerated feedstuffs. The quality of the feedstuffs, feed intake behaviour, and health should in addition all be monitored in order to ensure the efficacy of the ration. In many cases, it is more appropriate to investigate the daily ration, individual feed intake behaviour and horse health to provide more information on the suitability of the ration rather than relying on blood analyses, which may not truly reflect body status. A summary of the main dietary aspects that can be monitored to assess the efficacy of the diet are described in Table 6.

References

Bergero, D. and J. Nery, 2008. Hepatic diseases in horses. J. Anim. Physiol. Anim. Nutr. 92(3), 345-55.

Bochnia, M., I. Vervuert and M. Coenen, 2008. The impact of different feeding orders of hay or hay chaff combined with a concentrate meal on feed intake pattern in horses. Proc. Soc. Nutr. Physiol. Göttingen, 17, 91.

Bridges, C.H. and P.G. Moffitt, 1990. Influence of variable content of dietary zinc on copper metabolism of weanling foals. Am. J. Vet. Res. 51, 275-80.

Coenen, M. and S. Spitzlei, 1996. Zur Zusammensetzung des Hufhorns in Abhängigkeit von Alter, Rasse und Hufhornqualität. Pferdeheilkunde 12, 279-283.

Coenen, M. and I. Vervuert, 2001. Hinweise zur Rationsberechnung für Pferde. In: Dem Pferd auf's Maul geschaut. Eds. Coenen M. and I. Vervuert, ISBN 3-00-008728-1, pp. 97-100.

Ellis, A.D. and J. Hill, 2005. Nutritional physiology of the horse. Nottingham University Press, pp. 1-68.

GEH, 1994. Energie- und Nährstoffbedarf landwirtschaftlicher Nutztiere. Empfehlungen zur Energie- und Nährstoffversorgung der Pferde. DLG Verlag, Frankfurt (Main), pp. 41.

Hambleton, P.L., L.M. Slade, D.W. Hamar, E.W. Kienholz and L.D. Lewis, 1980. Dietary fat and exercise conditioning effect on metabolic parameters in the horse. J. Anim. Sci. 51(6), 1330-9.

Harris, P.A., J.D. Pagan, K.G. Crandell and N. Davidson, 1990. Effect of feeding thoroughbred horses a high unsaturated or saturated vegetable oil supplemented diet for 6 months following a 10 month fat acclimation. Equine Vet. J. Suppl. 30, 468-74.

INRA, 1990. Alimentation des Chevaux. W. Martin-Rosset (ed) INRA Publications, Route de St Cyr, 78000 Versailles

Jarvis, N.G., 2009. Nutrition of the aged horse. Vet. Clin. North Am. Equine Pract. 25(1), 155-66.

Kamphues, J, M. Coenen, C. Iben, E. Kienzle, J. Pallauf, O. Simon, M. Wanner and J. Zentek, 2009. Supplemente zu Vorlesungen und Übungen in der Tierernährung. M and H Schaper, Hannover, pp 71-156.

Kienzle, E, F. Möllmann, S. Nater, M. Wanner and B. Wichert, 2008. Mineral content of hay harvested in Bavarian and Swiss horse farms. Predictive value of cutting time, number of cut, botanical composition, origin and fertilization. J. Anim. Physiol. Anim. Nutr. 292, 712-7.

Kronfeld, D.S., P.L. Ferrante and D. Grandjean, 1994. Optimal nutrition for athletic performance, with emphasis on fat adaptation in dogs and horses. J. Nutr. 124 (12 Suppl), 2745S-2753S.

Martin-Rosset, W., 2008a. Energy requirements and allowances of excercising horses. In: Nutrition of the exercising horse. Eds. Saastamoinen, M.T. and W. Martin Rosset, 103-138.

Martin-Rosset, W., 2008b. Protein requirements and allowances of excercising horses. In: Nutrition of the exercising horse. Eds. Saastamoinen, M.T. and W. Martin Rosset, 183-206.

Matsui T., Y. Murakami, H. Yano, H. Fujikama, T. Osawa and Y. Asai, 1999. Phytate and phosphorus movements in the digestive tract of horses. Equine Vet. J. Suppl. 30, 71- 76.

McDowell, L., (ed.), 2003. Minerals in animal and human nutrition. Elsevier, Amsterdam, The Netherlands.

Meyer, H. and M. Coenen, (eds.), 2004. Pferdefütterung. Parey Verlag, Berlin, Germany.

Morris-Stoker, L.B., L.A. Baker, J.L. Pipkin, R.C. Bachman and J.C. Haliburton, 2001. The effect of supplemental phytase on nutrient digestibility on mature horses. Seventeenth symposium of the Equine Nutrition and Physiology Society, May 31-June 2, Lexington, KY, USA. 48-53.

National Research Council, 2005. Mineral tolerance of animals. 2nd ed. National Academies Press, Washington, DC, USA.

National Research Council, 2007. Nutrient requirements of horses. 6th ed. National Academies Press, Washington, DC, USA.

Richards, N., G.N. Hinchi and J.B. Rowe, 2006. The effect of current grain feeding practices on hindgut starch fermentation an acidosis in the Australian racing thoroughbred. Austr. Vet. J. 84(11), 402-407.

Smith, D.G., R.W. Mayes, T. Hollands, D. Cuddeford, H. Yule and E. Gillen, 2007. Validating the alkane pair technique to estimate dry matter intake in equids. J. Agric. Sci. 145(3), 273-281.

Schryver, H.F., H.F. Hintz, P.H. Craig, D.E. Hogue and J.E. Lowe, 1972. Site of phosphorous absorption from the intestine of the horse. J. Nutr. 102, 143-148.

Treiber, K.H, R.J. Geor, R.C. Boston, T.M. Hess, P.A. Harris and D.S. Kronfeld, 2008. Dietary energy source affects glucose kinetics in trained Arabian geldings at rest and during endurance exercise. J Nutr. 138(5), 964-70.

Van Doorn, D.A., M.E. Van der Spek, H. Everts, H. Wouterse and A.C. Beynen, 2004a. The influence of calcium intake on phosphorus digestibility in mature ponies. J. Anim. Physiol. A. Anim. Nutr. 88, 412-418.

Van Doorn, D.A., H. Everts, H. Wouterse and A.C. Beynen, 2004b. The apparent digestibility of phytate phosphorus and the influence of supplemental phytase in horses. J. Anim. Sci., 82, 1756-1763.

Vervuert, I., 2009. Food properties affecting starch digestion by healthy horses as measured by glycaemic and insulinaemic responses. Habilitationsschrift, University of Leipzig.

Vervuert, I. and M. Coenen, 2008. Feeding practice in horses competing for the Olympics. Unpublished data.

Vervuert, I., 2008. Feeding schedule during competition day in performance horses. In: Nutrition of the exercising horse. Eds. Saastamoinen MT, Martin Rosset W. 295-302.

Williamson, A., C.W. Rogers and E.C. Firth, 2007. A survey of feeding, management and faecal pH of Thoroughbred racehorses in the North Island of New Zealand. N. Z. Vet. J. 55(6), 337-41.

Wyse, C.A., K.A. McNie, V.J. Tannahill, J.K. Murray and S. Love, 2008. Prevalence of obesity in riding horses in Scotland. Vet. Rec. 162(18), 590-1.

Zeyner, A., J. Bessert and J. M, Gropp, 2002. Effect of feeding exercised horses on high-starch or high-fat diets for 390 days. Equine Vet. J. Suppl. 34, 50-7.

Metabolic actions of intermittent exogenous equine parathyroid hormone (ePTH 1-37) in horses

K. Weisrock[1], W. Martin-Rosset[2], B. Remy[3], W.G. Forssmann[4], N. Parvizi[5], M. Coenen[1] and I. Vervuert[1]
[1]Institute of Animal Nutrition, Nutrition Diseases and Dietetics, Faculty of Veterinary Medicine, University of Leipzig, Germany
[2]INRA Center Clermont- Ferrand, 63122 Saint-Genés Champanelle, France
[3]Department of Physiology of Reproduction, Faculty of Veterinary Medicine, University of Liège, 4000, Belgium
[4]IPF PharmaCeuticals, 30625 Hannover, Germany
[5]Friedrich Loeffler Institute (FLI), Institute of Animal Breeding, Functional Genetics and Bioregulation, 31535 Neustadt, Germany

Abstract

In humans, intermittent administration of parathyroid hormone (PTH) and its fragments have been shown to have anabolic skeletal effects and reduce fracture rates in osteoporotic stages. In horses, there is no information on the effects of exogenous intermittent application of PTH or PTH fragments on calcium and bone metabolism. Intermittent ePTH (1-37) subcutaneous application of 0.5 µg ePTH/kg BW (n=6), or of a placebo (n=6) was given in healthy horses for 120 days. Blood (ionized Ca^{2+}, total Ca_T, inorganic P_i, creatinine, intact PTH (1-84)) and urine samples (Ca, P, creatinine) were taken every 30 days 5 h after ePTH or placebo application. Bone mineral density (BMD) and bone mineral content (BMC) was performed with dual X-ray absorptiometry (DXA) on metacarpus and calcaneus before and after 120 days of ePTH or placebo administration. A significant temporary increase in blood Ca^{2+} and plasma Ca_T occurred 5 h after application of ePTH with a return to baseline levels 24 h post inj. Levels of endogenous intact PTH and fractional Ca and P excretion showed no differences between ePTH treatment and placebo group. During observation period, cancellous BMD of calcaneus increased significantly in both groups, but without any treatment related effect. Long term application of ePTH had no adverse effects in healthy horses. The response of blood Ca reflected the responsiveness of the target cells to PTH, while BMD and BMC showed no significant differences between placebo and exogenous ePTH.

Keywords: bone mineral density, bone strength, resorption, blood calcium

Introduction

Bone is a dynamic tissue which is characterized by continuous bone formation and bone resorption. There is a very strong coupling between the two processes. Bone resorption and bone formation are physiologically in a tightly regulated balance (Compston, 2007). In horses bone metabolism can be evaluated with bone resorption and formation markers in blood, bone mineral density (BMD), bone mineral contend (BMC) and strengh. On bone metabolism, clinical studies have proven influences of exercise (Kiley-Worthington, 1990), supplements as calcium and phosphorus (Vervuert *et al.*, 2002; Schryver *et al.*, 1970) and pharmacologically active substances as glucocorticoides (Geor *et al.*, 1995) and estrogen (Fox *et al.*, 2007).

In humans and animal species, intermittent administration of parathyroid hormone (PTH) and its fragments have been shown to have anabolic skeletal effects and reduce fracture rates in osteoporotic stages (Compston, 2007; Lindsay *et al.*, 2007). However, in horses there is almost no information on the effects of exogenous intermittent application of PTH or PTH fragments on calcium and bone metabolism (Vervuert *et al.*, 2005). Therefore, the aim of the study was to evaluate the effects of long-term daily administration of equine PTH fragment (ePTH 1-37) on calcium and bone metabolism during a typical pasturing period in healthy adult horses.

Material and methods

Long-term effects were investigated by 120 days of daily, intermittent ePTH (1-37) subcutaneous application of 0.5 µg ePTH/kg BW (n=6), or of a placebo (n=6) in healthy adult horses (geldings, mean BW: 598±41.3 kg, mean age: 10.4±0.8 years) which were kept on pasture. Blood samples and urine samples were taken every 30 days 5 h after ePTH or placebo application. Additionally, at days 90 and 120 a basal blood sample was obtained prior to ePTH or placebo administration (24 h post injection). Blood was analysed for ionized Ca^{2+}, total plasma Ca_T, inorganic P_i, creatinine, serum intact PTH (1-84) and the bone markers serum equine osteocalcin (eOC, Carstanjen *et al.*, 2003), carboxyterminal telopeptide of type I collagen (ICTP), bone-specific alkaline phosphatase (BAP, Jackson *et al.*, 1996) and carboxy-terminal cross-linked telopeptide of type I collagen (CTX-1, Carstanjen *et al.*, 2004). Urine was analysed for Ca, P and creatinine. Bone mineral density (BMD) and bone mineral content (BMC) was performed with dual X-ray absorptiometry (DXA) on metacarpus and calcaneus before and after 120 days of ePTH or placebo administration (Donabedian *et al.*, 2005).

Results and discussion

A temporary increase in mean blood Ca^{2+} and plasma Ca_T occurred 5 h after application of ePTH with a return to baseline levels 24 h post injection (treatment $P<0.05$). Mean levels of endogenous intact PTH and fractional Ca and P excretion showed no differences between ePTH treatment and placebo group. During observation period, cancellous BMD of calcaneus increased significantly in both groups, but BMD and BMC for the calcaneus and metacarpus showed no differences between ePTH treatment and placebo group. Bone formation markers (eOC and BAP) showed no differences between the beginning and the end of treatment period in both groups. A decrease in ICTP, a marker of bone resorption, was monitored during pastureing time in both treatment groups. These changes were not detected for CTX-1, another marker of bone resorption.

The results indicated the long term intermittent application of 0.5 µg ePTH/kg BW generally had no negative effects in healthy horses. The response of Ca in blood reflected the responsiveness of the target cells to exogenous PTH application, while bone markers and DXA bone analysis showed no significant differences between placebo and exogenous ePTH group during long term application (Table 1). The increase in cancellous BMD is likely an effect of summer pasturing period (Hoekstra *et al.*, 1999), may be related to a reduced bone resorption rather than an increase in bone formation, reflected by the drop in ICTP, a reliable indicator of bone resorption (Table 2). As expected, bone formation seemed to be more distinct in healthy horses when compared to those results obtained in osteoporotic stages in humans or other animal species.

Table 1. Bone mineral density (BMD) (mean ± SD, g/cm²) and bone mineral content (BMC) (mean ± SD, g/cm³) before and after 120 days of placebo (n=6) or ePTH treatment (n=6).

Localization	Treatment	BMD		BMC	
		Before	After	Before	After
Calcaneus	Placebo	1.76 ± 0.09^a	1.90 ± 0.14^b	12.6 ± 1.38^a	13.2 ± 0.77^a
	ePTH	2.03 ± 0.33^a	2.16 ± 0.40^b	14.4 ± 2.57^a	14.2 ± 2.86^a
Metacarpus	Placebo	2.53 ± 0.38^a	2.66 ± 0.32^a	16.9 ± 3.10^a	20.6 ± 2.53^a
	ePTH	2.63 ± 0.33^a	2.77 ± 0.22^a	14.6 ± 4.17^a	17.0 ± 7.36^a

Means in the same row with different superscripts are significantly different ($P<0.05$)

Table 2. Mean concentrations (±SD) of bone-specific alkaline phosphatase (BAP) (ng/ml), eOC (ng/ml), carboxyterminal telopeptide of type I collagen (ICTP) (ng/ml) and carboxy-terminal cross-linked telopeptide of type I collagen (CTX-1) (ng/ml) before or during experimental period of placebo (n=6) or ePTH treatment (n=6).

Parameter	Treatment	Experimental period (days)				
		0	30	60	90	120
BAP	Placebo	34.6±9.72^a	38.4±8.41^a	37.3±8.16^a	41.3±7.81^b	30.5±9.40^a
	ePTH	26.8±6.55^a	35.8±13.9^b	38.7±11.9^b	28.2±8.43^a	28.0±7.80^a
eOC	Placebo	9.72±13.94	9.19±5.14	5.28±4.36	8.15±6.31	8.45±4.44
	ePTH	12.9±7.73	15.1±3.60	11.1±2.66	17.2±3.66	17.5±2.47
ICTP	Placebo	5.40±0.22^a	6.43±0.47^b	4.85±0.28^b	5.05±0.48^a	4.10±0.64^b
	ePTH	5.92±1.30^a	7.12±0.67^b	5.82±0.77^a	6.03±0.88^a	5.25±0.89^b
CTX-1	Placebo	0.14±0.04	0.24±0.17	0.25±0.16	0.20±0.04	0.15±0.05
	ePTH	0.18±0.10^a	0.31±0.15^b	0.30±0.18^a	0.41±0.09^b	0.27±0.13^a

Means in the same row with different superscripts are significantly different to day 0 ($P<0.05$).

Acknowledgements

The study was funded by the German Research Foundation (DFG).

References

Carstanjen, B., J. Sulon, H. Banga-Mboko, J.F. Beckers and B. Remy, 2003. Development and validation of a specific radioimmunoassay for equine osteocalcin. Domest Anim Endocrinol. 24(1), 31-41.

Carstanjen, B., N.R. Hoyle, A. Gabriel, O. Hars, C. Sandersen, H. Amory and B. Remy, 2004. Evaluation of plasma carboxy-terminal cross-linking telopeptide of type I collagen concentration in horses. Am J Vet Res. 65(1), 104-109.

Compston, J.E., 2007. Skeletal actions of intermittent parathyroid hormone: effects on bone remodelling and structure. Bone 40(6), 1447-1452.

Donabedian, M., C. Delguste, G. Perona, P. Lebecque, F. Duboeuf, O. Lepage and W. Martin-Rosset W., 2005. Third metacarpal bone mineral density assessment in the standing horse by dual X-ray absorptiometry: suitability, precision and accuracy. Vet Comp Orthop Traumatol. 18(1), 26-30.

Fox, J., M.A. Miller, M.K. Newman, C.H. Turner, R.R. Recker and S.Y. Smith, 2007. Treatment of skeletally mature ovariectomized rhesus monkeys with PTH(1-84) for 16 months increases bone formation and density and improves trabecular architecture and biomechanical properties at the lumbar spine. J Bone Miner Res. 22(2), 260-273.

Geor, R., E. Hope, L. Lauper, S. Piela, J. Klassen, V. King and M. Murphy, 1995. Effect of glucocorticoids on serum osteocalcin concentration in horses. Am J Vet Res. 56(9), 1201-1205.

Hoekstra, K.E., B.D. Nielsen, M.W. Orth, D.S. Rosenstein, H.C. Schott and J.E. Shelle, 1999. Comparison of bone mineral content and biochemical markers of bone metabolism in stall- vs. pasture-reared horses. Equine Vet J Suppl. 30, 601-604.

Jackson, B., R. Eastell, R.G. Russell, L.E. Lanyon and J.S. Price, 1996. Measurement of bone specific alkaline phosphatase in the horse: a comparison of two techniques. Res Vet Sci. 61(2), 160-164.

Kiley-Worthington, M., 1990. The Behavior of Horses in Relation to Management and Training – towards ethologically sound Environments. Journal of Equine Veterinary Science. 10(1), 62-75.

Lindsay, R., H. Zhou, F. Cosman, J. Nieves, D.W. Dempster and A.B. Hodsman, 2007. Effects of a One-Month Treatment with Parathyroid Hormone (1-34) on Bone Formation on Cancellous, Endocortical and Periosteal Surfaces of the Human Ilium. J Bone Miner Res. 22(4), 495-502.

Schryver, H.F., P.H. Craig and H.F. Hintz, 1970. Calcium metabolism in ponies fed varying levels of calcium. J Nutr. 100(8), 955-964.

Vervuert, I., K. von Scheidt, S. Winkelsett, W.G. Forssmann and M. Coenen, 2005. Intermittent administration of human parathyroid fragment (hPTH 1-37) on calcium and phosphorus homeostasis and bone markers in resting horses: a preliminary study. Equine Comp Exercise Physiol. 2(4), 245-251.

Vervuert, I., M. Coenen, U. Wedemeyer, C. Chrobok, J. Harmeyer and H.P. Sporleder, 2002. Calcium homeostasis and intact plasma parathyroid hormone during exercise and training in young Standardbred horses. Equine Vet J. 34(7), 713-718.

Assessment of bone mineral density by dual X-ray absorptiometry in the horse: a methodological approach

I. Vervuert[1], J. Junge[1], W. Martin-Rosset[2] and M. Coenen[1]
[1]Institute of Animal Nutrition, Nutrition Diseases and Dietetics, Faculty of Veterinary Medicine, University of Leipzig, Germany
[2]INRA, Department of Animal Science, St. Genes Champanelle, France

Abstract

The aim of the present study was to evaluate a transportable device system to assess bone mineral density in horses. Bone mineral density (BMD) of the third metacarpus and metatarsus was assessed in 103 slaughter horses (range of age: 2-27 years, 7 stallions, 40 mares, 56 geldings, different breeds) using the transportable device system Lunar PIXI® (30200 MDL, GE Healthcare, Farfield, US). After BMD analysis, samples of the third metacarpal or metatarsal region were prepared for analysis of bone ash, Ca, P and Mg. Mean BMD (± SD) was significant higher in the third metatarsus (3.7±0.2 g/cm^2) in comparison to the third metacarpus (3.0±0.3 g/cm^2), but without any gender or age related effects. In the bone ash, mean Ca, P and Mg content did not differ between location, gender and age. As the main finding, there was no correlation between BMD and bone ash (r=0.06), Ca (r=0.017), P (r=0.09) or Mg (r=0.08). From the present data, it seemed that the transportable device system Lunar PIXI® did not precisely reflect bone mineral amount as DXA evaluates only mineral content of the surface.

Keywords: bone mineral density, dual x-ray absorptiometry

Introduction

In horses, monitoring bone metabolism for early detection of bone mineral losses during growth or strenuous training periods is of particular interest. In general, training in horses is known to increase bone mineral density on the dorsal aspect of the third carpal bone after several month of training (McCarthy and Jeffcott, 1992). These bone remodelling processes are necessary for the skeleton to adapt to mechanical stresses. Failure of this adaptive response can lead to injuries to the skeleton, which is a particular problem in young racehorses, because these horses start their rigorous training when they are still growing (Nielsen *et al.*, 1997). Quantitative ultrasound as well as urine and blood markers of bone formation and resorption are suitable tools for the investigation of bone metabolism and bone´s responsiveness to changes in its mechanical properties in horses during race training (Carstanjen *et al.*, 2003, Lepage *et al.*, 2001); however imaging techniques like dual X-ray absorptiometry might extend the tool of diagnosis.

Dual X-ray absorptiometry (DXA) is one of the preferred methods to assess bone mineral density (BMD) in humans. However, DXA (e.g. transportable device Lunar PIXI®) for the analysis of BMD of the third metacarpus or calcaneus is rarely used in horses and validation was performed only in a very limited number of horses (Carstanjen *et al.*, 2003, Donabedian *et al.*, 2005).

Material and methods

BMD of the third metacarpus and metatarsus was assessed in 103 slaughter horses (range of age: 2-27 years, 7 stallions, 40 mares, 56 geldings, different breeds) using the transportable osteodensitometer Lunar PIXI® (dual-energy x-ray, 30200 MDL, GE Healthcare, Farfield, US). The mid region of the metacarpus or metatarsus was defined by anatomical points, BMD was also analysed 1, 2 or 3 cm proximal or distal apart from the mid region. After BMD analysis, samples of the third metacarpal or metatarsal region were prepared for analysis of bone ash, Ca, P and Mg.

Results

Mean BMD was significant higher in the third metatarsus in comparison to the third metacarpus, but without any gender or age related differences (Table 1). BMD analysis medial or lateral (1, 2 or 3 cm) apart from the mid point of the metacarpus or metatarsus resulted in precision CVs (coefficient of variation) ranging from 6.8% (1 cm) and 9.5% (3 cm). In the bone ash, mean Ca, P and Mg content did not differ between location, gender and age (Table 1). As the main finding, there was no correlation between BMD and bone ash (r=0.06), Ca (r=0.017), P (r=0.09) or Mg (r=0.08).

Table 1. Mean (± SD) bone mineral density (BMD) (g/cm²), bone ash (%), Ca, P and Mg in bone ash (%) in the metacarpus and metatarsus.

Site	BMD g/cm²	Bone ash %	Ca	P	Mg
			in bone ash (%)		
Metacarpus	3.0 ± 0.3^a	69.7 ± 0.8^a	38.4 ± 24^a	17.7 ± 7.6^a	3.5 ± 0.4^a
Metatarsus	3.7 ± 0.2^b	69.8 ± 6.3^a	38.5 ± 21^a	17.7 ± 6.8^a	3.5 ± 0.3^a

Means in the same column with different superscripts are significantly different ($P<0.05$).

Discussion

From the present data, it seems that the transportable device system Lunar PIXI® did not precisely reflect bone mineral content as DXA evaluates only mineral content of the surface area. The higher BMD of the metatarsus compared to the metacarpus is possibly related to differences in the distance of the course of beam as girth and geometry varies between those bones.

In horses, DXA was predominately used to compare BMD density in growing horses (Donabedian *et al.*, 2006) or to evaluate equine parathyroid hormone treatment on bone metabolism in healthy horses (Weisrock *et al.*, 2010). Weisrock *et al.* (2010) used the DXA method under general anaesthetic conditions hereby defining anatomical points of the third metacarpus for standardization the region of interest. This procedure is urgently required as 1, 2 or 3 cm (distal or proximal) apart from the defined midpoint of the metacarpus or metatarsus remarkable variations up to 10% were observed in the present study. Under practical conditions like routine control of bone density on a race track, such a procedure is not feasible. In addition, we used disarticulated fore or hind limbs from slaughter horses, standardized positioning of the limbs under *in vivo* conditions without general anaesthesia is unequally challenging and needs still to be evaluated.

References

Carstanjen, B., F. Duboeuf, J. Detilleux and O.M. Lepage, 2003. Equine third metacarpal bone assessment by quantitative ultrasound and dual energy X-ray absorptiometry: an *ex vivo* study. J Vet Med A Physiol Pathol Clin Med. 50(1): 42-47.

Donabedian, M., G. Fleurance, G. Perona, C. Robert, O. Lepage, C. Trillaud-Geyl, S. Leger, A. Ricard, D. Bergero and W. Martin-Rosset, 2006. Effect of fast vs. moderate growth rate related to nutrient intake on development orthopaedic disease in the horse. Animal Research 55, 471-486.

Donabedian, M, C. Delguste, G. Perona, P. Lebecque, F. Duboeuf, O. Lepaga and W. Martin-Rosset, 2005. Third metacarpal bone mineral density assessment in the standing horse by dual X-ray absorptiometry: suitability, precision and accuracy. Vet Comp Orthop Traumatol. 18(1) :26-30.

Lepage, O.M., B. Carstanjen and D. Uebelhart, 2001. Non-invasive assessment of equine bone: an update. Vet J. 161(1): 10-22.

McCarthy, R.N. and L.B. Jeffcott, 1992. Effects of treadmill exercise on cortical bone in the third metcarpus of young horses. Res Vet Sci 52: 28-37.

Nielsen, B.D., G.D. Potter and E.L. Morris, 1997. Changes in the third metacarpal bone and frequency of bone injuries in young Quarter horses during race training – Observations and theoretical considerations. J Equine Vet Sci 17: 541-549.

Weisrock, K., W. Martin-Rosset, B. Remy, W.G. Forssmann, N. Parvizi, M. Coenen and I. Vervuert, 2010. Metabolic actions of intermittent exogenous equine parathyroid hormone (ePTH 1-37) in horses. In: Ellis, A.D., A.C. Longland, M. Coenen and N. Miragli (eds.), The impact of nutrition on the health and welfare of horses. EAAP publication no. 128, Wageningen Academic Publishers, Wageningen, the Netherlands, p. 116-119.

Using faecal pH to predict gut health in horses

J.A. Næsset and D. Austbø
Department of Animal and Aquacultural Sciences, Norwegian University of Life Sciences, Box 5003, 1432 Ås, Norway

Abstract

Four caecally cannulated horses (Norwegian Cold-blooded Trotter, bodyweight 540±10 kg, age 8-13 years) were fed four isoenergic rations in a Latin square design: (H) 14.5 kg timothy hay; (B) 6 kg timothy hay and 3.8 kg pelleted barley; (O) 6 kg timothy hay and 4.3 kg whole oats; (F): 7.5 kg timothy hay, 2.2 kg pelleted sugar beet pulp (SBP) and 0.4 kg sunflower oil The morning meal provided different amounts of starch (g/kg BW): (H) 0.3; (B) 2.2; (O) 1.7; (F) 0.2. The horses were adapted to each diet for 17 days prior measuring caecal pH. On test days pH was measured recordings taken every minute for 9 hours post feeding. Microsoft Excel (2007) was used to calculate: (a) the average pH in the first 30 minutes, (b) the average pH for every 15 minute period, (c) the average of the 10 lowest pH values, (d) the number of periods the average pH was lower than 7.6. Faecal samples were collected directly from the rectum. The hay diet (H) resulted in a caecal ph between 6.7 and 6.8 without postprandial variation. The caecal pH showed an enlarged scatter between 6.6 and 6.8 for the the fibrous diet F. both starch providing diets were associated with a pronounced postprandial decrease; the nadir was reached at 5-7 h after the barley containing meal (diet B) while the ingestion of oats (diet O) was reflected by the lowest pH at 3-4 h postprandially. The pH values in faeces did not sufficiently correspond to the caecal pH profile and are obviously not suitable to monitor caecal fermentation characteristic after a starchy meal.

Keywords: fibre, starch levels, caecal pH, hindgut environment

Introduction

In several studies faecal pH has been measured to predict gut health after feeding different rations (Van den Berg *et al.*, 2008; Williamson *et al.*, 2007; Hussein *et al.*, 2004). The horse is a monogastric herbivore depending largely on the cellulolytic microbial activity in caecum and colon for the fermentation of fibrous material to volatile fatty acid (VFA). This microbial fermentation of fibre is most efficient at pH 6-7 (Sjaastad *et al.*, 2003), and produces mainly acetate, propionate and butyrate. Caecal pH decreases markedly after feeding a meal containing more starch than the enzymatic digestion in the small intestine is able to utilize. Austbø (2005) reports caecal pH values falling to 6.1 for a period following a meal containing 2 grams of starch from barley/kg BW. The microbial fermentation of roughage can be markedly reduced because of the changes in the microbial flora with reduced cellulolytic activity and increased amylolytic activity. A caecal pH below 6 can be classified as a sub-clinical acidosis increasing the risk of gut related conditions such as colic and laminitis. The aim of this study was to compare the pH values in faecal samples with continuous logging of the caecal pH for 9 hours post feeding on horses fed either grass hay only, two different starch rich rations or a ration with high fat and fibre content.

Material and methods

Four caecally cannulated 8-13 years old geldings of Norwegian Cold -blooded Trotter (bodyweight 540±10 kg) were fed four isoenergic rations in a Latin square design. The horses were fed at 06:00, 16:00 and 22:00 every day. The rations consisted of; (H): 14.5 kg timothy hay, (B): 6 kg timothy hay and 3.8 kg pelleted barley, (O): 6 kg timothy hay and 4.3 kg whole oats, (F): 7.5 kg timothy hay, 2.2 kg pelleted sugar beet pulp (SBP) and 0.4 kg sunflower oil (the SBP were soaked in 6 litres of water before feeding). The morning meal consisted of; (H): 4.5 kg timothy hay (equivalent to 0.3 g starch / kg BW), (B): 2 kg timothy hay and 2 kg pelleted barley (equivalent to 2.2 g starch/kg BW), (O): 2 kg timothy hay and 2 kg whole oats (equivalent to 1.7 g starch/kg BW), (F): 2.5 kg timothy hay, 0.8 kg pelleted sugar beet pulp and 0.2 kg sunflower oil (equivalent to 0.2 g starch/kg BW). All rations included added salt and a commercial vitamin/mineral supplement to cover the nutrient requirements stated in NRC, 2007. The supplement and the rest of the feed were divided equally between the two other meals. The horses were housed in individual boxes of 9 square meters bedded with sawdust during the sampling days. They had *ad libitum* access to water from automatic bowls. During sampling periods the horses conducted a 40-60 minutes standardised exercise test on a high speed treadmill with variable speed between 1.5 -4.2 m/s, 3% slope and up to 16 kg pull through lines attached to the breast band harness. Between the sampling periods (17 days adaptation to new rations) the horses were daily turned out in a paddock as a group for 6 -7 hours.

In addition the horses were exercised for 1 hour in an automatic rotary exerciser with variable speed between 1.1 -4.75 m/s. The body weight remained unchanged during the experiment, and therefore the exercise level was suitable to the feed rations which were calculated to be 50% above maintenance level. All horses remained healthy during the entire experimental period. On sampling days a pH electrode (Hammilton, Polyplast DIN 60) was inserted into the caecum through the cannula and pH was measured using a pH meter (WTW, pH 340i) attached to the horse with recordings taken every minute for 9 hours post feeding. Microsoft Excel (2007) was used to calculate: (a) the average pH in the first 30 minutes, (b) the average pH for every 15 minute period, (c) the average of the 10 lowest pH values, (d) the number of periods the average pH was lower than 7.6 to express how many minutes the caecal pH was lower than 7.6. Faecal samples were collected either directly from rectum at 0, 150, 300 and 450 minutes post feeding or picked up from the floor if dropping was detected the last 10 -15 minutes before sampling time. 15 grams of the mixed faecal sample were added an equal amount of ion exchanged water and shaken. pH was measured in the liquid phase with a pH meter (WTW, pH340i) and a pH electrode (Hamilton, WTW SenTix41). The average faecal pH for each sampling time and rations was calculated with Microsoft Excel (2007). The calculated values from the recorded caecal pH and the average faecal pH were evaluated by analysis of variance using the PROC GLM procedure in SAS (9.1) with ration and horses as fixed effect in the model. The significance level was set to $P<0.05$. The model used was: $Y_{ij} = \mu + \alpha_i + \beta_j + \varepsilon_{ij}$ where Y is response, μ = overall mean, α is ration, β is horses and ε is residuals; i = 1, 2, 3, 4; j = 1, 2, 3, 4.

Results and discussion

A typical pattern of the recorded caecal pH for different rations is shown in Figure 1. The pH values show only small variation between 6.7 and 6.8 when the horses were fed roughage only (ration H). When fed starch rich rations, barley (B) and oats (O), the pH values started decreasing 60-120 minutes post feeding, and were decreased for a longer period when barley was fed. When the horses were fed the SBP and fat ration (F) the pattern showed a variation between 6.6 and 6.8, with no markedly decrease.

A variation in pH values between rectal samples and samples picked up from the floor shortly after dropping was detected. Therefore only rectal samples were used in this study. The average faecal pH is shown in Figure 2. The experimental rations did not show any significant difference in faecal pH (Table 1). Figure 2 shows that the lowest faecal pH value was observed when the horses were fed the barley ration (B), and the highest faecal pH value was observed when the horses were fed the oats ration (O). This does not reflect the pattern seen in the caecal pH (Figure 1), where a drop in pH is seen when the horses were fed the oats ration (O), which is a potentially harmful environment for the cellulolytic microbial activity in caecum.

Rectal pH values for different sampling times post feeding showed no variation compared to the pattern seen when recording pH directly in the caecum (Figure 3). The small variation between the

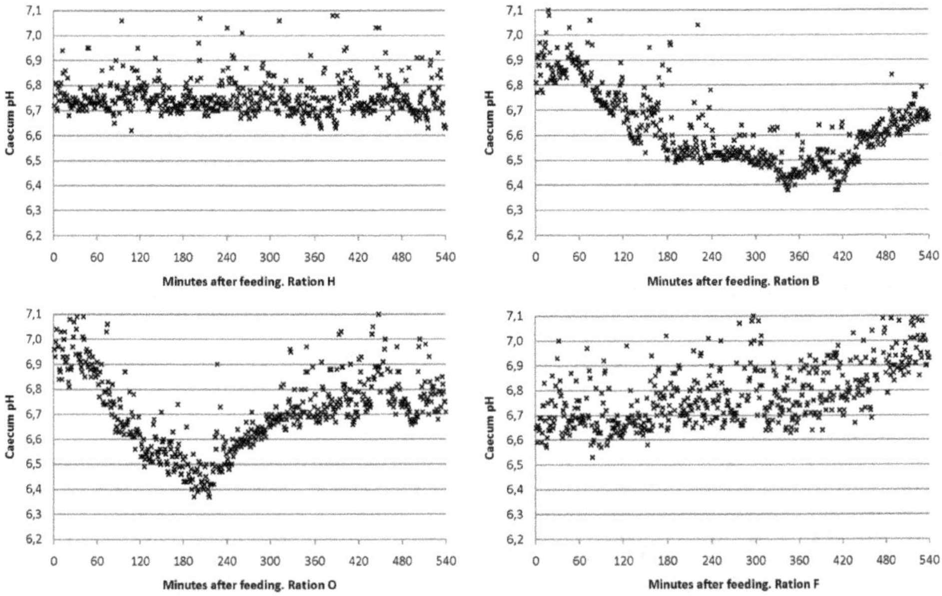

Figure 1. Example of pattern in recorded caecum pH for the different rations.
Ration H: 4.5 kg timothy hay (equivalent to 0.3 g starch/kg BW), ration B: 2 kg timothy hay and 2 kg pelleted barley (equivalent to 2.2 g starch/kg BW), ration O: 2 kg timothy hay and 2 kg whole oats (equivalent to 1.7 g starch/kg BW), ration F: 2.5 kg timothy hay, 0.8 kg pelleted sugar beet pulp and 0.2 kg sunflower oil (equivalent to 0.2 g starch/kg BW).

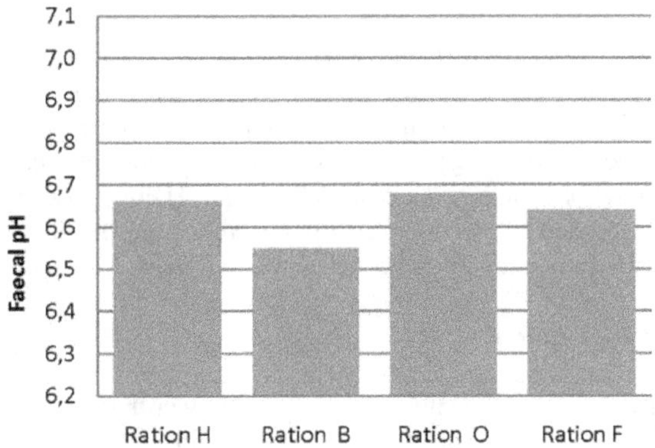

Figure 2. Average faecal pH for different rations.
Ration H: 4.5 kg timothy hay (equivalent to 0.3 g starch/kg BW), ration B: 2 kg timothy hay and 2 kg pelleted barley (equivalent to 2.2 g starch/kg BW), ration O: 2 kg timothy hay and 2 kg whole oats (equivalent to 1.7 g starch/kg BW), ration F: 2.5 kg timothy hay, 0.8 kg pelleted sugar beet pulp and 0.2 kg sunflower oil (equivalent to 0.2 g starch/kg BW).

rations indicates that the faecal pH do not reflect the short time variation in the caecum caused by a single meal. It is observed that the fibre rich rations result in a small decrease in faecal pH for each sample time. The starch rich ration results in a small increase in faecal pH for the two (ration O) or three (ration B) first sampling times, and then a drop for the samples taken at 300 and 450 minutes (ration O) or at 450 minutes (ration B) post feeding. This is probably a small effect of the increased amylolytic fermentation after feeding a starch rich meal, but this study cannot prove that these rations affect the faecal pH enough to predict gut health.

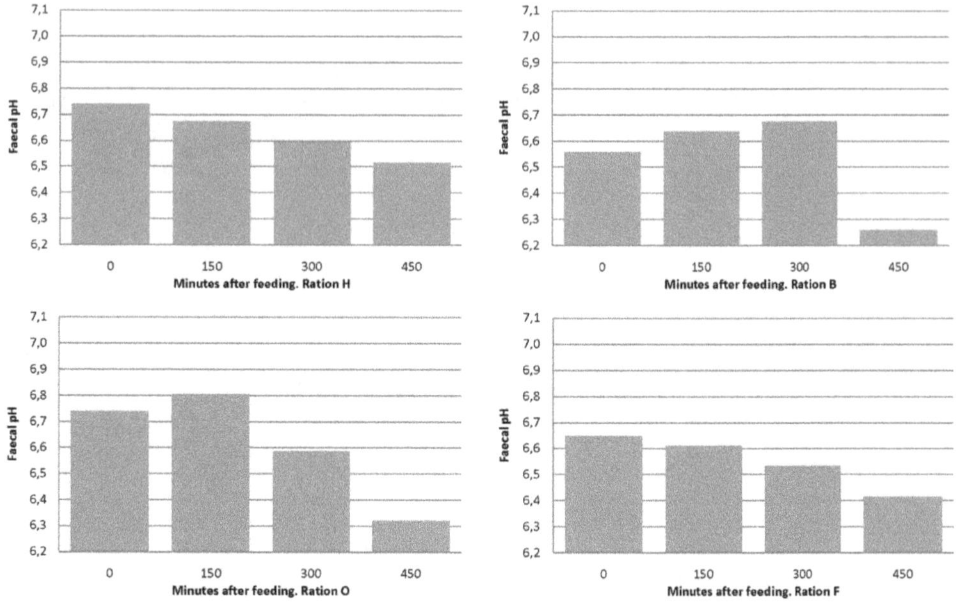

Figure 3. Average faecal pH for each ration's samples collected at different times post feeding. Ration H: 4.5 kg timothy hay (equivalent to 0.3 g starch/kg BW), ration B: 2 kg timothy hay and 2 kg pelleted barley (equivalent to 2.2 g starch/kg BW), ration O: 2 kg timothy hay and 2 kg whole oats (equivalent to 1.7 g starch/kg BW), ration F: 2.5 kg timothy hay, 0.8 kg pelleted sugar beet pulp and 0.2 kg sunflower oil (equivalent to 0.2 g starch/kg BW).

The statistic analysis shows individual differences when analysing the faecal pH, but no difference between horses when analysing the parameters related to the caecum pH. Table 1 shows no clear difference between the fibre rich rations or the starch rich rations when evaluating the average pH the first 30 minutes post feeding (morning pH), but there is a difference between ration O and F. This indicates that the caecum environment is still able to stabilise the pH during the night with these ration sizes. The minimum pH recorded in caecum is clearly different from the fibre rich rations and the starch rich rations, also with a tendency for lower pH on ration B than O (P=0.087). When analysing

Table 1. Response for different rations.

Ration	Morning pH[1]	Min pH[2]	Lower than 6.7[3] (min)	Faecal[4] pH
H	6.89	6.61_a	94.7_a	6.71
B	6.89	6.42_b	283.1_b	6.55
O	7.00_a	6.49_b	175.3_a	6.68
F	6.76_b	6.61_a	85.3_a	6.64

Ration H: 4.5 kg timothy hay (equivalent to 0.3 g starch/kg BW), ration B: 2 kg timothy hay and 2 kg pelleted barley (equivalent to 2.2 g starch/kg BW), ration O: 2 kg timothy hay and 2 kg whole oats (equivalent to 1.7 g starch/kg BW), ration F: 2.5 kg timothy hay, 0.8 kg pelleted sugar beet pulp and 0.2 kg sunflower oil (equivalent to 0.2 g starch/kg BW). [1] Caecum pH in the morning measured as the average of the first 30 minutes post feeding, [2] Minimum pH measured as the average of the 10 lowest values of 9 hours logging in caecum, [3] Average of a 15 minute period the caecum pH is lower than 6.7, [4] Average faecal pH for different rations. All values are means. Within a column, means with a different common subscript letter differ significantly (P<0.05).

how many minutes the caecum pH is lower than 6.7 after one single meal we found that ration B differ from the others. There is also a tendency for a difference between ration O and F (P=0.081).

Conclusion

Caecal pH values reflect the fermentative conditions as influenced by the actual meal. Faecal pH varies between rectal grab samples and samples picked up from the floor shortly after dropping was detected. Faecal samples will not show the short term variations caused by a single meal. Faecal pH values cannot be used as indicator of meal induced variations in caecal pH and hence equine gut health.

References

Austbø, D., 2005. pH-verdien i blindtarmen hos hest, effekt av rasjon og tid. Husdyrforsøksmøtet 325-327.

Hussein, H.S., L.A. Vogedes, G.C. Fernandez and R.L. Frankeny, 2004. Effects of cereal grain supplementation on apparent digestibility of nutrients and concentrations of fermentation endproducts in the feces and serum of horses consuming alfalfa cubes. Journal of Animal Science, 82: 1986-1996.

Sjaastad, Ø.V., K. Hove and O. Sand, 2003. Physiology of Domestic Animals. Oslo, Scandinavian Veterinary Press. 735 pp.

Van Den Berg, M., S.O. Hoskin, C.W. Roger, A. Grinberg, and N.D. Walker, 2008. The effect of dietary transition on faecal pH and microbiology of thoroughbred horses. In. 4th European Equine Health and Nutrition Congress, p. 123-127, The Netherlands.

Williamson, A., C.W. Rogers and E.C. Firth, 2007. A survey of feeding, management and faecal pH of Thorougbred racehorses in the NOrth Island of New Zealand. New Zealand Veterinary Journal, 55 (6): 337-341.

The effect of steam treatment on the total viable count, mould and yeast numbers in hay using the Haygain hay steamer

R. James and M. Moore-Colyer
Royal Agricultural College, Cirencester, Gloucestershire, GL 7 6JS, United Kingdom

Abstract

Five bales of English rye grass mix meadow small baled hay were steamed in the Haygain hay steamer using its unique method to introduce steam through a manifold system for 50 minutes. Microbiology test results showed that steaming in this way was very effective at reducing total viable count of bacteria, 99.08% and 98.84%, respectively and 100% reduction in fungi and yeast of the hay. This drastic improvement in the hygiene quality of whole bales of hay makes it a safe and suitable fodder for all types of horses.

Keywords: hay, soaking hay, steaming hay, hay steamer, stable environment, dust, respirable particles, fungal spores, Aspergillus, respiratory disorders, RAO, IAD

Introduction

Domestication of the horse has resulted in the combination of eating conserved forage in a stable environment, which has subjected the horse to dust (bacteria, fungal spores, plant particles and insect fragments) which can produce respiratory disorders (Clark, 1992). It is widely regarded that even good quality hay contains mould and fungal spores along with bacteria and other organic dust and this is due to the nature in which hay is made in the field. Moulds such as *Aspergillus fumigatus* are linked to the development of conditions such as recurrent airway obstruction (RAO) formerly known as COPD in horses (McGorum *et al.*, 1993) and farmer's lung in humans (Kotimaa *et al.*, 1991).

The incidence and significance of the resultant respiratory disorders is notably high for all types of horses. It was reported that 1 in 6 horses in the UK are diagnosed with recurrent airway obstruction and 80% of horses stabled part of the time suffer from some degree of airway inflammation (Horse and Hound, 2006). It is also a particularly large problem in the racing industry, an epidemiology study by Rossdale *et al.* (1985) found respiratory problems in race horses to be the second highest reason for lost training days after lameness. In concurrence with this finding, it is also reported that up to 80% of race horses are affected at some point in the first years of training in the UK (Wood *et al.*, 1999) and Australia (Christly *et al.*, 2001). In addition a study investigating the occurrence of inflammatory airway disease (IAD) in Japanese Thoroughbred race horses found 73.3% of horses who presented with a cough or poor performance were diagnosed with IAD.

While dust extracted shavings and cardboard can be used to minimise dust from bedding, it is more difficult to reduce the dust levels in fodder without negatively affecting the nutrient quality of the feed. Hay is an ideal conserved forage to feed stabled horses, however, hay contains significant levels of respirable mould and fungal spores especially when made in less than ideal conditions due to climatic limitations. In fact, hay has the potential to contain the highest concentration of all sources in the stable environment (Webster *et al.*, 1987; Raymond *et al.*, 1994; Robinson *et al.*, 2001).

Many owners soak hay to reduce airborne particles, but soaking hay nets leaches nutrients (Moore-Colyer, 1996) is laborious and heavy to handle and results in post-soak liquor that has a very high Biological Oxygen Demand (BOD) (War and Petch, 1992). Art *et al.* (2002) suggested that soaking bales of hay is only efficient when done for several hours and with the strings cut, thus permitting water penetration to the centre of the bale. This is neither suitable nor practical for larger establishments or racing yards.

Steaming hay has previously been shown by Blackman and Moore-Colyer (1998) to have none of the above disadvantages and is very effective at reducing the airborne respirable particles. Horse owners have attempted to steam with home-made steamers using plastic containers with boilers, and the equivalent is also currently sold in the UK as hay steamers. However, there has been no published data on their efficacy and they are thus not widely used by horse owners. In addition this method is only capable of steaming loose-hay or hay-nets which is not practical or economical for the larger yards.

The objective of the current study was to test how well the Propress Equine Haygain hay steamer reduced bacteria, mould and yeast numbers when complete bales of hay were steamed for 50 minutes.

Materials and methods

The Haygain hay steamer was made up of two components, firstly a steam generator; which heated the water to boiling point to produce steam. The steam escaped from the heating vessel through the elbow, specifically designed to create a small amount of back pressure (75-100 mb). The steam travelled down a multi layer construction hose with braided reinforcing and due to the back-pressure, distributed the steam evenly through the manifold system. The second component, the insulated chest contained the steam distribution manifold inside, referred to as the hay chest, which infused the steam into the hay shown by Figure 1.

Operation of the Haygain hay steamer

A fully strung bale was placed onto the steam manifolds inside the hay chest and pushed down evenly and firmly so the steam lances pierced the hay to their full length. The lid containing a thermometer which read the atmospheric temperature inside the hay chest was secured. The filler cap on the steam generator was then removed and filled to the maximum level as directed by the side-glass (approximately seven litres) with clean water, plugged into an electrical supply and switched on. The hay steamer was then left to steam the hay for 50 minutes.

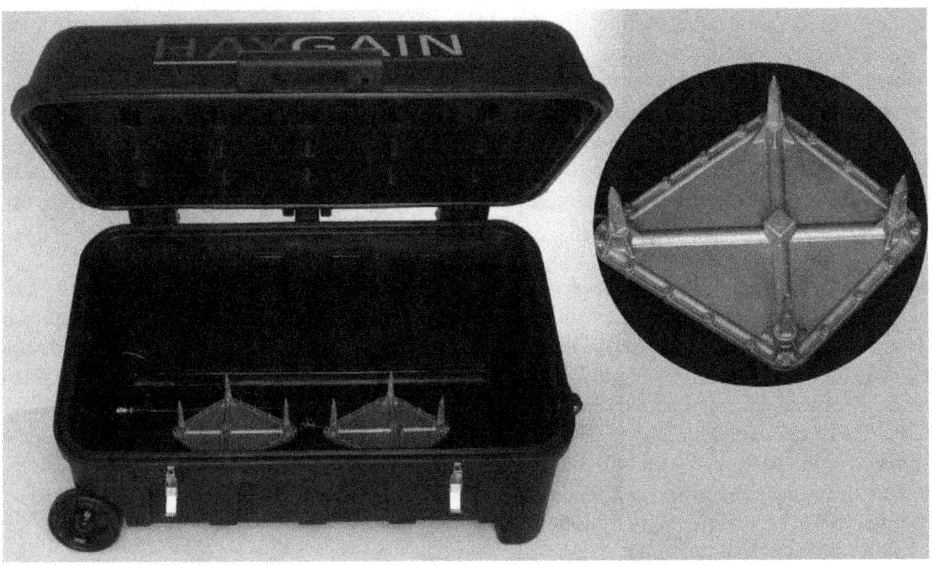

Figure 1. Component of the Haygain hay steamer.

Methodology

Five bales of good quality rye grass mix meadow hay were randomly selected from hay purchased from a 'horse hay' merchant in Gloucestershire. Each bale was subjected to the following treatment. Dry samples were taken from five areas of the bale using large tongs and temporarily stored in a sterile glass beaker. The bale was then placed onto the spiked manifold in the steamer and the boiler (containing approx 7 litres of water) turned on. The bale was left in the steamer for 50 minutes, where upon the boiler was switched off and the lid was carefully removed. Tongs were then used to take samples from another five areas of the bale and also placed into a sterile glass beaker. The dry and steamed samples were then separately prepared using the following procedure.

One gram hay was roughly chopped with sterilized scissors into a stomacher bag and 79 ml of maximum recovery dilutent (MRD) added. The mixture was then 'stomached' for 2 minutes. 1 ml of the sample was then taken and put into 9 ml of MRD in a sterile screw-top tube. Sequential dilutions were then prepared down to 10^{-4}. 1 ml was taken from each of 10^{-1}, 10^{-2}, 10^{-3}, 10^{-4} dilutions and placed onto 3M™ petrifilms for total viable count (TVC), mould and yeast, two replicates were prepared for each dilution. The 3M™ Petrifilm™ Aerobic Count (AC) Plate, a sample-ready-culture-medium system which contains nutrients, a cold-water-soluble gelling agent, and a tetrazolium indicator that facilitates colony enumeration of aerobic bacteria was used for TVC. The dilutions were plated out, the top film was lifted and 1 ml of the dilutent was dispensed using a Volac high precision Ultra micropipette, the top film was dropped onto the sample and then the plastic spreader was used to distribute the sample evenly over the entire Petrifilm plate growth area before the gel was formed. These were then incubated at 33 °C for 3 days. The 3M™ Petrifilm Yeast and Mold (YM) count plate, a sample-ready culture medium system which contains nutrients supplemented with antibiotics, a cold-water-soluble gelling agent and an indicator that facilitates yeast and mould enumeration was used to determine yeast and mould counts using the same preparation as described for TVC. The mould and yeast Petrifilms were then incubated at 20 °C for 5 days.

Data analysis

Colony numbers were counted following the 3M™ interpretation guide and using a standard colony counter, an illuminated magnifier. For TVC all red colonies regardless of size or intensity were counted. The circular growth area was approximately 20 cm^2, on plates that contained more than 300 colonies, three representative squares were counted and used to determine the average number per square. The average was then multiplied by 20 to determine the estimated count per plate. High concentrations of colonies on the Petrifilm are known to cause the entire growth area to become red or pink. These are recorded as too numerous to count (TNTC) and the higher dilutions were used for actual counts.

Mould and yeast counts were also determined using a standard colony counter and the 3M™ interpretation guide was used to distinguish between the yeast and mould based on their characteristics. Yeast colonies were small, had defined edges, were pink-tan to blue-green in colour, appeared raised and had a uniform colour. The mould colonies were large with diffuse edges, they were variable in colour, appeared flat and had a dark centre. The circular growth area was approximately 30 cm^2, on plates that contained more than 150 colonies, representative squares were used to determine the estimated count per plate by multiplying the average number by 30. High numbers of yeast colonies are known to cause the entire area to turn blue and mould to turn the area black or yellow, this was recorded as TNTC and the higher dilutions were used to determine actual counts.

All counts were recorded into an excel spreadsheet and for every bale, each dilution and repetition was recorded alongside each treatment (dry or steamed). Differences between dry and steamed hay were determined using the non-parametric Wilcoxon signed rank test.

Results

Table 1 shows the microbiological test results of dry and hay steamed in the production model of the Haygain hay steamer for 50 minutes.

Table 1. Total colony forming units (CFU/g) from dry hay and hay steamed in the Haygain hay steamer production model (HG1000) for 50 minutes.

Parameter	Dry	Steamed	Probability
TVC	381,573	4,453	0.008
Fungi	1.85×10^8	0	0.008
Yeast	6,893,333	0	0.008

TVC: total variable count.

Discussion and conclusions

The results of this experiment clearly show that steaming for 50 minutes in the Haygain hay steamer produced hay devoid of fungi or yeast (100% reduction) and a 98.84% reduction in bacterial contamination. Steamed bales were fed to horses within 12 hours of being steamed and the palatability of the hay was described as very good.

Steaming hay in the Haygain hay steamer drastically improved the hygiene quality of whole bales of hay. The microbial contamination was reduced to zero for fungi and yeasts and by 98.84% for bacteria. Minimising exposure to these respirable particles is hugely beneficial to the health of the horse and in particular the respiratory system. Pathogenic challenge to both the respiratory and digestive systems is therefore all but eliminated making the steamed hay an excellent feed for all performance horses.

References

Art, T., B.C. McGorum and P. Lekeux, 2002. Environmental control of respiratory disease. In: Lekeux, P. (ed.), Equine Respiratory Diseases. International Veterinary Information Service, New York, NY, USA.

Blackman, M. and M.J.S. Moore-Colyer, 1998. Hay for horses: the effects of three different wetting treatments on dust and mineral content. Animal Science. 66, 745-750.

Christley, R.M., D.R. Hodgson, R.J. Rose, J.L.N. Wood, S.W.J. Reid and J.L. Hodgson, 2001. A case-control study of respiratory disease in Thoroughbred racehorses in Sydney, Australia. Equine Veterinary Journal. 33: 256-264.

Clarke, A., 1992. Environmental monitoring in relation to equine respiratory disease. In: Robinson, N.E. (ed.), Current therapy in equine medicine, third edition. W.B. Saunders, Philadelphia, USA, p. 310-315.

Horse and Hound, 2006. Breathing problems more common than ever. Horse and Hound magazine. IPC media. Available online: http://www.horseandhound.co.uk/horsecare/397/73480.html [date accessed 15/04/10].

Kotimaa, M.H., L. Oksanen and P. Koskela, 1991. Feeding and bedding materials as sources of microbial exposure on dairy farms. Scandinavian Journal of Work, Environmental & Health 17: 117-122.

McGoram, B.C., P.M. Dixon and R.E.W. Halliwell, 1993. Responses of horses affected with chronic obstructive pulomonary disease to inhalation challenges with mould antigens. Equine Veterinary Journal 25: 261-267.

Moore-Colyer, M., 1996. Effects of soaking hay fodder for horses on dust and mineral content. Animal Science 63: 337-342.

Raymond, S.L., E.F. Curtis and A.F. Clarke, 1994. Compariative dust challenges faced by horses when fed alfalfa hay cubes or hay. Equine Practice. 16: 42-47.

Robinson, N.E., F.J. Derksen, C.A. Jackson, D. Peroni and V. Gerber, 2001. Management of heaves. Equine Veterinary Education. 13: 247-259.

Rossdale, P.D., R. Hopes, N.J. Digby and K. Offord, 1985. Epidemiological study of wastage among racehorses 1982 and 1983. Veterinary Record 116 (3): 66-69.

Warr, E.M. and J.L. Petch, 1992. Effects of soaking hay on its nutritional quality. Equine Veterinary Education. 5: 169-171

Webster, A.J., A.F. Clarke, T.M. Madelin and C.M. Wathes, 1987. Air hygiene in stables. 1: Effects of stable design, ventilation and management on the concentration of respirable dust. Equine Veterinary Journal. 19 (5): 448-53.

Wood, J.L.N., J.R. Newton, N. Chanter, H.G.G. Townsend, K.H. Lakhani, J.A. Mumford, R. Sinclair, M.H. Burrell, R.C. Pilsworth, M. Shepherd, D. Dugdale, B.M.B. Herinckx, J.P.M. Main, H. Windsor and D. Windsor, 1999. A longitudinal epidemiological study of respiratory disease in racehorses: disease definitions, prevalence and incidence. In: Wernery, U., J.F. Wade, J.A. Mumford and O. Kaaden (eds.), Proceedings of the 8th International Conference of Equine Infectious Disease VIII. R&W Publications, Newmarket, United Kingdom, p. 64-70.

Validation of long chain fatty acids as markers to estimate diet composition of equines fed on grassland-heathland vegetation components

L.M.M. Ferreira[1], A.S. Santos[1], R. Celaya[2], V. Falco[3], M.A.M. Rodrigues[1] and K. Osoro[2]
[1]CECAV-Univeristy of Trás-os-Montes and Alto Douro, Animal Production Group, P.O. Box 1013, 5001-901 Vila Real, Portugal
[2]SERIDA- Servicio Regional de Investigación y Desarrollo Agroalimentario, Apartado 13, 33300 Villaviciosa, Asturias, Spain
[3]Departamento de Agronomia, Universidade de Trás-os-Montes e Alto Douro, P.O. Box 1013, 5001-801 Vila Real, Portugal

Abstract

The main objective of this study was to validate the use of long-chain fatty acids (LCFA) as markers to estimate diet composition of equines fed with diets composed of herbaceous and woody species. Six crossbreed mares (434±59 kg LW) were divided in 2 groups and each group was fed one of two different diets: D1 100% *Lolium perenne*; D2 70% *L. perenne* + 30% heather (*Erica* spp., *Calluna vulgaris*). Diet composition was estimated from LCFA concentrations (C_{22} to C_{34}) in diet and faeces, with (i.e. mean recovery rate of the dietary treatment that the animal belonged to) or without correction for incomplete faecal recoveries using least-squares procedures. Results showed clear differences in the LCFA profiles among diet components. LCFA faecal recovery was in general incomplete and tended to decrease with carbon chain length in a curvilinear fashion ($P<0.001$) and was affected by diet composition ($P<0.001$). Results also showed that diet composition was accurately estimated with or without faecal recovery correction in both diets, confirming the usefulness of LCFA as diet composition markers for studying diet selection of equines in diverse plant communities.

Keywords: long-chain fatty acids, diet composition, heathlands

Introduction

N-alkanes have been used with success as markers to estimate diet composition of different herbivore species. One major limitation of these markers is their very low concentrations in herbaceous species, which could limit the accuracy of diet composition estimates. Also, similarities in the alkane profiles observed between several herbaceous species could lead to less precise estimates of diet composition (Oliván *et al.*, 2007). Long chain fatty acids (LCFA) have been suggested as an alternative or in addition to the alkanes in order to overcome these constrains (Ferreira *et al.*, 2009, 2010). The main objective of this study was to validate LCFA as markers to estimate diet composition of equines fed with simple diets.

Material and methods

Six crossbreed mares (434±59 kg LW) were divided in 2 groups and were housed in metabolism pens to allow total faecal collection. Each group was fed one of two different diets: D1 100% *Lolium perenne*; D2 70% *L. perenne* + 30% heather (*Erica* spp., *Calluna vulgaris*). Diet composition reflected the grazing behaviour of equines on heathland areas with improved pastures. Animals were offered daily a total of 1.0 kg of DM/100 kg LW. The low feeding level used in this trial (1 kg DM/100 kg LW) was selected to ensure that all diet components were completely consumed by the animals, preventing the existence of refusals (the botanical separation of the diet components in the refusals would be difficult to accomplish) and diet selection, and ensuring a uniform intake of a diet during the trial.

The experimental period lasted 11 days and included a 7-day adjustment period followed by 4-day period for collecting samples of diet components and animal faeces. Samples of faeces and diet components were freeze-dried and ground prior to LCFA analysis. Long-chain fatty acid concentrations of diet components and faeces were analysed in duplicate according to the methods of Dove and Mayes (2006).

Diet composition was estimated from LCFA concentrations (C_{22} to C_{34}) in diet and faeces, with or without correction for incomplete faecal recoveries using least-squares procedures which minimise the sum of squared discrepancies between the actual LCFA proportions in faeces and the calculated proportions (different combinations of diet components) (Salt *et al.*, 1994). LCFA faecal concentrations were corrected using mean recovery rate of the dietary treatment that the animal belonged to. Besides *L. perenne* and heather, another woody species (*Ulex gallii*), that is characteristic of these marginal vegetation communities and can be selected by equines in situations where the grass availability is low, was considered as a possible diet component to test the capability of LCFA markers for estimating feeds which are not part of the diet but are available for consumption. This was done to test the capability of LCFA markers in estimating feeds which are not part of the diet but are available for consumption. Accuracy of diet composition estimates was assessed by the Kulcyznski similarity index (KSI). $KSI = (\sum 2c_i)/(\sum(a_i + b_i))*100$, where c_i is the lesser percentage of i component in the two diets and $(a_i + b_i)$ is the sum of the percentages of each plant component in both diets.

Effects of diet composition and carbon chain length on faecal recovery of each LCFA in faeces, and diet composition and faecal recovery correction on KSI, were examined by ANOVA.

Results and discussion

Results showed differences in the LCFA profiles between plant components (Table 1) indicating their clear discrimination when using these markers as previously observed (Dove and Mayes, 2006).

Even-chain prevailed over the odd-chain ones, representing over 83% of total LCFA content. LCFA faecal recovery was in general incomplete and tended to decrease with carbon chain length in a curvilinear fashion (P<0.001) and was affected by diet composition (P<0.001). These results contrast with the positive relationship between carbon chain length of LCFA and faecal recovery

Table 1. Long-chain fatty acid (LCFA) content (mg/kg DM) of the plant components used in the diets of equines.

	Lolium perenne	Heather	Ulex gallii
C_{22}	570±100.8	801±142.9	603±68.6
C_{23}	148±29.6	103±18.0	141±20.6
C_{24}	447±54.3	530±108.4	217±25.3
C_{25}	143±32.9	249±70.3	77±15.4
C_{26}	1,002±132.1	405±57.7	92±6.0
C_{27}	115±37.7	127±52.9	30±4.0
C_{28}	479±68.1	456±149.2	287±37.0
C_{29}	40±7.3	65±8.7	102±11.6
C_{30}	458±98	747±86.7	531±73.4
C_{32}	247±59.8	718±147.7	38±6.6
C_{34}	96±10.2	254±38.4	1±0.6
Total	3,745±366.3	4,455±456.5	2,119±209.1
Total even-chain	3,299±269.6	3,911±451.3	1,769±175.3

observed in goats (Ferreira *et al.*, 2009) and sheep (Ferreira *et al.*, 2010) fed with the same type of plant components.

Accurate estimates of diet composition (Table 2) were observed for either diets when applying or not applying faecal recovery correction. The inclusion of *U. gallii* in the calculations did not affect the estimation of diet composition. These results highlight that correction procedures are unnecessary when LCFA faecal recovery is unaffected by chain length due to the importance of the relative concentration of LCFA, rather than the absolute recoveries (Dove and Mayes, 2006). Similar results were obtained in our laboratory for the alkane markers (Ferreira *et al.*, 2007).

In conclusion, the results obtained in this study confirm that LCFA can be useful markers to estimate diet composition, at least in simple diets. Moreover, for this animal species, diet composition can be accurately estimated even without a previous correction of the LCFA faecal concentrations. However, more research is needed to assess the usefulness of these markers to estimate the composition of more complex diets, in addition or as an alternative to the alkanes. These faecal markers can be an important tool for studying diet selection of equines in diverse plant communities.

Table 2. Comparison of known proportions (± standard deviations) of dietary components and those estimated using the LCFA markers corrected or not for incomplete faecal recovery.

Diet		Lolium perenne	Heather	Ulex gallii	KSI
D1	Known	1.000	0.000	0.000	
	Estimated[1]	1.000±0.000	0.000±0.000	0.000±0.000	100.0±0.0
	Estimated[2]	0.964±0.032	0.013±0.017	0.023±0.029	96.4±3.2
D2	Known	0.700	0.300	0.000	
	Estimated[1]	0.733±0.033	0.262±0.026	0.005±0.008	96.0±2.3
	Estimated[2]	0.692±0.043	0.287±0.008	0.021±0.036	96.7±2.7

[1] Using LCFA faecal concentrations without recovery correction; [2] using LCFA faecal concentrations corrected with treatment mean faecal recoveries.

References

Dove, H. and R.W. Mayes, 2006. Protocol for the analysis of n-alkanes and other plant-wax compounds and for their use as markers for quantifying the nutrient supply of large mammalian herbivores. Nat. Protoc. 4, 1680-1697.

Ferreira, L.M.M., S. Carvalho, V. Falco, R. Celaya, U. García, A.S. Santos, M.A.M. Rodrigues and K. Osoro, 2009. Assessment of very long-chain fatty acids as complementary or alternative natural fecal markers to n-alkanes for estimating diet composition of goats feeding on mixed diets. J. Anim. Sci. 87, 2732-2745.

Ferreira, L.M.M., R. Celaya, V. Falco, M. Oliván, A.S. Santos, C. Guedes, M.A.M. Rodrigues and K. Osoro, 2010. Evaluation of very long-chain fatty acids and n-alkane epicuticular compounds as markers for estimating diet composition of sheep fed heathland vegetation species. Anim. Feed Sci. Technol. 156, 75-88.

Ferreira, L.M.M., U. García, M.A.M. Rodrigues, R. Celaya, A. Dias-da-Silva and K. Osoro, 2007. The application of the *n*-alkane technique for estimating the composition of diets consumed by equines and cattle feeding on upland vegetation communities. Anim. Feed Sci. Technol. 138, 47-60.

Oliván, M., L.M.M. Ferreira, U. García, R. Celaya and K. Osoro, 2007. Application of *n*-alkanes as diet compositionmarkers in grazing/browsing goats and sheep: effect of using different faecal recovery corrections and plant species grouping approaches. Aust. J. Agr. Res. 58, 1013-1022.

Salt, C.A., R.W. Mayes, P.M. Colgrove and C.S. Lamb, 1994. The effects of season and diet composition on the radiocaesium intake by sheep grazing on heather moorland. J. Appl. Ecol. 31, 125-136.

Steaming hay for horses: the effect of three different treatments on the respirable particle numbers in hay treated in a Haygain steamer

C. Stockdale and M.J.S. Moore-Colyer
Royal Agricultural College, Cirencester, Gloucestershire GL7 6JS, United Kingdom

Abstract

Respirable particle numbers were measured in four different hays subjected to a 50 minute steam treatment in a Haygain steamer. Dry hay ranged from 51,254 to 10,711 particles/litre air/kg hay, and reflect the wide variation of UK hay fed to horses. Steaming and immediate shaking reduced the respirable particle numbers from 25,699 to 1,586 particles/litre air/kg hay, a reduction of 94%, while steaming + 24-hour rest period reduced particle numbers to 5,398, a mean reduction of 80%. These results show that steaming is a highly successful way to reduce respirable particle numbers in hay.

Keywords: hay, steam, haygain, respirable particles

Introduction

It is known that hay and straw are the primary sources of potentially allergenic respirable particles in the stable environment (McGorum, 2001). These particles can initiate the debilitating condition recurrent airways obstruction (RAO) that causes respiratory distress, coughing and nasal discharge. Replacing the straw bedding with low dust or dust extracted shavings can solve part of the problem, but replacing the hay with another forage such as haylage, is not always a viable alternative. Blackman and Moore-Colyer (1998) reported positive results (95% reduction in respirable particle numbers) when steaming 5 kg hay nets in a home-made steamer, but to-date no other work has been published on steaming hay. This study sought to determine the efficacy of a commercially available steamer, known as the Haygain, at reducing respirable particle numbers in complete-strung bales of 4 different hays, representing typical hay fed to horse across the UK.

Method

Four different hays of varying species and quality (determined by visual assessment of the number of seed heads, leaf:stem ratio, colour and smell), but typical of hay fed to horses across the UK were selected for testing. All hays were made using the conventional field-drying system, involving air drying accompanied by 3-5 turnings with a tractor mounted hay turner, before baling and immediate storage in a stack in a covered barn. Two replicate bales of each hay type were then subjected to 3 treatments. Treatment 1 was dry hay shaken under the cyclone as detailed below, treatment 2 was steamed for 50 minutes in the Haygain and shaken immediately, and treatment 3 was steamed for 50 minutes in the Haygain and left for 24 hours before shaking. In each steaming treatment the temperature inside the bale was between 98 and 104 °C, as measured by irreversible plastic temperature strips placed onto rulers and inserted into 3 areas (middle, left and right ends) of each bale. Respirable particle (RP) numbers were measured from a 5 kg sample taken from the middle of each bale. The hay was placed onto a clean plastic sheet in the laboratory and shaken vigorously with a four-pronged fork under a cyclone air sampler for 3 minutes. Particles were captured on cellulose filter papers, which were then mounted in triacetate, left to 'clear' for a minimum of 48 hours before being counted using an eyepiece graticule under x 40 magnification. Differences in respirable particle numbers between the three treatments were determined using analysis of variance and lsd test = $t_{(error\ df)}$ x s.e.d.

Results

Table 1 shows that significant differences ($P<0.005$) existed between respirable particle numbers in hays from different areas of the UK. Hay 1 had 4.7 times the number of particles than hay 4, the latter being similar to hay 3. Hay 2 was intermediate with over double the number of particles of hay 3 and just under half that found in hay 1. These differences reflect the variable nature of hays made across the UK, and can be attributed to weather conditions during harvesting and subsequent storage conditions, although grass species differences and stage of maturity may also play a part.

Table 2 demonstrates that steaming 4 different hay for 50 minutes in the Haygain steamer significantly ($P<0.001$) reduced the respirable particle numbers by 94% compared with dry hay. Moreover steaming the hay and leaving it to rest for 24 hours reduced the respirable particle numbers by 79%.

Discussion

Steaming is an effective method for reducing respirable particles numbers in all hays whether only slightly dusty (hay 4) or highly contaminated (hay 1). Table 3 shows that steaming and immediate shaking reduced the respirable particle numbers in all hays by 91% (hay 3) to 96% (hay4). Steaming and leaving to rest for 24 hours reduced the respirable particle numbers in these two hays by 58% (hay 3) and 75% (hay 4), with hays 1 and 2 showing reductions of 83% and 82% respectively. Steaming a hay bale (average cost of the dry hay bale 4.7 euro) for 50 minutes costs approximately 0.58 euro (based on a unit cost of 16.9 kWh, Scottish Power, 2010) and so steamed hay is considerably

Table 1. Mean respirable particle (RP) numbers (/kg hay/litre of air) from 2 replicates of four different dry hays.

	Hay 1	Hay 2	Hay 3	Hay 4	s.e.d	Sig
RP numbers	5,1254^{c}	28,506^{b}	12,327^{a}	10,711^{a}	3,897.3	0.005

Values in the same row not sharing common superscripts differ significantly ($P<0.05$).

Table 2. Mean respirable particle numbers from 2 replicates of four different hays (/kg hay/litre of air) detected in dry, steamed and steamed + 24 hours.

	Dry hay	Steamed hay	Steamed hay + 24 hours	s.e.d	Sig
RP numbers	25,699^{a}	1,586^{b}	5,398^{b}	1,937.5	0.001

Values in the same row not sharing common superscripts differ significantly ($P<0.001$).

Table 3 Mean respirable particle numbers (/kg hay/litre of air) from four different hays (2 replicate bales of each hay) detected in dry, steamed and steamed + 24 hours.

Treatment	Dry	Steamed	Steamed + 24 hours	s.e.d	Sig
Hay 1	51,254^{e}	2,458^{ab}	8,476^{abc}		
Hay 2	28,506^{d}	2,166^{a}	5,141^{abc}		
Hay 3	12,327^{c}	1,203^{a}	5,278^{abc}		
Hay 4	10,711^{bc}	515^{a}	2,699^{ab}	3,875.0	0.001

^{abcde} Values in the table not sharing common superscripts differ significantly ($P<0.001$)

cheaper than haylage at 7.70 euro for an equivalent sized bale. These results show that steaming is a highly successful way to reduce respirable particle numbers in hay. Feeding steamed hay as part of a 'clean-air regime' to stabled horses should reduce the air contamination level to well below the $10^8/m^3$ respirable particles (Lacey, 1990) which is thought to be the threshold for induction of RAO.

References

Blackman, M and M.J.S. Moore-Colyer, 1998. Hay for horses: the effects of three different wetting treatments on dust and mineral content. Animal Science 66, 745-750.

Lacey, J., 1990. Aeriobiology and health: The role of airborne fungal spores in respiratory disease. In: Hawksworth, D. (ed.), Frontiers in Mycology. CAB International, United Kingdom.

The effect of a binding agent on occurrence of the mycotoxin Zearalenon (ZON) in equine faeces

H.J. Walton[1], A.D. Ellis[1], K. Meynen[2] and P. Harris[3]
[1]Nottingham Trent University, School of Animal, Rural and Environmental Sciences, Brackenhurst, Southwell, Nottinghamshire, NG25 0QF, United Kingdom
[2]Kemin Nutrisurance, Atealaan 4N, 2200 Herentals, Belgium
[3]Equine Studies Group, Waltham Centre for Pet Nutrition, Freeby Lane, Waltham on the Wolds, LE14 4RT Leicester, United Kingdom

Introduction

Mycotoxins are secondary metabolites produced by fungi and can be found in cereals as well as forages. The mycotoxin zearalenon (ZON) can affect fertility and has led to abortion in pigs (Raymond *et al.*, 2005, Juhasz *et al.*, 2001). Garcia *et al.* (2003) found that a binder prevented absorption of mycotoxins from the digestive tract in broilers. However, unabsorbed mycotoxins may have possible negative consequences for micro-biota in the equine hindgut. The aim of this study was to investigate the effectiveness of a novel binding agent produced by Kemin Nutrisurance in binding the *Fusarium* mycotoxin ZON *in vivo* and to measure the effect on the hindgut environment *in vitro*.

Material and methods

Six geldings (629 kg±67 kg, 7-18 years) were split into 2 treatment groups according to weight and type in a 2 phase (16 days each, 10 days washout) crossover design (control: pellets and hay or binder: added at 8 g per day to pellets). Diets remained constant at maintenance requirements (as fed hay to low energy pellets ratio 4:1) and horses were exercised daily for 45 minutes. Faecal sub-samples (4 days of full faecal collection using nappies) were analysed for ZON (HPLC analysis,) and samples were used as the inoculum in an *in vitro* gas fermentation study (Loweman *et al.*, 1999). Data were analysed using ANOVA incorporating treatment and phase effects and interactions; means given with standard error (SPSS).

Results and discussion

The pellets contained low natural background levels of ZON (22-84 ppb). Faecal output (17.2±1.2 kg) and faecal pH (6.87±0.1) were not affected by treatment or phase and showed no interactions. There was no effect of treatment on *in vitro* gas production. These findings indicate no adverse effects on the hindgut environment. There was a trend towards increased faecal ZON recovery (+3.2 ppb) when horses were fed the binder (P=0.073) and a significant increase in ZON recovery for phase 2 (+6.1 ppb; P=0.001) which may reflect an increase in pellet feed ZON levels for phase 2 (+62 ppb), so for ZON levels there was a slight treatment x phase interaction. This study showed a trend for the binder to reduce absorption of naturally occurring ZON from the digestive tract but further work is needed.

References

Garcia, A.R., E. Avila, R. Rosiles and V.M. Petrone, 2003. Evaluation of two mycotoxin binders to reduce toxicity of broiler diets containing Ochratoxin A and T-2 Toxin contaminated grain. Avian Diseases 47, 691-699.

Juhasz, J., M.E. Nagy, M. Kulcsar, G. Szigeti, J. Reiczigel and G. Huszenicza, 2001. Effect of low dose zearalenone exposure on luteal function, follicular activity and uterine oedema in cycling mares. Acta. Vet. Hungary 49, 211-222.

Loweman, R.S., M.K. Theodorou, J.J. Hyslop, M.S. Dhanoa and D. Cuddeford, 1999. Evaluation of an *in vitro* batch culture technique for estimating the *in vivo* digestibility and digestible energy content of equine feeds using equine faeces as the source of microbial inoculum, Animal Feed Science Technology 80, 11-27.

Raymond S.L., T.K. Smith and H.V.L.N. Swamy, 2005. Effects of feeding a blend of grains naturally contaminated with Fusarium mycotoxins on feed intake, metabolism, and indices of athletic performance of exercised horses, Journal of Animal Science 83, 1267-1273.

Feeding working donkeys in Mexico; success in simplicity

F. Burden and A. Trawford
The Donkey Sanctuary, Sidmouth, Devon, EX10 0NU, United Kingdom

Introduction

Mexico has approximately 3.3 million donkeys which provide draught power for rural families. In order to design future interventions for problems associated with nutrition for the working donkey in Mexico a baseline survey was carried out to determine feeding practices in two sites.

Material and methods

All donkeys (n=216) presented for veterinary treatment at a DS-UNAM clinic were examined and a survey completed using observation of the animal and questioning of the owner. Details of body condition score (BCS) on a 1-5 scale, feed availability and feeding practices were recorded along with other health and management data. Kruskal-Wallis analysis was used to determine association between feeds and BCS and grazing and BCS. Descriptive statistics were performed for other variables.

Results and discussion

The majority of the donkeys examined had access to pasture outside working hours (>16 hours), 45% of donkeys had access to poor-quality, sparse pasture whilst 55% had access to 'lush' pasture. BCS was not significantly different between the two grazing areas (P=0.90). 77% of donkeys received supplementary feeding, the provision of most supplementary feeds (maize stover, cracked maize, bran, oats, sorghum and tortillas) did not have a significant effect (P=0.35) on BCS which had a median of 2.5/5. However, results indicated that provision of alfalfa did positively affect BCS (P=0.010).

The results of this study showed that nutrition for donkeys is satisfactory for the majority of animals in the areas surveyed with little apparent disease related to malnutrition. Contrary to some preconceptions most donkeys appeared to have a satisfactory BCS suggesting that when given access to pasture and supplementary feeds as deemed appropriate by their owners, they are able to maintain a healthy BCS. These results suggest that nutritional intervention by charities to improve donkey welfare may not be needed in these regions. Further analysis of survey data may suggest other areas for intervention that may lead to significant welfare improvements.

Part 4. Nutrition and gastro-intestinal health

Conditions affecting gastrointestinal tract health

A.E. Durham
Liphook Equine Hospital, Forest Mere, Liphook, Hampshire, GU30 7JG, United Kingdom

Abstract

Given the close physical relationship between ingested nutrients and the gastrointestinal tract, it is no surprise that various aspects of equine nutrition are closely associated with risk of several forms of colic, diarrhoea and gastric ulceration. This review discusses the recognised dietary risk factors for gastrointestinal diseases and the probable causal association with common feeds and feeding regimes for which the horse is poorly adapted.

Keywords: gastrointestinal health, factors affecting

Introduction

The frequent necessity to manipulate equine diets in order to satisfy our requirements for equestrian activity and to facilitate equine management are often not in the best interests of gastrointestinal health and consequent intolerance of such practices may then manifest as gastrointestinal disease. Many aspects of equine nutrition are known to markedly influence the gastrointestinal health of horses in both an immediate and a long term manner. Dietary quality, quantity and the pattern of ingestive behaviour permitted or restricted by nutritional management are all implicated in the risk of conditions such as colic, diarrhoea and gastric ulceration. Clearly many other non-dietary factors such as exercise, parasitism, stable management, stereotypic behaviours and infectious agents (Archer *et al.*, 2006, Mair *et al.*, 1990, Videla and Andrews, 2009) are also important in all of these conditions and these should always be considered alongside diet when problems arise.

Nutritional aspects of equine colic

Horses appear somewhat predisposed to suffering abdominal pain and colic is a condition very familiar to equine veterinarians with approximately 5 cases per 100 horses per year being typically expected (Archer and Proudman, 2006). Far higher colic incidence may occasionally be encountered in certain premises and populations (Uhlinger, 1992; Tinker *et al.*, 1997a) which should ideally trigger application of preventative measures based on epidemiologic studies so that the problem may be effectively controlled. Colic has enormous negative consequences in terms of the welfare of the afflicted individuals, loss of training and competition days, treatment costs and, sometimes, death of the horse (Tinker *et al.*, 1997a; Traub-Dargatz *et al.*, 2001).

In a minority of colic cases surgical treatment may be required and a specific causal lesion is then frequently identified, as is also the case in those horses unfortunate enough to be subject to *post mortem* examination. In some non-surgical cases such as large colon impactions, a confident specific diagnosis can also be made on the basis of clinical examination. However, in many cases the precise cause of colic in an individual horse remains unknown or speculative (Cohen *et al.*, 1999, Kaneene *et al.*, 1997, Tinker *et al.*, 1997b). It is likely that for many cases of colic, pain is caused fundamentally by intestinal distension and stretching of the mesentery and visceral peritoneum. This may be due to a physical obstruction of the intestine (e.g. food impaction, strangulation, displacement) or a functional obstruction resulting from dysmotility (e.g. post operative ileus, equine grass sickness, 'spasmodic' colic). Both the quality and quantity of dietary intake have marked influences on the nature of ingesta as a result of alterations in hydration, pH and osmolarity as well as gas and froth generation. These changes may result in intestinal obstructions as described above either directly or via mucosal inflammation that then leads to dysmotility. The relative balance between dietary content of slowly fermented fibre (e.g. hays, haylages, sugar beet pulp) versus rapidly fermented non-structural carbohydrates (NSC) (e.g. grass fructan or cereal starch) is key to the generation of diet-related colic.

Colic risks associated with cereal and concentrate feeds

Several studies have found that the risk of colic, irrespective of type, significantly increases when horses are fed amounts of cereals or concentrates typical of performance horse diets (Hudson *et al.*, 2001, Kaya *et al.*, 2009, Tinker *et al.*, 1997b). In studies not including multivariate analyses, it should be considered that any apparent association between high levels of concentrate feeding and colic could be partly explained by other non-dietary factors associated with high concentrate feeding that might also predispose to colic such as breed (e.g. Thoroughbred (Hudson *et al.*, 2001)) and a more intensive exercise and competition programme (Cohen *et al.*, 1995, 1999; Durham and White, 2008, Kaneene *et al.*, 1997). Hillyer *et al.* (2001) found that the incidence of colic in UK racehorses appeared synchronised with their competition season when they were presumably receiving higher concentrate feeds although this finding might have other, non-dietary explanations. Nevertheless, increased colic risk has been demonstrated in association with increased cereal or concentrate feeding even after controlling for other variables (Table 1). Tinker *et al.* (1997b) found that horses consuming moderate quantities (2.5-5 kg/day) of concentrates were at an almost 5 times increased risk of colic, and those consuming larger quantities (>5 kg/day) were at a greater than 6 times increased risk in comparison to grazing horses. A further study found that feeding >2.7 kg oats per day was associated with an almost 6-fold increased risk of colic (Hudson *et al.*, 2001). Given that racehorses, for example, may commonly receive in excess of 7 kg concentrates daily (Richards *et al.*, 2006, Southwood *et al.*, 1993) they are likely to be at increased risk of colic in comparison to horses receiving less starch-rich diets. Some studies have suggested processed and pelleted feeds pose a greater risk of colic than wholegrain cereals (Morris *et al.*, 1989; Tinker *et al.*, 1997b) but this has not been found by others (Cohen *et al.*, 1999; Little and Blikslager, 2002). Starch fed in excess of the limited digestive capacity of the equine small intestine is likely to have adverse consequences on caeco-colonic microbial stability that has been the subject of many studies (Goodson *et al.*, 1988; Clarke *et al.*, 1990; Potter *et al.*, 1992; De Fombelle *et al.*, 2001; Drogoul *et al.*, 2001; Julliand *et al.*, 2001; Hussein *et al.*, 2004; Lopes *et al.*, 2004) and is covered in greater detail elsewhere in these proceedings (Veronique Julliand). Additionally, gastric and small intestinal transit may be hastened by the more voluminous chyme associated with high starch feeds, further limiting pre-caecal digestibility and increasing hindgut delivery of NSC (Clarke *et al.*, 1990; Drogoul *et al.*, 2001; Metayer *et al.*, 2004).

In addition to the increased risk of general colic in association with concentrate feeding as described above, some studies have indicated association between concentrates and certain specific types of colic (Table 1). For example, duodenitis-proximal jejunitis (DPJ) may be particularly associated with

Table 1. Colic risk factors associated with concentrate/cereal feeding (all data derived from multivariate analyses).

Colic type	Variable	OR	95% CI	P	Reference
General	Fed >5 kg/day concentrate	6.3	1.8-22.0	0.004	Tinker *et al.* 1997b
	Fed >2.7 kg/day oats	5.9	1.6-22.0	0.009	Hudson *et al.* 2001
	Fed 2.5-5 kg/day concentrate	4.8	1.4-16.6	0.01	Tinker *et al.* 1997b
	No whole grains fed	2.5	1.25-5.0	0.01	Tinker *et al.* 1997b
DPJ	Concentrate fed (per kg)	1.3	1.1-1.6	0.001	Cohen *et al.* 2006
Colon impaction	Concentrates fed (donkeys)	5.2	1.6-16.4	0.005	Cox *et al.* 2009

DPJ: duodenitis proximal jejunitis.

concentrate feeding. Cohen *et al.* (2006) found that horses with DPJ were fed significantly more concentrates than horses with colic due to other causes or lame control horses (4.1 vs. 2.7 kg/day). Colon impactions and displacements have also been associated with higher levels of concentrate feeding (Cox *et al.*, 2009, Hillyer *et al.*, 2002) and this relationship remained significant after controlling for other variables in a study of impaction colic in donkeys. High starch and low fibre diets may be associated with dehydration of colonic ingesta and explain this association (Lopes *et al.*, 2004).

Colic risks associated with grazing

When compared with stable confinement, pasture turnout has generally been found to be associated with a reduced risk of colic although this finding might sometimes be partly explained by other factors such as increased hay or concentrate feeding, closer observation of stabled horses, stereotypic behaviours, weather patterns, exercise, disease and injuries and effects of drugs used to medicate sick or injured horses, (Cohen *et al.*, 1999; Hillyer *et al.*, 2002; Archer *et al.*, 2006). Nevertheless increased grazing has remained significantly associated with decreased risk of colic in studies that have controlled for other potentially confounding factors (Table 2). Hudson *et al.* (2001) found that fully stabled horses or those with a recent reduction in grazing were three times more likely to have colic as those at pasture full time. Colon impactions appear especially associated with decreased access to grazing with a greater than 30 times increased risk of this type of colic in horses that were not grazed in one study (Hillyer *et al.*, 2002). A similar, but less strong relationship between lack of pasture access and impaction colic was also found in donkeys (Cox *et al.*, 2009). Fructans contained in herbage are indigestible and most will reach the hindgut where fermentation will tend to reduce pH as previously discussed for starches. The risk of enterolithiasis is significantly reduced in horses with access to pasture even after controlling for other risk factors (Cohen *et al.*, 2000, Hassel *et al.*, 2004, 2008), perhaps as a result of mild colonic acidification that would deter enterolith formation (Hassel *et al.*, 2004, 2008).

Several studies have found no significant association between pasture access and colic risk, perhaps due to possible confounding effects of pooling all colic cases together (Cohen *et al.*, 1995, Cohen and Peloso, 1996, Kaya *et al.*, 2009); whereas grazing may actually increase the risk of certain colic subtypes. Sand impactions and grass sickness cases are seen only in grazing horses (Ragle *et al.*, 1989, Wood *et al.*, 1998). Soil types and pasture management practices that might promote soil

Table 2. Colic risk factors associated with grazing (all data derived from multivariate analyses).

Colic type	Variable	OR	95% CI	P	Reference
General	No grazing or recent decrease in grazing	3.0	1.4-6.6	0.007	Hudson *et al.* 2001
DPJ	Grazing pasture	3.5	1.8-6.8	0.0002	Cohen *et al.* 2006
Enterolithiasis	No pasture access	6.7	1.4-33.3	<0.05	Hassel *et al.* 2004
	<50% pasture turnout	4.5	1.4-13.9	<0.01	Cohen *et al.* 2000
	No pasture access	2.8	1.1-7.6	0.04	Hassel *et al.* 2008
Colon impaction/ displacement	No pasture access (horses)	30.2	25.7-35.5	0.005	Hillyer *et al.* 2002
	No pasture access (donkeys)	3.4	1.3-8.8	0.04	Cox *et al.* 2009
Grass sickness	No co-grazing ruminants[1]	9.1	4.2-20.0	0.001	Newton *et al.* 2004
	No grass cutting[1]	8.3	4.5-16.7	<0.001	Newton *et al.* 2004
	Previous cases on premises	2.2	1.4-3.6	0.002	Wood *et al.* 1998

DPJ: duodenitis proximal jejunitis; [1] combining co-grazing and grass cutting may neutralise the potential benefit of each strategy alone.

disturbance (e.g. mechanical pasture sweepers) appear to increase the risks of grass sickness further whereas removal of grass by co-grazing ruminants or by grass-cutting may offer some protection against the disease in grazing horses (Newton *et al.*, 2004). The proposed causal association between grass sickness and soil borne toxicoinfectious botulism may explain these associations (Hunter *et al.*, 1999). Increased risk of grass sickness recurrence on the same premises (Wood *et al.*, 1998) also suggests specific local factors associated with grass/pasture that predispose to the condition. Interestingly DPJ, another one of the few causes of colic that has been associated with increased grazing activity (Cohen *et al.*, 2006) has also been aetiologically linked with enteric toxico-infection (Feary and Hassel, 2006). One study found that grazing horses were at more than three times the risk of DPJ in comparison with stabled horses.

Colic risks associated with forage

Preserved forages tend to be rich in slowly fermentable carbohydrate, reflected by high neutral detergent fibre (NDF) and acid detergent fibre (ADF) and are qualitatively similar to the dietary intake for which the equine gastrointestinal tract is theoretically best suited. However, forages of various types have been shown to have positive, neutral or negative impacts on gastrointestinal health in horses (Table 3) (Cohen *et al.*, 1995, Cohen *et al.*, 1999). Poor quality hay especially high in ADF has been shown to be associated with increased colic risk in studies examining the effects of feeding coastal grass hay or hay fed from round bales (Cohen and Peloso, 1996, Hudson *et al.*, 2001). Poor hygienic quality of hay may also lead to colic. In one recent study, hays fed to horses that developed colic were far more likely to be of low hygienic quality compared with hays fed to a control group (Kaya *et al.*, 2009).

Intestinal impactions are associated with feeding poor quality hays and there is a strong association between feeding coastal Bermuda grass hay and development of ileal impaction (Little and Blikslager, 2002). Although the vast majority of cases of colon impaction and displacements are seen in hay fed horses, this observation may be confounded by related variables such as reduced grass intake and exercise (Hillyer *et al.*, 2002). There is a very strong association between alfalfa feeding and the harmful effect on the intestine of cantharidin when blister beetles infest the hay (Helman and Edwards, 1997); furthermore alfalfa hay represents the strongest known risk factor for development of enterolithiasis in horses (Cohen *et al.*, 2000; Hassel *et al.*, 2004, 2008; Morris *et al.*, 1989). One study found a mixed legume-grass hay was more likely to be associated with DPJ cases than other forms of colic (Morris *et al.*, 1989). In contrast, certain hay types have been associated with reduced colic risk; feeding oat and grass hays appears to offer significant protection against enterolith formation even after controlling for reduced intake of alfalfa (Hassel *et al.*, 2008, Morris *et al.*,

Table 3. Colic risk factors associated with forage (all data derived from multivariate analyses).

Colic type	Variable	OR	95% CI	P	Reference
General	Round bale hay	2.5	1.1-5.6	0.028	Hudson *et al.* 2001
	Multiple changes in hay in last year	2.1	1.2-3.8	0.01	Tinker *et al.* 1997b
	Coastal grass hay	1.7	1.01-2.70	0.0451	Cohen and Peloso, 1996
Enterolithiasis	Alfalfa >70% of diet	10.8	2.6-44.0	<0.05	Hassel *et al.* 2004
	Oat hay <50% diet	5.0	1.6-14.3	<0.01	Hassel *et al.* 2008
	Alfalfa >50% of diet	4.7	1.4-15.6	0.01	Hassel *et al.* 2008
	Grass hay <50% diet	4.6	1.6-12.5	<0.01	Hassel *et al.* 2008
	Fed any alfalfa hay	4.2	1.3-12.9	0.01	Cohen *et al.* 2000
Ileal impaction	Coastal Bermuda grass	4.4	2.1-9.1	<0.05	Little and Blikslager 2002

1989). Alfalfa feeding was associated with decreased risk of small intestinal strangulations in one study although possible confounding variables were not examined (Morris *et al.*, 1989). Similarly, although Bermuda grass hay was associated with decreased risk of developing DPJ, this may have been an indirect result of decreased pasture access and increased concentrate feeding (Cohen *et al.*, 2006, Morris *et al.*, 1989).

Colic risks associated with dietary changes

Dietary factors relating to quality and quantity of cereal, grazing and forage have a significant influence on colic risk as described above. However, it appears that dietary changes probably pose the greatest risk of nutrition-associated colic in horses (Table 4) presumably due to induced dynamism and instability in gastrointestinal microflora populations and consequent dysfermentation, pH variability, mucosal inflammation and dysmotility. De Fombelle *et al.* (2001) found that changing the diet from 100% hay to 70% hay/30% rolled barley was associated with significant changes in hindgut flora and volatile fatty acid (VFA) content. Changes in the source of hay or concentrates, changing pasture, changes in the amounts or frequency of feeding and changing the usual feeding times significantly increase the risk of colic (Tinker *et al.*, 1997b; Cohen *et al.*, 1999; Hudson *et al.*, 2001, Wood *et al.*, 1998) emphasising the requirement for gradual implementation of any necessary dietary changes in horses. The greatest colic risk following diet changes appears to last for approximately two weeks in most studies (Cohen *et al.*, 1995; Cohen and Peloso, 1996; Mehdi and Mohammed, 2006) although the risk of grass sickness and EFE following changes in grazing may extend for longer than this (Archer *et al.*, 2008, Wood *et al.*, 1998). Hillyer *et al.*, (2002) found that the risks of simple colon obstruction and distension were greatest within 7 days of a dietary change (forage change OR=22.00; concentrate change OR=12.03) but were still significantly increased between 8-14 days of the change (forage change OR=4.88; concentrate change OR=3.01). Dietary changes between 15 and 28 days previously were not significantly associated with colic. A change of hay in the previous 2 weeks was the strongest diet-related risk factor for colic in one study, increasing colic risk 10 fold (Cohen *et al.*, 1999). Hudson *et al.* (2001) found that a change in hay posed almost double the risk of colic when compared with a change in concentrates fed. It is perhaps surprising that several studies have indicated that a change in forage poses a greater colic risk than a change in concentrate or cereal feeding (Cohen *et al.*, 1999; Hudson *et al.*, 2001; Hillyer *et al.*, 2002, Little and Blikslager, 2002).

Table 4. Colic risk factors associated with diet changes (all data derived from multivariate analyses).

Colic type	Variable	OR	95% CI	P	Reference
General	Change of hay within previous 2 weeks	9.8	1.2-81.5	0.035	Cohen *et al.* 1999
	Change of hay within previous 2 weeks	4.9	2.1-11.4	<0.001	Hudson *et al.* 2001
	Single change in concentrate in last year	3.6	1.6-5.4	<0.001	Tinker *et al.* 1997b
	No grazing or recent decrease in grazing	3.0	1.4-6.6	0.007	Hudson *et al.* 2001
	Change in concentrate within previous 2 weeks	2.6	0.9-7.2	0.064	Hudson *et al.* 2001
	Diet change within previous 2 weeks	2.2	1.7-2.8	<0.001	Cohen *et al.* 1995
	Mulitple changes in concentrate in last year	2.2	1.2-4.1	0.02	Tinker *et al.* 1997b
Grass sickness	Change of grazing within previous 2 weeks	29.7	6.7-130	<0.001	Wood *et al.* 1998
	Change of grazing within previous 2 -3 months	4.1	1.0-16.6	0.048	Wood *et al.* 1998
EFE	Reduction in grazing in previous 28 days	3.7	1.4-9.7	<0.01	Archer *et al.* 2008

EFE: epiploic foramen entrapment.

One study indicated that rates of chewing new forages may take 2 weeks to completely adapt and therefore horses may swallow larger particles when moving from lower to relatively higher structural fibre forages (Ellis and Hill, 2002). A change in pasture represents one of the greatest risk factors for grass sickness and remains significant for up to 3 months following the change (Wood *et al.*, 1998). Fructan content of grasses and herbage can be markedly variable between and within grass species (Longland and Cairns, 2000) creating the possibility for hindgut disturbance when grazing is changed. Both temperature and light exposure can also have marked and acute effects on fructan content of grass with potential acute destabilisation of intestinal microflora (Longland and Byrd, 2006). A recent decrease in grazing significantly increases the risk of EFE even after controlling for other important risk factors such as stereotypic behaviour (Archer *et al.*, 2008).

Colic risks associated with other aspects of nutrition

Horses that were fed carrots as part of their daily ration were found to be at significantly greater risk of developing enteroliths in one study (Table 5) although the reason for this finding is not known (Hassel *et al.*, 2008). EFE appears to have strong associations with presumed psychologic stress and anxiety as stereotypic behaviours represent the strongest risk factor for this disease. Consistent with this it has been found that risk of EFE is increased when horses are fed at the same time as close companions (Archer *et al.*, 2008) as this practice may frequently lead to a general increase in levels of excitement, anticipation of feeding and perhaps anxiety. It has also been found that the risk of EFE is significantly reduced in horses that are offered mineral or salt licks (Table 5) although the explanation of this finding is not known (Archer *et al.*, 2008). A similar protective non-significant trend of offering mineral/vitamin supplementation was found in a study of enterolithiasis cases but again remains unexplained (Hassel *et al.*, 2008).

Table 5. Colic risk factors associated with other aspects of nutrition (all data derived from multivariate analyses).

Dietary category	Variable	OR	95% CI	P	Reference
Enterolithiasis	Carrots fed	2.7	0.98-7.64	0.05	Hassel *et al.* 2008
EFE	Fed at same time as close companions	4.0	1.3-14.3	0.01	Archer *et al.* 2008
	No access to mineral/salt lick	2.9	1.1-8.3	0.03	Archer *et al.* 2008

EFE: epiploic foramen entrapment.

Nutritional aspects of equine diarrhoea

The study of diarrhoea in horses is somewhat impaired by a generally low diagnostic rate in clinical cases. Specific causes were found ante mortem in less than 25% of cases in 2 series of diarrhoea cases reported in adult horses (Love *et al.*, 1992, Mair *et al.*, 1990) and in 45% of foals in another study (Frederick *et al.*, 2009). In none of these 3 studies were dietary factors implicated in the causes of diarrhoea although given the large number of cases without a specific diagnosis, it is still possible that aspects of nutrition were significant or even prominent. Although ingestion of certain toxic plants and chemicals is known to cause diarrhoea in horses (Cohen, 2002), the general putative association between dietary factors and diarrhoea are largely anecdotal and based on intuitive principles rather than evidence-based studies. Unsurprisingly, observations from experimental models for laminitis induction clearly indicate that ingestion of large quantities of cereal starch or oligofructose have the capacity for provoking diarrhoea (Rowe *et al.*, 1994, van Eps and Pollitt, 2006) and it appears reasonable to assume that similar dietary factors may increase the risk of diarrhoea in horses receiving diets high in cereals or concentrates or when pasture fructan intake is high. The potentially anti-microbial effects of high levels of dietary oils (Janssen *et al.*, 2002) should also be considered although horses appear relatively tolerant of feeding fat in comparison to high NSC diets. Dietary changes are also likely to increase the risk of diarrhoea and have been shown to increase *Salmonella* shedding in hospitalised horses (Traub-Dargatz *et al.*, 1990).

Nutritional aspects of equine gastric ulceration

Equine gastric ulceration is a disease of high prevalence, especially in performance horse populations (Videla and Andrews, 2009) and diet has long been implicated as a risk factor for ulcerogenesis along with intense and/or frequent exercise (Hammond *et al.*, 1986). The equine stomach has a distinctive division of upper squamous mucosa and lower glandular mucosa and each of these two regions demonstrates significantly different properties relevant to the links between diet and gastric ulceration (Merritt, 1999). Horses evolved as grazers that would consume high-fibre, low NSC grasses over very prolonged feeding periods (Houpt, 1990). In keeping with these normal feeding patterns, gastrointestinal secretions in the horse may be constant rather than meal-induced in contrast to the physiologic characteristics of carnivores. It has been shown that the equine stomach secretes hydrochloric acid almost continually, leading to prolonged exposure of glandular mucosa to low pH in the normal horse (Campbell-Thompson and Merritt, 1990; Murray, 1997). Strong antacid defence is essential to withstand this constantly very low pH environment and several prostaglandin-dependent protective mechanisms are recognised in the glandular mucosa comprising secretion of bicarbonate-rich mucus, high rates of mucosal perfusion (to remove any penetrating protons and maintenance of potential for a high metabolic rate) and rapid epithelial coverage and regrowth (restitution) following epithelial cell death (Murray, 1999). In contrast to the well defended glandular mucosa, the squamous mucosa dorsal to the *margo plicatus* offers less resistance to acid exposure. The squamous epithelium is covered by a thin phospholipid barrier (Ethell *et al.*, 2000) and also defensive keratinisation may develop in the *stratum corneum* (Martineau *et al.*, 2009). However, the main resistance to acidic challenge to the squamous mucosa arises from glycoconjugate substances containing bicarbonate secreted by the cells in the *stratum spinosum* (Andrews *et al.*, 2005). There is nevertheless remarkable contrast in the inherent acid-resisting properties of the two gastric divisions. The most logical explanation for the relative weakness of antacid defence in the equine gastric squamous mucosa is that evolutionary pressures have not been present to lead to its development. Observations of the pH and gross appearance of solid ingesta within the grass- or forage-fed equine stomach support this supposition indicating that the squamous mucosa is normally in contact with alkaline saliva-soaked fibrous matter. This solid and pH-stratified gastric ingesta serves to limit mixing of gastric contents and protects the squamous mucosa from the highly acidic gastric fluid in the more dependent glandular region (Figure 1, Merritt, 2003, Murray and Grodinsky, 1989).

Gastric ulcer risks associated with cereal and concentrate feeds

Associations between cereal feeding and ulceration of the gastric squamous mucosa are well recognised. Coenen (1990) found that 10/27 horses fed a mixed concentrate/hay feed developed gastric squamous mucosal ulcers compared with 0/21 control horses fed hay only. An *in vitro* study of the effects of short-chain fatty acids (SCFAs) on gastric mucosa concluded that horses fed >0.5 kg grain/100 kg bodyweight every 6 to 8 hours were likely to be at higher risk for the development of squamous mucosal ulcers (Andrews *et al.*, 2006). Consistent with this finding, a recent epidemiologic study found that horses consuming more than 2 g starch per kg bodyweight per meal (e.g. 2-3 kg concentrates or cereal for a 500 kg horse) were at more than double the risk of having significant squamous mucosal ulcers compared with horses receiving less starch (OR 2.6, 95% CI 1.2-5.4, P=0.01) (Luthersson *et al.*, 2009).

Diets high in concentrates and cereals (and therefore low in forage) will inevitably reduce the normal pH stratification described above and promote fluidity and mixing of gastric contents owing to the

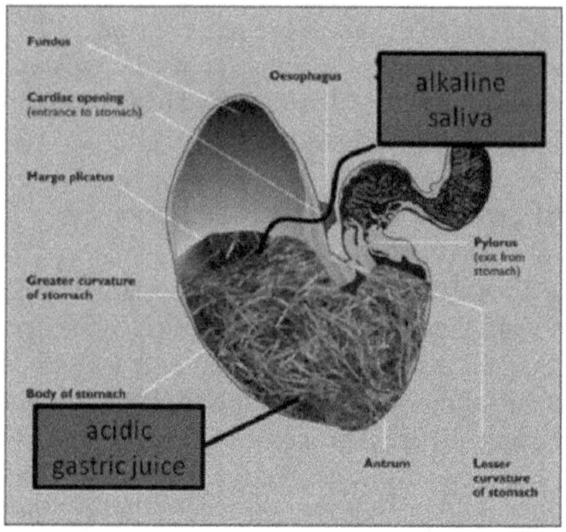

Figure 1. Diagrammatic representation of the stomach of a grazing or forage-fed horse demonstrating pH stratification of solid intragastric contents.

absence of the high-volume, solid, fibrous mass provided only by forage and grass-based intakes (Figure 2; Argenzio, 1999). Consequently exposure of the poorly defended squamous mucosa to highly acidic gastric secretions is more likely increasing the chances for erosive injury. High cereal diets may also stimulate a more prolonged gastrin secretory response than forage diets leading to enhanced acid secretion and even lower pH (Smyth *et al.*, 1989; Sandin *et al.*, 1998). This is also compounded by the reduced salivary buffer secreted in response to rapid cereal meal consumption in contrast to slow and prolonged forage mastication (Meyer *et al.*, 1985) and also by slower gastric emptying associated with large, high-starch meals (Metayer *et al.*, 2004).

Despite the poor acid-defence mechanisms within the squamous mucosa described above, studies have shown remarkable resistance to damage when exposed to hydrochloric acid alone (Nadeau *et al.*, 2003a,b), suggesting that further ulcerogenic challenges must exist. In addition to hydrochloric acid, many weaker SCFAs are present in the equine stomach and these may be important in ulcerogenesis. Significant bacterial fermentation of NSCs occurs in the stomach and small bowel of the horse as well as the hindgut (Al Jassim and Andrews; 2009; Mackie and Wilkins, 1988). Indeed, bacterial populations within the equine stomach may well be even more numerous than those found in the caecum and colons, and include *Streptococcus, Lactobacillus, Clostridium, Prevotella, Pseudomonas*

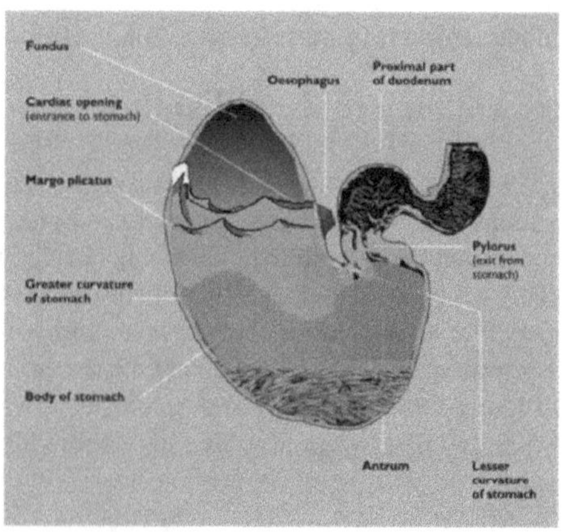

Figure 2. Diagrammatic representation of the stomach of a cereal-fed, forage-deprived, exercising horse demonstrating loss of pH stratification of gastric contents and potential movement of acidic fluid and contact with the poorly defended upper squamous mucosa.

and *Propionibacterium* (Al Jassim and Andrews, 2009). Undoubtedly the neutral or mildly acidic pH of the squamous mucosal area is most favourable to bacterial colonisation although gastric bacterial populations appear remarkably tolerant to acid exposure (Al Jassim and Andrews, 2009). When gastric bacteria are presented with dietary NSC they produce many different SCFAs including acetic, propionic, butyric, isobutyric, valeric, isovaleric and lactic acids, especially in the squamous region of the stomach (Al Jassim and Andrews; 2009; Nadeau *et al.*, 2000, 2003a,b). Lactic and acetic acids are found in the greatest concentrations and are frequently present in the stomach of cereal fed horses at >10 mmol/l (Al Jassim, 2006; Nadeau *et al.*, 2000). One study showed that the addition of sorghum grains to the diet of horses previously fed hay only, lead to a significant increase in lactate concentration throughout the gastrointestinal tract, but most especially in the non-glandular area of the stomach where concentrations >40 mmol/l were found (Al Jassim, 2006).

SCFAs produced by gastric bacterial fermentation have been implicated in the pathogenesis of equine gastric squamous mucosal ulceration (Nadeau *et al.*, 2003a,b). These are all relatively weak acids and when exposed to pH <4.5 they become almost totally undissociated as a result of accepting protons from the much stronger and prevalent gastric hydrochloric acid. When protonated, SCFAs become lipophilic and able to diffuse into the relatively unprotected gastric squamous epithelial cells. The bicarbonate-rich mucus barrier and high perfusion rate of the gastric glandular mucosa largely protects this area from intracellular SCFA uptake. When SCFA molecules arrive in the near neutral pH of the squamous epithelial cell cytosol, they will tend to dissociate and release protons leading to a marked decrease in cytosolic pH, cellular dysfunction and death, thus promoting squamous mucosal ulceration (Figure 3, Andrews *et al.*, 2006; Nadeau *et al.*, 2003a,b).

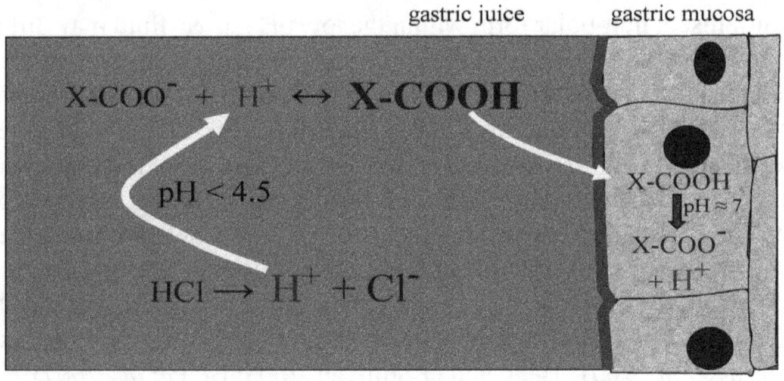

Figure 3. Illustration of intracellular transport of short chain fatty acids (X-COOH) in a low pH medium (see text for further explanation).

Gastric ulcer risks associated with forage

Forage quality, quantity and availability have a significant influence of the likelihood of gastric ulceration. Feeding a diet of alfalfa and grain results in higher gastric SCFA concentrations, but reduced acidity and less severe gastric ulceration than when a diet of bromegrass hay or coastal Bermuda grass hay is fed (Nadeau *et al.*, 2000; Lybbert *et al.*, 2007). The higher SCFA production with the former diet is not surprising in view of inclusion of grain and the higher gastric pH may have resulted from the buffering effect of higher protein and/or calcium content of the alfalfa/grain diet (Nadeau *et al.*, 2000). As described above, SCFAs are likely to be less injurious to the gastric squamous mucosa at pH>4.5. A recent study indicated that forage of even lower quality was also more ulcerogenic. Horses fed straw were at more than 4 times the risk of having significant squamous mucosal ulcers when compared to those receiving hay or haylage (OR 4.2, 95% CI 1.3-13.8, P=0.02) (Luthersson *et al.*, 2009). In addition to the proposed explanation of calcium and/or protein buffering of gastric acidity, the authors of the latter study also proposed that the highly lignified and silicated nature of straw might be irritating to the gastric mucosa or perhaps disturb the stability of gastric ingesta stratification.

Consistent with an evolutionary adaptation to almost continual feed intake, feed deprivation and intermittent feeding programmes have been shown to be an effective means of decreasing gastric pH and inducing squamous mucosal ulcers in horses (Murray and Eichorn, 1996; Murray and Schusser, 1993). A recent study suggests that daytime hay deprivation is likely to be more harmful than overnight fasting as nocturnal periods of low pH in the non-glandular area appear to be common even in hay fed horses owing to reduced nocturnal ingestive behaviour (Husted *et al.,* 2009). The latter study found that when less than 0.3 kg hay was consumed over any 4 hour period, the pH in the non-glandular area was consistently <4, creating significant ulcerogenic risk. Voluntary fasting in healthy horses never extends beyond 3 to 5 hours (Ralston, 1984) and a significant decrease in gastric pH is seen approximately 6 hours after feeding alfalfa and grain, suggesting the loss of the buffering effect of the gastric ingesta (Nadeau *et al.,* 2000). Consistent with these findings, a recent epidemiologic study found that forage feeding intervals greater than 6 hours represented the strongest dietary risk factor for ulceration of the gastric squamous mucosa (OR 5.3, 95% CI 1.4-20.0, $P=0.01$) (Luthersson *et al.,* 2009).

When the equine stomach is continually filled with a physically stable, high-fibre mat of forage soaked in saliva there are beneficial consequences beyond pH buffering and physical protection of the squamous mucosa from acidic gastric fluid. The equine pylorus offers little resistance to retrograde flow from the duodenum and frequent boluses of pancreatic and biliary secretions will enter the stomach when not prevented from doing so by solid ingesta (Merritt, 1999). Although there may be significant buffering of gastric acid by pancreatic bicarbonate (Merritt, 1999), the presence of bile acids and also SCFAs produced from duodenal fermentation of NSCs may be harmful to the gastric mucosa. As described above in the context of SCFAs associated with cereal feeding, bile acids in their non-ionised, non-polar form within the low pH gastric fluid may diffuse into mucosal cells leading to cell death and necrosis (Argenzio, 1999; Berschneider *et al.,* 1999; Murray, 1999). It is possible that bile acids might further disrupt the protective phospholipid barrier that normally offers some protection to the squamous mucosal surface (Geor, 2000). Although potentially injurious to the poorly defended and high-risk gastric squamous mucosa, it is the pyloric mucosa that likely experiences the highest concentrations of bile acids in duodenal reflux prior to dilution in gastric fluid and this might potentially overwhelm the normally strong mucosal defence of this area of the stomach leading to pyloric ulceration. This theory is supported by this author's experience of encountering especially severe pyloric ulcers located in the path of duodenal reflux streams and also the relatively high prevalence of ulcers in the pyloric area in competition horses that might receive relatively low forage diets (Murray *et al.,* 2001; Begg and O'Sullivan, 2003; Bell *et al.,* 2007).

Gastric ulcer risks associated with grazing

It is generally accepted that grazing reduces the risks of gastric ulceration in horses (Reese and Andrews, 2009) presumably due to similar mechanisms as those described above for forage feeding. However, grazing horses are occasionally encountered in practice that have severe squamous mucosal and/or pyloric ulceration without any further obvious risk factors for gastric ulceration (AED personal observations). In a study in racehorses, Bell *et al.* (2007), found that the prevalence of gastric ulcers in stabled, partly grazed and fully grazed horses to be 94%, 89% and 100% respectively indicating no protective effect of grazing. Similarly a high prevalence of gastric ulcers has been reported in pastured pregnant (67%) and non-pregnant (76%) broodmares in another study (Le Jeune *et al.,* 2006). This suggests that either other ulcerogenic factors may be capable of overwhelming the putative protective effect of grazing (e.g. exercise, cereals, an equine gastric Helicobacter species) or perhaps that grazing itself might be ulcerogenic under certain circumstances. It has been found that gastric pH falls significantly during the night even in horses fed ad libitum hay presumably due to reduced nocturnal feeding activity (Husted *et al.*, 2009) and this might also be true of grazing horses, thus maintaining a potential ulcerogenic challenge. It might also be that an ingested mass of grass within the stomach may sometimes be insufficiently alkaline, stable or pH-stratified to protect the gastric mucosa. Furthermore, as many grasses contain high levels of NSC, this may also serve as a significant source of injurious gastric concentrations of SCFAs that might lead to ulceration as previously described for cereal and concentrate feeds. Studies comparing grass quality and ingestion patterns in grazing horses with and without gastric ulcers might prove interesting in these respects.

Gastric ulcer risks and other dietary factors

Theoretically beneficial mechanisms of action of dietary oils in high-fat feeds might include provision of substrate for prostaglandin synthesis (likely to primarily aid gastric *glandular* mucosal defence) or possibly by binding free bile acids within the gastric fluid. Possible increases in gastric emptying rates may also be found in horses fed high fat versus high-starch diets thereby potentially limiting SCFA accumulation (Lorenzo-Figueras *et al.*, 2005). Although one study found reduced gastric acid secretion and higher levels of potentially gastroprotective prostaglandins in horses fed small amounts (45 ml) of corn oil (Cargile *et al.*, 2004), a further study that involved feeding higher levels of dietary fat found no benefit on ulcer severity (Frank *et al.*, 2005).

The administration of hypertonic electrolyte solutions and pastes is frequently practiced in competition horses (Rose and Lloyd, 1992; Auer *et al.*, 1993; Nyman *et al.*, 1996). This treatment may be significantly harmful to the gastric squamous mucosa and has been associated with significant squamous ulcerogenesis presumably due to osmotic and irritant effects (Holbrook *et al.*, 2005). Luthersson *et al.*, (2009) found that horses without free access to water in the paddock were more likely to have gastric ulcers regardless of location within the stomach.

Conclusion

The equid gastrointestinal tract most likely evolved in association with a diet comprising grasses, rushes and sedges high in fibre and low in NSC that was gradually ingested in a so-called 'trickle-feeding' pattern over perhaps 16 hours each day. It is generally the case that gastrointestinal dietary intolerances manifesting as colic, diarrhoea or gastric ulceration are largely explainable in terms of deviations from this natural feeding regimen. It is logical to assume therefore that gastrointestinal health will benefit from attempts to mimic such a diet and ingestive patterns. Indeed, despite the common practice of feeding cereal and grazing high quality managed pastures, there is evidence that the equine gastrointestinal tract is very poorly adapted to high rates of NSC consumption associated with these practices and gastrointestinal disease is a common consequence. Additionally, uniformity of dietary ingredients appears to be beneficial to the gastrointestinal tract and this supports the stability of bacterial populations with the gastrointestinal tract. The equine gastrointestinal tract does not appear to tolerate dietary changes very well. Although dietary changes clearly do occur naturally in feral grazing horses (McInnes and Vavra, 1987, Stuska *et al.*, 2008) this is in a gradual fashion with changes in the seasons and weather slowly influencing quality and quantity of ingested herbage. Some intra-day variability in patterns of consumption might also be normal in domesticated horses with free access to hay (Husted *et al.*, 2009) although tend to be more marked and potentially more harmful in horses offered cereal meals with limited access to forage. Furthermore, many differing choices and sources of preserved forage and concentrated feed are available for the modern domesticated horse creating a greater likelihood of abrupt dietary changes. Entero-bacterial stability may be disturbed to the detriment of gastrointestinal health both when blanket dietary changes are implemented, such as when grazing horses are stabled and fed hay instead of grass, and with the less obvious but potentially significant intra-day dietary variability resulting from high-starch meals interspersed by forage consumption. In practice it will frequently prove difficult (or impossible) to exactly mimic a natural diet and dietary behaviour owing to the increased nutritional requirements of working horses and the relatively high nutrient density of available feeds, forages and pasture in comparison to the typical feral diet. Nevertheless, potential dietary risk factors can still be identified and moderated in horses that are considered to be at risk of, or found to be suffering from, gastrointestinal diseases.

References

Al Jassim, R.A.M., 2006. Supplementary feeding of horses with processed sorghum grains and oats. Animal Feed Science and Technolology 125: 33-44.

Al Jassim R.A. and F.M. Andrews, 2009. The bacterial community of the horse gastrointestinal tract and its relation to fermentative acidosis, laminitis, colic, and stomach ulcers. Veterinary Clinics of North America Equine Practice 25: 199-215.

Andrews, F.M., B.R. Buchanan, S.B. Elliot, N.A. Clariday and L.H. Edwards, 2005. Gastric ulcers in horses. Journal of Animal Science 83: E18-21.

Andrews, F.M., B.R. Buchanan, S.H. Smith, S.B. Elliott and A.M. Saxton, 2006. *In vitro* effects of hydrochloric acid and various concentrations of acetic, propionic, butyric or valeric acid on bioelectric properties of equine gastric squamous mucosa. American Journal of Veterinary Research 67: 1873-1882.

Archer, D.C., G.L. Pinchbeck, N.P. French and C.J. Proudman, 2008. Risk factors for epiploic foramen entrapment colic in a UK horse population: a prospective case-control study. Equine Veterinary Journal 40: 405-410.

Archer, D.C., G.L. Pinchbeck, C.J. Proudman and H.E. Clough, 2006. Is equine colic seasonal? Novel application of a model based approach. BMC Veterinary Research 24: 27-37.

Archer, D.C. and C.J. Proudman, 2006. Epidemiological clues to preventing colic. Veterinary Journal 172: 29-39.

Argenzio, R.A., 1999. Comparative pathophysiology of nonglandular ulcer disease: a review of experimental studies. Equine Veterinary Journal (suppl.) 29: 19-23.

Auer, D.E., K.V. Skelton, S. Tay and F.C. Baldock, 1993. Detection of bicarbonate administration (milkshake) in standardbred horses. Australian Veterinary Journal 70: 336-340.

Begg, L.M. and C.B. O'Sullivan, 2003. The prevalence and distribution of gastric ulceration in 345 racehorses. Australian Veterinary Journal 81: 199-201.

Bell, R.J.W., J.K. Kingston, T.D. Mogg and N.R. Perkins, 2007. The prevalence of gastric ulceration in racehorses in New Zealand. New Zealand Veterinary Journal 55: 13-18.

Berschneider, H.M., A.T. Blikslager and M.C. Roberts, 1999 Role of duodenal reflux in nonglandular gastric ulcer disease of the mature horse. Equine Veterinary Journal(suppl.) 29: 24-29.

Campbell-Thompson, M.L. and A.M. Merritt, 1990. Basal and pentagastrin-stimulated gastric secretion in young horses. American Journal of Physiology 259: R1259-R1266.

Cargile, J.L., J.A. Burrow, I. Kim, N.D. Cohen and A.M. Merritt, 2004. Effect of dietary corn oil supplementation on equine gastric fluid acid, sodium, and prostaglandin E2 content before and during pentagastrin infusion. Journal of Veterinary Internal Medicine 18: 545-549.

Clarke, L.L., M.C. Roberts and R.A. Argenzio, 1990. Feeding and digestive problems in horses. Veterinary Clinics of North America Equine Practice 6: 319-337.

Coenen, M., 1990. The occurrence of feed-induced stomach ulcers in horses. Schweizer Archiv fur Tierheilkunde 132: 121-126.

Cohen, N.D., 2002. Toxic colitides. In: Divers T., T. Mair, N. Ducharme (eds.) Manual of Equine Gastroenterology. WB Saunders, London, pp 417-21

Cohen, N.D., P.G. Gibbs and A.M. Woods, 1999. Dietary and other management factors associated with colic in horses. Journal of the American Veterinary Medical Association 215: 53-60.

Cohen, N.D., P.L. Matejka, C.M. Honnas and R.N. Hooper, 1995. Case-control study of the association between various management factors and development of colic in horses. Texas Equine Colic Study Group. Journal of the American Veterinary Medical Association 206: 667-673.

Cohen, N.D. and J.G. Peloso, 1996. Risk factors for history of previous colic and for chronic, intermittent colic in a population of horses. Journal of the American Veterinary Medical Association 208: 697-703.

Cohen, N.D., E. Toby, A.J. Roussel, E.L. Murphey and N. Wang, 2006. Are feeding practices associated with duodenitis-proximal jejunitis? Equine Veterinary Journal 38: 526-531.

Cohen, N.D., C.A. Vontur, and P.C. Rakestraw, 2000. Risk factors for enterolithiasis among horses in Texas. Journal of the American Veterinary Medical Association 216: 1787-1794.

Cox, R., F. Burden, L. Gosden, C. Proudman, A. Trawford and G. Pinchbeck, 2009. Case control study to investigate risk factors for impaction colic in donkeys in the UK. Preventive Veterinary Medicine 92:179-187.

De Fombelle, A., V. Julliand, C. Drogoul and E. Jacotot, 2001. Feeding and microbial disorders in horses: 1-Effects of an abrupt incorporation of two levels of barley in a hay diet on microbial profile and activities. Journal of Equine Veterinary Science 26: 439-445.

Drogoul, C., A. de Fombelle and V. Julliand, 2001. Feeding and microbial disorders in horses: 2: Effect of three hay:grain ratios on digesta passage rate and digestibility in ponies. Journal of Equine Veterinary Science 26: 487-491.

Durham, A.E. and N.A. White, 2008. The gastrointestinal tract of performance horses: medical, nutritional and surgical considerations. In: Lindner, A. (ed.) The Acutely Poorly Performing Sport Horse. Wageningen Academic Publishers, Wageningen, the Netherlands, pp 39-68.

Ellis, A.D. and J. Hill, 2002. Feed factors affecting intake behaviour and water intake in horses. In: Proceedings of the 53rd Annual Meeting of the European Association of Animal Science. September 1-4, 2002. Cairo, pp. 259.

Ethell, M.T., D.R. Hodgson and B.A. Hills, 2000. Evidence for surfactant contributing to the gastric mucosal barrier of the horse. Equine Veterinary Journal 32: 470-474.

Feary, D.J. and D.M. Hassel, 2006. Enteritis and colitis in horses. Veterinary Clinics of North America Equine Practice 22: 437-479.

Frank, N., F.M. Andrews, S.B. Elliott and J. Lew, 2005. Effects of dietary oils on the development of gastric ulcers in mares. American Journal of Veterinary Research 66: 2006-2011.

Frederick, J., S. Giguère and L.C. Sanchez, 2009. Infectious agents detected in the feces of diarrheic foals: a retrospective study of 233 cases (2003-2008). Journal of Veterinary Internal Medicine 23: 1254-1260.

Geor, R.J., 2000. Gastric surface active phospholipid – a role in protection of the squamous epithelial mucosa? Equine Veterinary Journal 32: 458-459.

Goodson, J., W.J. Tyznik, J.H. Cline and B.A. Dehority, 1988. Effects of an abrupt diet change from hay to concentrate on microbial numbers and physical environment in the cecum of the pony. Applied Environmental Microbiology 54: 1946-1950.

Hammond, C.J., D.K. Mason and K.L. Watkins, 1986. Gastric ulceration in mature thoroughbred horses. Equine Veterinary Journal 18: 284-287.

Hassel, D.M., B.M. Aldridge, C.M. Drake and J.R. Snyder, 2008. Evaluation of dietary and management risk factors for enterolithiasis among horses in California. Research in Veterinary Science 85: 476-480.

Hassel, D.M., P.C. Rakestraw, I.A. Gardner, S.J. Spier and J.R Snyder, 2004. Dietary risk factors and colonic pH and mineral concentrations in horses with enterolithiasis. Journal of Veterinary Internal Medicine 18: 346-349.

Helman, R.G. and W.C. Edwards, 1997. Clinical features of blister beetle poisoning in equids: 70 cases (1983-1996). Journal of the American Veterinary Medical Association 211: 1018-1021.

Hillyer, M.H., F.G. Taylor and N.P. French, 2001. A cross-sectional study of colic in horses on thoroughbred training premises in the British Isles in 1997. Equine Veterinary Journal 33: 380-385.

Hillyer, M.H., F.G. Taylor, C.J. Proudman, G.B. Edwards, J.E. Smith and N.P. French, 2002. Case control study to identify risk factors for simple colonic obstruction and distension colic in horses. Equine Veterinary Journal 34: 455-463.

Holbrook,T.C., R.D. Simmons, M.E. Payton and C.G. MacAllister, 2005. Effect of repeated oral administration of hypertonic electrolyte solution on equine gastric mucosa. Equine Veterinary Journal 37: 501-504.

Houpt, K.A., 1990. Ingestive behaviour. Veterinary Clinics of North America Equine Practice 6: 319-337.

Hudson, J.M., N.D. Cohen, P.G. Gibbs and J.A. Thompson, 2001. Feeding practices associated with colic in horses. Journal of the American Veterinary Medical Association 219: 1419-1425.

Hunter, L.C., J.K. Miller and I.R. Poxton, 1999. The association of Clostridium botulinum type C with equine grass sickness: a toxicoinfection? Equine Veterinary Journal 31: 492-499.

Hussein, H.S., L.A.Vogedes, G.C. Fernandez, and R.L. Frankeny, 2004. Effects of cereal grain supplementation on apparent digestibility of nutrients and concentrations of fermentation end-products in the feces and serum of horses consuming alfalfa cubes. Journal of Animal Science 82: 1986-1996.

Husted, L., L.C. Sanchez, K.E. Baptiste and S.N. Olsen, 2009. Effect of a feed/fast protocol on pH in the proximal equine stomach. Equine Veterinary Journal 41: 658-662.

Jansen, W.L., S.N.J. Geleen, J. van der Kuilen and A.C. Beynen, 2002. Dietary soyabean oil depresses the apparent digestibility of fibre in trotters when substituted for an iso-energetic amount of corn starch or glucose. Equine Veterinary Journal 34: 302-305.

Julliand, V., A. de Fombelle, C. Drogoul and E. Jacotot, 2001. Feeding and microbial disorders in horses: Part 3: Effects of three hay:grain ratios on microbial profile and activities. Journal of Equine Veterinary Science 26: 543-546.

Kaneene, J.B., R.A. Miller, W.A. Ross, K. Gallagher, J. Marteniuk and J. Rook, 1997. Risk factors for colic in the Michigan (USA) equine population. Preventive Veterinary Medicine 30: 23-36.

Kaya, G., I. Sommerfeld-Stur and C. Iben, 2009. Risk factors of colic in horses in Austria. Journal of Animal Physiology and Animal Nutrition 93: 339-349.

Le Jeune, S.S., J.E. Nieto, J.E. Dechant and J.R. Snyder, 2006. Prevalence of gastric ulcers in Thoroughbred broodmares in pasture. In: Proceedings of the American Association of Equine Practitioners, December 2-6, 2006. San Antonio, TX, USA, p 264.

Little, D. and A.T. Blikslager, 2002. Factors associated with development of ileal impaction in horses with surgical colic: 78 cases (1986-2000). Equine Veterinary Journal 34: 464-468.

Longland, A.C. and B.M. Byrd, 2006. Pasture nonstructural carbohydrates and equine laminitis. Journal of Nutrition 136: 2099S-2102S.

Longland, A.C. and A.J. Cairns, 2000. Fructans and their implications in the aetiology of laminitis. In: Proceedings 3rd International Conference on Feeding horses, April 3-7, 2000. Stoneleigh, Warwickshire, UK, pp. 52-55.

Lopes, M.A., N.A. White, M.V. Crisman and D.L. Ward, 2004. Effects of feeding large amounts of grain on colonic contents and feces in horses. American Journal of Veterinary Research 65: 687-694.

Lorenzo-Figueras, M., T. Preston, E.A. Ott and A.M. Merritt, 2005. Meal-induced gastric relaxation and emptying in horses after ingestion of high-fat versus high-carbohydrate diets. American Journal of Veterinary Research 66: 897-906.

Love, S., T.S. Mair and M.H. Hillyer, 1992. Chronic diarrhoea in adult horses: a review of 51 referred cases. Veterinary Record 130: 217-219.

Luthersson, N., K.H. Nielsen P. Harris and T.D. Parkin, 2009. Risk factors associated with equine gastric ulceration syndrome (EGUS) in 201 horses in Denmark. Equine Veterinary Journal 41: 625-630.

Lybbert, T., P. Gibbs, N. Cohen, B. Scott and D. Sigler, 2007. Feeding alfalfa hay to exercising horses reduces the severity of gastric squamous mucosal ulceration. Proceedings of the American Association of Equine Practitioners, December 1-5, 2007. Orlando, FL, USA, pp. 525-526.

Mackie, R.I. and C.A. Wilkins, 1988. Enumeration of anaerobic bacterial microflora of the equine gastrointestinal tract. Applied Environmental Microbiology 54: 2155-2160.

Mair, T.S., L.V. de Westerlaken, P.J. Cripps, S. Love, 1990. Diarrhoea in adult horses: a survey of clinical cases and an assessment of some prognostic indices. Veterinary Record 126: 479-481.

Martineau, H., H. Thompson and D. Taylor, 2009. Pathology of gastritis and gastric ulceration in the horse. Part 1: range of lesions present in 21 mature individuals. Equine Veterinary Journal 41: 638-644.

McInnis, M.L. and M. Vavra, 1987. Dietary relationships among feral horses, cattle, and Pronghorn in southeastern Oregon. Journal of Range Management 40: 60-66.

Mehdi, S. and V. Mohammed, 2006. A farm-based prospective study of equine colic incidence and associated risk factors. Journal of Equine Veterinary Science 26: 171-174.

Merritt, A.M., 1999 Normal equine gastroduodenal secretion and motility. Equine Veterinary Journal (suppl.) 29: 7-13.

Merritt, A.M., 2003. The equine stomach: a personal perspective (1963-2003). Proceedings of the American Association of Equine Practitioners, November 21-25, 2003. New Orleans, LA, USA, pp. 75-102.

Metayer, N., M. Lhote, A. Bahr, N.D. Cohen, I. Kim, A.J. Roussel and V. Julliand, 2004. Meal size and starch content affect gastric emptying in horses. Equine Veterinary Journal 36: 436-440.

Meyer, H., M. Coenen and C. Gurer, 1985. Investigations on saliva production and chewing effects in horses fed various feeds Proceedings of the Equine Nutrition and Physiology Society, May 23-25, 1985. Michigan State University, MI, USA, pp 38-41.

Morris, D.D., J.N. Moore and S. Ward, 1989. Comparison of age, sex, breed, history and management in 229 horses with colic. Equine Veterinary Journal (suppl.) 7: 129-132.

Murray, M.J., 1997 Suppression of gastric acidity in horses. Journal of the American Veterinary Medical Association 211: 37-40.

Murray, M.J., 1999. Pathophysiology of peptic disorders in foals and horses: a review. Equine Veterinary Journal (suppl.) 29: 14-18.

Murray, M.J. and E.S. Eichorn, 1996. Effects of intermittent feed deprivation, intermittent feed deprivation with ranitidine administration, and stall confinement with ad libitum access to hay on gastric ulceration in horses. American Journal of Veterinary Research 57: 1599-1603.

Murray, M.J., C. Grodinsky, C.W. Anderson, P.F. Radue and G.R. Schmidt, 1989. Gastric ulcers in horses: a comparison of endoscopic findings in horses with and without clinical signs. Equine Veterinary Journal S(suppl.) 7: 68-72.

Murray, M.J., Y.S. Nout and D.L. Ward, 2001. Endoscopic findings of the gastric antrum and pylorus in horses: 162 cases (1996-2000). Journal of Veterinary Internal Medicine15: 401-406.

Murray, M.J. and G.F. Schusser, 1993. Measurement of 24-h gastric pH using an indwelling pH electrode in horses unfed, fed and treated with ranitidine. Equine Veterinary Journal 25: 417-421.

Nadeau, J.A., F.M. Andrews, A.G. Mathew, R.A. Argenzio, J.T. Blackford, M. Sohtell and A.M. Saxton, 2000. Evaluation of diet as a cause of gastric ulcers in horses. American Journal of Veterinary Research 61: 784-790.

Nadeau, J.A., F.M. Andrews, C.S. Patton, R.A. Argenzio, A.G. Mathew and A.M. Saxton, 2003a. Effects of hydrochloric, acetic, butyric, and propionic acids on pathogenesis of ulcers in the nonglandular portion of the stomach of horses. American Journal of Veterinary Research 64: 404-412.

Nadeau, J.A., F.M. Andrews, C.S. Patton, R.A. Argenzio, A.G. Mathew and A.M. Saxton, 2003b. Effects of hydrochloric, valeric and other volatile fatty acids on pathogenesis of ulcers in the nonglandular portion of the stomach of horses. American Journal of Veterinary Research 64: 413-417.

Newton, J.R., E.J. Hedderson, V.J. Adams, B.C. McGorum, C.J. Proudman, J.L. Wood, 2004. An epidemiological study of risk factors associated with the recurrence of equine grass sickness (dysautonomia) on previously affected premises. Equine Veterinary Journal 36: 105-112.

Nyman, S., A. Jansson, K. Dahlborn and A. Lindholm, 1996. Strategies for voluntary rehydration in horses during endurance exercise. Equine Veterinary Journal (suppl.) 22: 99-106.

Potter, G.D., F.F. Arnold, D.D. Householder, D.H. Hansen and K.M. Brown, 1992. Digestion of starch in the small or large intestine of the equine. Europische Konferenz uber die Ernahrung des Pferdes. PferdeHeilkunde Sonderheft 1: 107-111.

Ragle, C.A., D.M. Meagher, C.A. Lacroix and C.M. Honnas, 1989. Surgical treatment of sand colic. Results in 40 horses. Veterinary Surgery 18: 48-51.

Ralston, S.L., 1984. Controls of feeding in horses. Journal of Animal Science 59:1354-1361.

Reese, R.E. and F.M. Andrews, 2009. Nutrition and dietary management of equine gastric ulcer syndrome. Veterinary Clinics of North America Equine Practice 25: 79-92.

Richards, N., G. Hinch and J. Rowe, 2006. The effect of current grain feeding practices on hindgut starch fermentation and acidosis in the Australian racing Thoroughbred. Australian Veterinary Journal 84: 402-407.

Rose, R.J. and D.R. Lloyd, 1992. Sodium bicarbonate: more than just a 'milkshake'? Equine Veterinary Journal 24: 75-76.

Rowe, J.B., M.J. Lees and D.W. Pethick, 1994. Prevention of acidosis and laminitis associated with grain feeding in horses. Journal of Nutrition 124(12 suppl.): 2742S-2744S.

Sandin, A., K. Girma, B. Sjöholm, A. Lindholm and G. Nilsson, 1998. Effects of differently composed feeds and physical stress on plasma gastrin concentration in horses. Acta Veterinaria Scandinavica 39: 265-272.

Smyth, G.B., D.W. Young and L.S. Hammond,1989. Effects of diet and feeding on postprandial serum gastrin and insulin concentration in adult horses. Equine Veterinary Journal (suppl.) 7: 56-59.

Southwood, L.L., D.L. Evans, W.L. Bryden and R.J. Rose, 1993. Nutrient intake of horses in thoroughbred and standardbred stables. Australian Veterinary Journal 70: 164-168.

Stuska, S., S.E. Pratt, H.L. Beveridge and M. Yoder, 2008. Nutrient composition and selection preferences of forages by feral horses: the horses of Shackleford Banks, North Carolina. Available at http://www.nps.gov/calo/parkmgmt/upload/Nutrient%20Composition%20and%20Selection%20Preferences.pdf. Accessed 17 January 2010.

Tinker, M.K., N.A. White, P. Lessard, C.D. Thatcher, K.D. Pelzer, B. Davis and D.K. Carmel, 1997a. A prospective study of equine colic incidence and mortality. Equine Veterinary Journal 29: 448-453.

Tinker, M.K., N.A. White, P. Lessard, C.D. Thatcher, K.D. Pelzer, B. Davis and D.K. Carmel, 1997b. A prospective study of equine colic risk factors. Equine Veterinary Journal 29: 454-458.

Traub-Dargatz, J.L., C.A. Kopral, A.H. Seitzinger, L.P. Garber, K. Forde and N.A. White, 2001. Estimate of the national incidence of and operation-level risk factors for colic among horses in the United State, spring 1998 to spring 1999. Journal of the American Veterinary Medical Association 219: 67-71.

Traub-Dargatz, J.L., M.D. Salman and R.L. Jones, 1990. Epidemiologic study of salmonellae shedding in the feces of horses and potential risk factors for development of the infection in hospitalized horses. Journal of the American Veterinary Medical Association 196: 1617-1622.

Uhlinger, C., 1992. Investigations into the incidence of field colic. Equine Veterinary Journal. (suppl.) 13: 16-18.

Van Eps, A.W. and C.C. Pollitt, 2006. Equine laminitis induced with oligofructose. Equine Veterinary Journal 38: 203-208.

Videla, R. and F.M. Andrews, 2009. New perspectives in equine gastric ulcer syndrome. Veterinary Clinics of North America Equine Practice 25: 283-301.

Wood, J.L.N., E.M. Milne and D.L. Doxey, 1998. A case-control study of grass sickness (equine dysautonomia) in the United Kingdom. Veterinary Journal 156: 7-14.

Equine microbial gastro-intestinal health

S. Sadet-Bourgeteau and V. Julliand
URANIE-USC 'Nutrition du cheval athlète', AgroSup Dijon, 26 Bd Dr Petitjean, 21079 Dijon Cedex, France

Abstract

High numbers of a diverse array of micro-organisms inhabit the equine gastro-intestinal tract (GIT), guaranteeing functionality and stability in the face of stresses and environmental fluctuations. The first part of this review reports on the number, diversity and identification of the autochthonous microbial communities present in the different anatomic segments. Although bacteria are present in great concentrations along the length of the GIT, viruses, protozoa and fungi have so far been reported only in the hindgut. There is a special focus on the microbiota of equine faeces. Despite recent advances on the microbial diversity in the different segments of the GI tract, there still is a profound lack of knowledge about these microorganisms. The adult optimal microbiota depends on the establishment in the young animal. The second part of this review is dedicated to the colonization by viral, fungal and bacterial communities of the foal GIT. Despite a rapid colonization following birth, the digestive population in its whole complexity is probably not fully established until weaning. There is a need to better understand the colonization of the bacterial communities in foals, measure correlations with the establishment of immunity and the impact on to the future adult digestion. The final part of this review firstly presents the potential role of the GI microbiota for the host in terms of gut health. Whereas little is known in horses, it could be accepted like in other mammals, that commensal gut bacteria play an essential role in angiogenesis, postnatal intestinal maturation, mucosal barrier fortification, and nutrient absorption. Secondly, the complex GIT microorganism interactions organized in trophic chains to degrade fibre, hydrolyze rapidly digestible carbohydrates, fat, and utilize protein are described. The nutritional role of the microbial communities is probably even more essential than previously thought and from the stomach to the colon the contribution of feed hydrolysis and fermentation has to be precisely evaluated for better comprehension.

Keywords: gut microbial ecosystem, nutrition

Introduction

It is accepted that biological ecosystems with a higher diversity of species are more stable in the face of stresses and environmental fluctuations (Elton, 1958; Loreau and Behera, 1999). This is mainly based on the existence of redundant species that, due to their genetic relationships, are capable of quickly compensating for each other in case of deletion of one of them. This general postulate can also be applied to gut microbial ecosystems.

As for other herbivores, horses possess a gut microbial ecosystem which is essential for efficient digestion, especially fibre degradation thus allowing them to survive on a forage-based diet. To achieve high intensity exercise, athletic horses need energy-dense diets often containing high-starch cereal grains, leading to changes in microbial populations inhabiting the gastrointestinal (GI) tract (Bailey *et al.*, 2003) that can have severe consequences on the horse health. High-starch diets, by modifying the gut microbial population, sometimes result in acidosis, colic and laminitis (Garner *et al.*, 1977; Goncalves *et al.*, 2002; Bailey *et al.*, 2003; Frape, 2004; Milinovich *et al.*, 2006; Milinovich *et al.*, 2007; Milinovich *et al.*, 2008). Better understanding of the microbial diversity is thus crucial to stabilise the equine gut ecosystem which is necessary not only for better function and resistance to nutritional stresses but also probably to environmental stresses. There is an optimal microbiota in horses, which depends upon good establishment in the young animal. Once established this optimal microbiota then contributes to the overall health of the animal.

The gut microbiota colonization begins at birth. The foetus develops in a sterile environment and its gastrointestinal tract is consequently devoid of micro-organisms until parturition. The physical process of foaling exposes the newborn to environmental bacteria from the dam's vagina, faeces, and saliva, provoking the beginning of the equine GI tract microbial colonization.

To improve horse performance and health, it is essential to know the composition of the gut microbiota, its establishment and its roles. Different tools are available to study this ecosystem. The current knowledge of equine GI microbiology and ecology has been almost exclusively obtained by classical cultivation of microbes, based on anaerobic techniques (Hungate, 1969) on specific media which have generally been modified (Grubb and Dehority, 1976) and adapted for horses (Baruc *et al.*, 1983; Julliand *et al.*, 1999). These culture-based techniques allow phenotype classification but underestimate biodiversity since they do not take into account the uncultured bacteria and do not discriminate between genetically close species. The advent of genetic techniques has revealed an extensive microbial diversity that was previously undetected (Stahl *et al.*, 1988; Pace, 1997).

In the present review, we will address a special focus on microbial diversity and data obtained by genetic techniques. Few studies have reported data using these tools thus, when necessary, information will be complemented with results obtained by culture-based techniques. The first part will describe the autochthonous microbial communities present in the different anatomic segments of the horse GIT. The second part will be dedicated to the establishment of the bacterial communities in foals. The final part will present the potential role of the GI microflora on the health and wellbeing of the host together with its nutritional role.

The GI tract microbiota: diversity, composition and quantities

Both quantitative and qualitative information will be reported about the identification and characterization of the microorganisms isolated to date in different segments of the equine GIT. The limited knowledge of some segments is related to methodological difficulties. Some studies have been conducted on slaughtered (Kern *et al.*, 1974) or anesthetized horses (De Fombelle *et al.*, 2003), which allow rapid and controlled removal of the required segment content but give selective information and prevent any repeatability. Moreover, the post-mortem or post-anesthesia impact is unknown and limits the extrapolation of these latter data to live horses. Surgical procedures where permanent cannulas have been fitted in the stomach (Merritt and Campbell-Thompson, 1987), the duodenum (Wolter *et al.*, 1979), the ileum (Roger *et al.*, 1990; Gerhards *et al.*, 1991; Peloso *et al.*, 1994; Leao *et al.*, 1999), the caecum (Pulse *et al.*, 1973; Ralston *et al.*, 1983; Drogoul, 2000) or the colon (Drogoul, 2000), offering the opportunity to repeat the collection of digesta from these different segments. This allows researchers to study the microfloral composition or the influence of various parameters thereon. A minimally invasive technique has been reported to collect repeatedly gastric contents (Varloud *et al.*, 2006). As for the rectum, its anatomical localization permits easy collection of their contents. However while numerous studies are derived from faecal observations they may not accurately represent the intestinal microflora, either quantitatively or qualitatively.

Equine gut microbiota are composed of fungi, protozoa, viruses, archaebacteria and bacteria. The presence of each of these micro-organisms have not been demonstrated in each segment of the GIT. According to one work based on DNA clone libraries, viruses appeared to be the most predominant micro-organisms in equine faeces, bacteria being second. Archae and eukaryotes (protozoa and fungi) are less numerous (63%, 20%, 7% and 6% respectively of clones sequenced) (Cann *et al.*, 2004). These data are in accordance with previous works based on conventional cultural techniques which found that viral particles and bacteria were the two predominant microbial types in equine faeces versus ciliates protozoa and fungi; these latter were the less important micro-organisms.

Viral diversity

Bacteriophages were described for the first time in the equine large intestine in 1970 (Alexander *et al.*,.1970). The number of phage particles was expected about 10^{10} to 10^{11} per gram of faeces (Table 1) (Cann *et al.*, 2004; Golomidova *et al.*, 2007). Out of the 63% of viruses quantified in the faecal microbial community, 52% were identified as *Siphoviridae*, 26% as unclassified phages, 17% as *Myoviridae*, 4% as *Podoviridae*, and one clone (2%) as a vertebrate *Ortopoxvirus* (Cann *et al.*, 2004). Another study based on classical microscopy highlighted more than sixty morphologically distinct phage types in faeces (Kulikov *et al.*, 2007).

Table 1. Microbial concentrations in the different parts of the digestive tract of horses.

Digestive segment	Bacteriophages (particles/g)	Fungi (zoospores/ml)	Protozoa (cells/ml)	Total anaerobic bacteria	References
Stomach	NR	NR	NR	10^6-10^8 CFU/ml	Varloud *et al.* 2007
Small intestine	NR	NR	NR	10^6-10^9 CFU/ml	De Fombelle *et al.* 2003
Duodenum	NR	NR	NR	2.9x10^6 CFU/ml	Mackie and Wilkins, 1988
Jejunum	NR	NR	NR	2,9x10^7 CFU/ml	Mackie and Wilkins, 1988
Ileum	NR	NR	NR	3,8x10^7 CFU/ml	Mackie and Wilkins, 1988
Caecum	NR	10^2 to 10^4	2.2 x 10^5	10^7-10^8 CFU/ml	Orpin, 1981; Julliand *et al.* 1997; Ozeki *et al.* 1973; Mackie and Wilkins, 1988; Moore and Dehority, 1993; Julliand *et al.* 2001; Medina *et al.* 2002; De Fombelle *et al.* 2003
Colon	NR	NR	7.7 x 10^5	10^7-10^9 CFU/ml	Ozeki *et al.* 1973; Mackie and Wilkins, 1988; Moore and Dehority, 1993; Julliand *et al.* 2001; Medina *et al.* 2002; De Fombelle *et al.* 2003
Right ventral colon	NR	NR	NR	10^8-10^9 CFU/ml	De Fombelle *et al.* 2003
Left dorsal colon	NR	NR	NR	10^8-10^9 CFU/ml	De Fombelle *et al.* 2003
Faeces	10^{10}-10^{11}	NR	10^4-10^6	10^{10}-10^{12} CFU/g	Cann *et al.* 2004; Golomidova *et al.* 2007; Imai *et al.* 1999; Gurelli *et al.* 2009

NR: no reference.

The role of equine bacteriophages has not been studied. As in the rumen they could exert a significant and continual lysis of bacteria (Klieve and Swain, 1993), including cellulolytic bacteria (Klieve *et al.*, 2004) thus having an impact on fibre digestion.

Fungal diversity

In 1961, yeast belonging to the *Candida* and *Torulopsis* genera and fungi identified as *Geotrichum candidum* were isolated from the large intestinal content of the horse (Batista *et al.*, 1961). Later, other strictly anaerobic fungi were isolated from the equine hindgut (Orpin, 1981; Gold *et al.*, 1988; Li *et al.*, 1989; Breton *et al.*, 1991; Teunissen *et al.*, 1991a; Teunissen *et al.*, 1991b; Gaillard-Martinie *et al.*, 1995; Julliand *et al.*, 1998). To our knowledge, there are few studies on equine gut fungal diversity. In the caecum from 10^2 to 10^4 zoospores/ml were counted (Table 1) (Orpin, 1981; Julliand *et al.*, 1998). As observed in the rumen, *in vivo* evaluation of fungal populations is difficult because the number of zoospores can be connected either to a number or a quantity of rhizoids. The determination of chitin specific constituents of fungal walls is reliable for pure cultures, but not *in vivo*.

Microscopic observations of equine caecal contents and use of cultural techniques has established that fungi were microscopic, anaerobes, and belonged to the class *Phycomycetes* (Orpin, 1981). The majority of the caecal fungal strains belonged to the genus Piromyces and four species were distinct in equines: *Piromyces mae* (Li *et al.*, 1989; Gaillard-Martinie *et al.*, 1992), *Piromyces rhizinflata* (Breton *et al.*, 1991), *Psoroptes equi* (Li and Heath, 1993) and *Piromyces citronii* (Gaillard-Martinie *et al.*, 1995). Some strains were classified as *Caecomyces equi* (Gold *et al.*, 1988). Certain uni-flagellates zoospores were identified as *Sphaeromonas communi, Piromonas communis* (Liebetanz, 1910), and other multi-flagellates organisms as *Neocallimastix equi* (Vavra and Joyon, 1966).

Equine strains of *P. citronii* not only showed specific morphological characteristics (Gaillard-Martinie *et al.*, 1995), but also metabolic and genetic traits (Julliand *et al.*, 1998). They also exhibited a more rapid *in vitro* growth rate compared to ruminal strains (Julliand *et al.*, 1998), which could be related to a physiological adaptation of the equine species to shorter retention time of feed particles in the equine hindgut compared to the rumen.

Protozoal diversity

Due to their large size, protozoa were described early in equine gut by Gruby and Delafond in 1843. In the large intestine, the quantity of protozoa was evaluated from 10^4 to 10^6 micro organisms/ml (Table 1) (Ozeki *et al.*, 1973). Colonic protozoal density (7.7×10^5) would appear to be greater

Figure 1. Uniflagellates zoospores of Pyromyces citronii *strains (Julliand, 1996).*

than in caecum (2.2 x 10^5) (Table 1) (Ozeki *et al.*, 1973). Protozoal diversity was first studied in the faeces as a model to investigate large colonic protozoal diversity (Adam, 1951). However, more recent studies have been conducted on colonic and cecal contents.

Microscopic observations showed that most equine intestinal protozoa were ciliated and close to those found in the rumen (Geyer and Drepper, 1973). Around thirty genera have been identified in the horse digestive tract (Ozeki *et al.*, 1973; Ike *et al.*, 1983; Ike *et al.*, 1985; Imai *et al.*, 1999; Gürelli and Göçmen, 2009) (Table 2). Protozoal diversity differs according to the segment of the gut, especially between the colon and caecum. *Blepharocorys*, *Cycloposthium* (Adam, 1953) and Paraisotricha (Ozeki *et al.*,1973) were considered to be the dominant genera in the caecum, whereas *Ampullacula, Blepharosphaera, Ditoxum, Prorodonopsis, Spridinium, Tetratoxum, Triadinium* and *Trpalmaria*, were absent in the caecum but specific for the colon (Ozeki *et al.*, 1973). Moore and Dehority (1993) completed these data and found 16.7% and 50.9%, *Buetschlia*, 19.9% and 15.9% *Didesmis*, 66.8% and 25.0% *Blepharoprosthium sp.* in the caecal and colonic contents respectively. In faeces, *Bundleia, Holophryoides* and *Cycloposthium* are the predominant genera (Imai *et al.*, 1999; Kobayashi *et al.*, 2006; Gürelli and Göçmen, 2009).

Table 2. Equine gut protozoal genera (Ozeki et al., *1973; Ike* et al., *1983; Ike* et al., *1985; Imai* et al., *1999; Gürelli and Göçmen, 2009).*

Genus	Caecum	Colon	Faeces
Allantosoma; Alloiozona; Blepharoconus; Blepharocorys; Blepharoprosthium; Bundleia; Cycloposthium; Didesmis; Holophryoides; Paraisotricha	x	x	x
Ampullacula; Blepharosphaera; Ditoxum; Prorodonopsis; Spirodinium; Tetratoxum; Triadinium; Tripalmaria		x	x
Chlamydobundleia; Charonnautes; Charonina; Circodinium; Cochliatoxum; Hemiprorodon; Ochoterenaia; Paraisotrichopsis; Polymorphella; Walskana			x

Bacterial diversity

Gastric bacterial diversity

In the stomach, the average concentration of total bacteria can reach 10^6-10^8 CFU/ml (Table 1) (Varloud *et al.*, 2007). Currently, phylogenic composition of the gastric bacterial community has not been described. Gastric concentrations of total anaerobic bacteria obtained post-mortem reported average values of 10^8 (Kern *et al.*, 1974) or 1.4 x 10^9 CFU/ml (De Fombelle *et al.*, 2003). In conscious horses, data confirmed an abundant microbial concentration in the chyme and provided complementary information about the postprandial changes of the microbial concentration: total anaerobes increased from 5.8x10^5 to 1.7x10^7 CFU/ml within the first hour and reached 3.7x10^8 CFU/ml, 3.5 h postprandially (Varloud *et al.*, 2004).

Bacteria implicated in the metabolism of starch and highly fermentable carbohydrates (lactate producers and lactate utilizers) constitute the predominant microflora and it is recognized that *Lactobacillus* and *Streptococcus* are the main genera (Varloud *et al.*, 2007). After distribution of the meal, post-mortem Lactobacilli, Streptococci and lactate utilisers concentration reached 1.5x10^8, 2.6x10^7 and 2.3x10^7 CFU/ml respectively (De Fombelle *et al.*, 2003). Cellulolytic bacteria were counted in the equine stomach in very low numbers (less than 10^2 CFU/ml) (Kern *et al.*, 1974; De Fombelle *et al.*, 2003).

Studies have been done to investigate the stomach mucosal bacterial diversity. Molecular biology tools (DNA-DNA hybridization or sequencing) highlighted that *Lactobacillus crispatus*, *Lactobacillus reuteri*, and *Lactobacillus agilis* were specific to the stomach wall and were not found in stomach contents (Yuki *et al.*, 2000). *Lactobacillus crispatus* and *L. agilis* were found to adhere to horse epithelial cells *in vitro* but not to those of rats suggesting an equine specificity (Yuki *et al.*, 2000). *Lactobacillus mucosae* and *Lactobacillus delbrueckii* would be specific of gastric juice (Al Jassim *et al.*, 2005). *Lactobacillus salivarius* has been observed both in stomach mucosa and chyme (Yuki *et al.*, 2000; Al Jassim *et al.*, 2005).

Small intestine bacterial diversity

Strictly anaerobic bacteria colonise the small intestine of the horse, with numbers ranging from 10^6-10^9 CFU/ml (Table 1) (De Fombelle *et al.*, 2003). More precisely bacterial counts showed a substantial bacterial population in the duodenum ($2.9\text{x}10^7$ per g (wet weight) of sample), and increased ten-fold in the jejunum and ileum (respectively $2.9\text{x}10^7$ and $3.8\text{x } 10^7$ (Table 1) (Mackie and Wilkins, 1988). The phylogenic bacterial community composition of the small intestine has not been described yet.

Pathogens such as *Clostridia sp.*, *Proteus sp.*, *Staphylococci sp.* and *Pseudomonas sp.* have been detected in the jejunal contents at an average of 10^3 CFU/ml (Kollarczik *et al.*, 1992). Bacteria implicated in the metabolism of starch and highly fermentable carbohydrates constitute the predominant microflora: *Lactobacillus, Enterobacteria, Enterococcus, Streptococcus*, lactate-utilizing bacteria. *Lactobacillus* concentrations ranged in average from 10^7 to 10^9 CFU/ml in the small intestinal contents (Alexander and Davies, 1963; Kollarczik *et al.*, 1992; De Fombelle *et al.*, 2003); *Enterobacteria* and *Enterococcus* counts varied from 10^7 to 10^8 CFU/ml and 10^5 to 10^8 CFU/ml respectively in the jejunal intestinal content (Kollarczik *et al.*, 1992) and Streptococcus counts from 10^7 to 10^8 CFU/ml in the small intestinal content (De Fombelle *et al.*, 2003). The concentrations of *Lactobacillus* were lower in the small intestine than in the stomach whereas those of Streptococci counts were higher. Streptococci were also greater in the small intestine than *Lactobacillus*; this was probably related to the higher value of pH in the ileal contents (De Fombelle *et al.*, 2003) that could be more propitious for *streptococcal* growth. The concentrations of lactate-utilizing bacteria ranged on average from 10^7 to 10^8 CFU/ml in the small intestinal contents (Alexander and Davies, 1963; De Fombelle *et al.*, 2003).

Cellulolytic bacteria were counted in the equine jejuno-ileum in very low numbers ($3.0 \text{ x } 10^2$ CFU/ml) (Kern *et al.*, 1974; De Fombelle *et al.*, 2003).

Proteolytic bacteria have been found in a high numbers (10^6-10^7 CFU/ml) in the small intestine (Mackie and Wilkins, 1988). In the duodenum, these bacteria would represent almost all cultivable bacteria (58%) whereas it would be 19 and 42% in jejunum and ileum respectively (Mackie and Wilkins, 1988). In horses, the small intestine seems to be an important place for microbial proteolysis in accordance with great proteolytic activity (30 fold greater in ileum of pony than in the colon or cæcum) (Kern *et al.*, 1974).

Large intestine bacterial diversity (caecum/colon)

In the different sections of the hindgut (the caecum, the right and left ventral colon, the left and right dorsal colon, the transverse colon and descending colon) a large density of strictly anaerobic micro-organisms has been reported: concentrations of total anaerobic bacteria vary from 10^7 up to 10^8 CFU/ml in the caecum and colon (Table 1) (Mackie and Wilkins, 1988; Moore and Dehority, 1993; Julliand *et al.*, 2001; Medina *et al.*, 2002; De Fombelle *et al.*, 2003).

Phylogenetic studies have shown that Firmicutes were the predominant phylum (72%) in the hindgut (Daly *et al.*, 2001; Daly and Shirazi-Beechey, 2003). The numerical prevalence of *Firmicutes* suggests that this is the greater functional group within intestinal ecosystems; and indeed it contains the majority of cellulolytic and fibrolytic organisms such as *Clostridium spp.*, *Ruminococcus spp.*, *Butyrivibrio spp.*, and *Eubacteria spp.* The second most important phyla (20%) was Bacteroidetes in accordance with other results found in humans (Wilson and Blitchington, 1996; Suau *et al.*, 1999), swine (Cotta *et al.*, 2003) or ruminants (Whitford *et al.*, 1998; Tajima *et al.*, 2000; Edwards *et al.*, 2004; Yu *et al.*, 2006). Other phyla have been found for equine hindgut bacterial diversity: *Spirochetaceae* (3%), *Verrucomicrobiales* (3%), High % G+C Gram-positive Bacteria (HGCGPB ; <1%) and *Proteobacteria* (<1%) (Daly *et al.*, 2001; Daly and Shirazi-Beechey, 2003). Work of Daly *et al.* (2001) highlighted that only 5% of the bacterial sequences corresponded to known organisms whose sequences are available in public databases. The vast majority (89%) was unknown suggesting that the anaerobiotic flora of the equine hindgut is severely under-represented in the public domain and that the equine flora may contain many novel bacterial species (Daly *et al.*, 2001).

Bacterial diversity seems to be different between colon and caecum and seems greater in the colon. Whereas it was shown that caecal bacteria were separated in three different phyla, colonic bacteria were separated into six. In the two segments *Firmicutes* and *Bacteroidetes* appeared to be predominant (Figure 2) (Daly *et al.*, 2001; Daly and Shirazi-Beechey, 2003).

Bacteria have been classified in different functional groups in the caecum and colon of the horse (Mackie and Wilkins, 1988) as glycolytic, amylolytic, lactate utilizing, proteolytic, hemicellulolytic and cellulolytic bacteria.

Streptococci, Lactobacilli and lactate-utilising bacteria are considered to be the main glycolytic and amylolytic bacteria. It has been observed that the concentration of Streptococci and Lactobacilli was less numerous in the caecum than in the colon (caecum : 10^7 CFU/ml of contents) (Julliand *et al.*, 2001; Medina *et al.*, 2002; De Fombelle *et al.*, 2003).

Lactate-utilizing bacteria averaged 10^7 CFU/ml of cecal (Goodson *et al.*, 1988; De Fombelle *et al.*, 2001; Julliand *et al.*, 2001; Medina *et al.*, 2002; De Fombelle *et al.*, 2003) and colonic contents (De Fombelle *et al.*, 2001; Julliand *et al.*, 2001; Medina *et al.*, 2002; De Fombelle *et al.*, 2003). Their concentration tended to be systematically higher in the colon than in the caecum. *Streptococcus bovis, S. equines* (Julliand and Goachet, 2005), *Lactobacillus salivarus, L. mucosae, L. delbrueckii* and *Mitsuokelal jalaludinii* (Al Jassim *et al.*, 2005) have been described as the main lactic acid producing bacteria of the hindgut. Using 16S rDNA clones libraries from the digestive content of the caecum and colon, Daly *et al.*, (2001) highlighted that Streptococcus bovis was in the minority

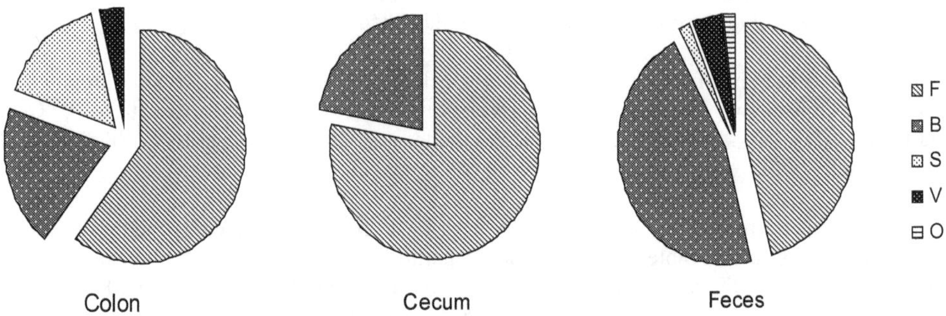

Figure 2. Hindgut phyla bacterial comparison (F: firmicutes, B: bacteroidetes, S: spirochaetes, P: proteobacteria, H: high % G+C Gram-positive Bacteria, V: verrucomicrobiales, O: other) (Daly et al., *2001; Daly and Shirazi-Beechey, 2003; Yamano* et al., *2008).*

in the colon (2% of clones) (Daly *et al.*, 2001). In the caecum this bacterial species was predominant (Bailey *et al.*, 2003). *Veillonella sp.* and *Megasphera* would be the main lactate-utilizing bacteria in equine hindgut (Baruc *et al.*, 1983; Maczulak *et al.*, 1985).

Conventional microbiology techniques highlighted that cellulolytic bacterial numbers vary from 10^4 to 10^7 bacteria/ml of intestinal contents (Kern *et al.*, 1973; Julliand *et al.*, 2001) with a higher abundance in the caecum than in the colon, indicating that the caecum is probably the main site of fibre digestion (Kern *et al.*, 1973; Julliand *et al.*, 2001). This is in contradiction with a recent work realised by real-time PCR (Hastie *et al.*, 2008) which showed that there were significantly fewer cellulolytic bacteria in the caecum than the dorsal colon and rectum. Several cellulolytic bacterial species have been identified in the caecum: *Ruminococcus flavefaciens*, *Ruminococcus albus*, *Fibrobacter succinogenes*, *Clostridium spp.*, *Butyrivibrio spp.*, and *Eubacterium spp.* (Julliand *et al.*, 1999; Daly *et al.*, 2001). *Ruminococcus flavefaciens* would be the predominant cellulolytic bacteria in the hindgut and caecum (Julliand *et al.*, 1999; Daly *et al.*, 2001; Hastie *et al.*, 2008). *F. succinogenes* would be the second most important cellulolytic species, less numerous in the colon than in the caecum (Lin and Stahl, 1995; Hastie *et al.*, 2008). Novel F. succinogenes caecal lineages have been identified (Lin and Stahl, 1995; Hastie *et al.*, 2008). The third main caecal cellulolytic bacteria was *Ruminococcus albus* (Julliand *et al.*, 1999). These results disagree with those of Daly *et al.*, (2001) which never identified *F. succinogenes* or *Ruminococcus albus* in the caecal contents. According to the authors, this could result from an unknown PCR bias or perhaps differences in diet resulting in low abundance of *F. succinogenes* and *Ruminococcus albus* that declined under the detection threshold.

The proteolytic bacteria form a high proportion of the total cultivable bacteria (Kern *et al.*, 1973; Baruc *et al.*, 1983; Maczulak *et al.*, 1985; Mackie and Wilkins, 1988) showed a higher number of colony counts of proteolytic bacteria in the caecum (10^8 bacteria/g of digestive content) than in the colon.

Fecal bacterial diversity

Total bacteria in faecal contents were determined as 10^{10} to 10^{12} bacteria/g. Equine faecal bacterial diversity is not well known. However, at least four different phyla have been highlighted for this last 'digestive segment'. These phyla were *Firmicutes* (46%), *Bacteroidetes* (46%), *Verrucomicrobia* (4%) and *Spirochaetes* (1.7%) (Figure 2) (Yamano *et al.*, 2008; Willing *et al.*, 2009). The proportion of faecal *Firmicutes* was less important than in the hindgut, contrary to *Bacteroidetes*, which was predominant in the faeces. The faecal bacterial community seems to be different and more diverse than those present in the colon and caecum. This is in accordance with recent results obtained by our research team. Using a fingerprint technique (Automated Ribosomal Intergenic Spacer Analysis) we confirmed that the bacterial structure of the equine faecal bacterial community was different from that of the hindgut (Figure 3; Bourgeteau-Sadet, unpublished).

Average concentrations of total anaerobic bacteria, cellulolytic bacteria, Lactobacilli, Streptococci and Lactate-utilisers were reported to approximate 10^8 CFU, 10^6 CFU, 10^6 to 10^7, 10^6 to 10^8 CFU and 10^7 to 10^8 CFU/g faeces respectively (De Fombelle *et al.*, 2003; Da Veiga *et al.*, 2005; Julliand and Goachet, 2005).

In equine faeces, the number of Lactobacilli vary from 10^7 to 10^9 cells/g of faeces (Endo *et al.*, 2007). In a recent work based on molecular technique (Denaturing Gradient gel Electrophoresis and real-Time PCR) Lactobacilli were predominant in horse faeces (Endo *et al.*, 2009). To our knowledge, thirteen bacterial species would belong to the *Lactobacillus* genera (Table 3) (Endo *et al.*, 2007; Morita *et al.*, 2007; Endo *et al.*, 2008; Endo *et al.*, 2009; Morita *et al.*, 2009). *Lactobacillus hayakitensis, Lactobacillus equi* (Endo *et al.*, 2009; Morita *et al.*, 2009), *Lactobacillus johnsonii* (Endo *et al.*, 2009) and *Lactobacillus equigenerosi* (Morita *et al.*, 2009) were predominant in the *Lactobacillus*

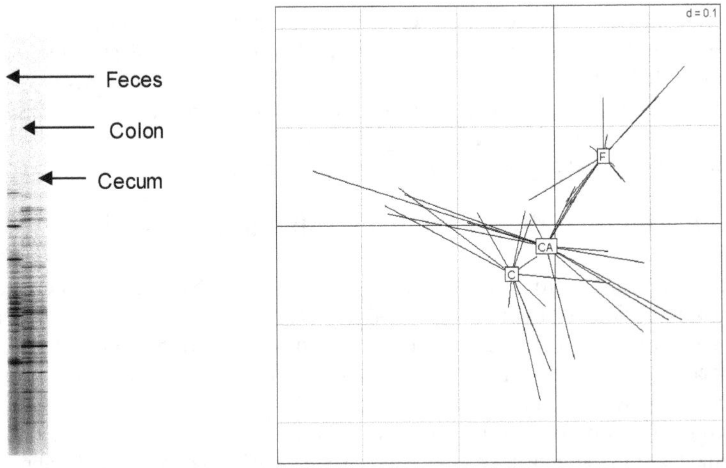

Figure 3. Comparison of bacterial community structure between equine colon (C), caecum (CA) and faeces (F) using ARISA and treated by Principal Component Analysis (data unpublished).

group. *L. equi* would be faeces specific (Morotomi *et al.*, 2002). Compared to Lactobacillus and *Bifidobacterium, Streptococcus* was the second predominant genera in faeces. *Streptococcus bovis* and *Streptococcus equinus* (Table 3) were the main species of this group (Endo *et al.*, 2007). Few studies have been done on equine faecal *Bifidobacteria*. In faeces, the number of *Bifidobacteria* varies from 10^3 to 10^5 CFU/g of faeces (Endo *et al.*, 2007) which is very low compared to other animal species (10^7 to 10^9 CFU/g for calves) (Orban *et al.*, 1997; Mentula *et al.*, 2005; Selim *et al.*, 2005; Rada *et al.*, 2006). Among *Bifidobacterium* species described in equine faeces, *Parascardovia denticolens* would be predominant (Endo *et al.*, 2007).

As observed for the hindgut, few identified sequences (3.8%) showed more than 97% similarity to known bacteria. This suggests that equine faecal bacterial community is highly specific to the host (Yamano *et al.*, 2008).

Table 3. Main equine faecal Lactobacilli, Streptococci and Bifidobacteria (Endo et al.*, 2007; Morita* et al.*, 2007; Endo* et al.*, 2008; Endo* et al.*, 2009; Morita* et al.*, 2009).*

	Bacterial genera		
	Lactobacillus	*Streptococcus*	*Bifidobacterium*
Bacterial species	*salivarius* *reuteri* *delbrueckii* *buchneri* *vitulinus* *equi* *hayakitensis* *equigenerosi sp* *gastricus* *johnsonii* *agilis*	*bovis* *equinus* *equi*	*boum*

In several studies of equine gut bacterial diversity, faeces are used and results extrapolated to the hindgut. This practise is questionable. In humans, Marteau *et al.* (2001), suggested that the microflora from the right hand side of the colon would be more appropriate than studying faeces for functions occurring in the caecum, such as fermentation of dietary fibre and endogenous substrates (Marteau *et al.*, 2001). For equines further investigations are necessary to know the relevance of using faeces to measure hindgut changes of the functional microflora.

Establishment of the GI tract microbiota in the foal

There is a dearth of information on this topic despite the fact that it is a critical area of research from both physiological and pathological aspects, and is important for a better understanding of the roles of the digestive ecosystem in adults. To understand the colonization, the succession and the mechanisms of interactions between intestinal microorganisms and their host, it is critical to distinguish between autochthonous (indigenous) and allochthonous (non-indigenous) microbes. Autochthonous microorganims are considered to colonize the habitat natively, whereas the allochthonous microorganisms cannot colonize the same habitat except under abnormal circumstances. The distinction between autochthonous and allochthonous microbes in studies of the acquisition and development of GI microbiota is difficult, especially in young animals or infants, in whom bacteria are acquired transiently during and immediately after birth, as well as from the surrounding environment during this important development phase (Mackie *et al.*, 1999).

To date the only material available for studying the GIT microbiology of the foal was meconium, (collected immediately after birth), and faeces, although the degree to which these represent colonic contents is not known (see earlier paragraph).

In foals, the meconium was reported to be free of micro-organisms (Sakaitani *et al.*, 1999), which confirmed that the GIT of neonates is sterile at birth. Following birth, the GIT of the foal is rapidly colonized by a variety of microorganisms.

Fungi could not be quantified in the faeces of foals but they were observed under the microscope from the dilutions 10^{-2} which demonstrated that fungi are established in the GIT of new-born horses from the second week of life (Julliand *et al.*, 1996).

Before the occurrence of heat diarrhoea at between eight and fourteen days of age, protozoa were not observed in the faeces of foals (Masri *et al.*, 1986). Another study confirmed that ciliates establish in the GIT in the eleventh or twelfth day after birth at very low levels (10^2 to 10^3 protozoa/ml) (Ike *et al.*, 1985). From the onset of diarrhoea, the presence of protozoa were reported unequally between foals but at 21 days of age their presence was systematic in all foals in moderate numbers (Masri *et al.*, 1986) and reached the mothers level at 35 day (10^2 to 10^3 protozoa/ml) (Ike *et al.*, 1985). The authors correlated the establishment of ciliates in the foal to their intestinal concentration in the mothers. Irrespective of age, ciliates of the genera *Blepharoprostium, Blepharospherae, Blepharocorys* and *Cycloposthium* were observed (Masri *et al.*, 1986). The ciliates belonging to *Buetschiilidae* and *Cycloposthiidae* appeared the earliest among all ciliates in the foal faeces, after 10 days. In contrast *Paraisotrichidae* ciliates were detected from the 30th day after birth (Ike *et al.*, 1985).

In the new-born foal, diverse maternal and environmental bacteria established within the first 24 hours (Julliand *et al.*, 1996), and populations of aerobes and facultative anaerobes were higher than strict anaerobes (10^9 and 10^8 CFU/g respectively). At 3-days, foal's faeces contained 10^9 total bacteria/g in average (Sakaitani *et al.*, 1999; Yuyama *et al.*, 2004). As the foal matured, the concentrations of facultative anaerobes declined and were replaced by strict anaerobes, reversing the ratio facultative to strict in the first week of life (Sakaitani *et al.*, 1999; Yuyama *et al.*, 2004): at 12 weeks of age the concentration of strictly anaerobic bacteria being 10^9 CFU/g of faeces (Julliand *et al.*, 1996). Total anaerobes concentrations above 10^9 CFU/g of faeces were even reported from the second (Yuyama *et al.*, 2004) or third day of life (Sakaitani *et al.*, 1999). The reversal in terms of dominance between total anaerobic flora and total aerobic flora has also been described in young ruminant herbivores but appears only at weaning in monogastric animals (except young rabbits) (Fonty *et al.*, 1995). The concentration of total anaerobic bacteria was close to that measured in adults (Julliand *et al.*, 1996). Thirty-eight to eighty percent of caecal bacteria in the hindgut are strict anaerobes (McCreery *et al.*, 1971; Kern *et al.*, 1973; Kern *et al.*, 1974).

In new-born foals, Julliand *et al.* (1996) found populations of Clostridia, Enterococci, Enterobacteriaceae, Lactobacilli, Streptococci and Staphylococci. Initially, the population of Enterococci and Enterobacteriaceae was predominant amongst the aerobic microflora (Sakaitani *et al.*, 1999; Yuyama *et al.*, 2004). In 3-day-old foals, Sakaitani *et al.* (1999) confirmed the presence of Bacteroidaceae, Clostridia, Enterobacteriaceae, Enterococci, Lactobacilli, *Bacillus sp.* and Staphylococci at the following detection rates: 100, 100, 100, 100, 81, 62 and 50% respectively. The concentrations of Bacteroidaceae and Enterobacteriaceae (2.10^8 CFU/g) were slightly superior to that of Enterococci (5.10^7 CFU/g). Both *Clostridium perfringens*, ά-toxin producing type, and *Clostridium difficile* were identified. *C. perfringens* were enumerated at the concentration of 5.10^7 CFU/g of faeces and were cultured from 64% in 8-12 hours-old neonates and from more than 90% in 3-day-old foals (Tillotson *et al.*, 2002). In a study on the prevalence of *C. difficile* in horses, it was reported that about 30% of healthy normal foals hosted *C. difficile* within the first 13 days of life but that all older foals (aged from 1 to 6 months) were negative (Baverud *et al.*, 2003).

After 6 weeks of age, the predominant populations were that of Bacteroidaceae and Lactobacilli, this latter having increased up to 10^8 CFU/g of faeces (Julliand *et al.*, 1996; Sakaitani *et al.*, 1999). The subdominant bacteria remained Enterobacteria, Enterococci and Staphylococci (around 10^5 CFU/g of faeces). After the first six weeks, clostridial concentrations decreased under 10^4 CFU/g faeces (Julliand *et al.*, 1996; Sakaitani *et al.*, 1999) and Clostridia were detected at a very low rate: 22% at 30 days of life and 0% at 60 days of life (Sakaitani *et al.*, 1999). As *Clostridia* were not detectable at 60 days, it was concluded that this bacteria only colonized the gut transiently from immediately after birth (Sakaitani *et al.*, 1999). Neither *Rotavirus* nor *Salmonella* or *Clostridium difficile*, common viral and bacterial organisms implicated in foals diarrhoea (Dunkel and Wilkins, 2004), was detected from birth to 60 days of age (Sakaitani *et al.*, 1999).

Cellulolytic bacteria were detected in the faeces between the 3rd and the 5th day of life, before foals were fed any solid food (Julliand, data unpublished). Within the first week, 2.10^2 cellulolytic bacteria/g of faeces were enumerated and the concentration increased as the animal matured reaching adult values at the age of two months (Julliand, data unpublished). The functioning of the horse digestive ecosystem appears to be similar to that of a pre-ruminant herbivore from a very early age. In young rabbits, cellulolytic bacteria only establish with solid feed ingestion. It is essential to ensure the normal GIT colonisation by cellulolytics and then to preserve it in order to prevent further troubles in the fibre degradation.

Interestingly, an abrupt decrease of total anaerobes and cellulolytic concentrations was reported around the 10th day of life followed by an increase and then stabilization. This marked decrease of bacteria in the foal digestive tract probably indicated a period where the commensal bacteria cannot play their protective effect against potential pathogens in the ecosystem of the foal (Julliand *et al.*, 1996). The digestive population in its whole complexity is probably not fully established until weaning.

Current knowledge about the contamination modes of the GIT by microorganisms is limited in foals. During and immediately after birth neonates are exposed to diverse microbial populations that originate from the mother and the surrounding environment. Active coprophagy of fresh maternal faeces were reported for each foal 7 days after birth by Ike *et al.*, (1985).

In horses like in other herbivores, it has been suggested that either 'contaminated' feed (indirect transmission) or mare-foal mouth contact (direct transmission) is the most likely mechanism of acquiring gut fungi in young animals (Li and Heath, 1993). Ike and co-workers strongly suggested that these are orally transmitted to the foal via the maternal faecal ciliates (Ike *et al.*, 1985). As for bacteria, the transmission would probably arise by the close contact between the new-born and its mother (vaginal, oral transmission) and other *equidae* (oral transmission) or by coprophagy. However, the neonate will not be populated with every microbial population it is exposed to. Only specific microorganisms will establish following a succession of different phases whatever the species. The process has been described in humans, chicks, pigs and young ruminants and remains remarkably similar. The microbial succession currently depicted in foals follows the same trends (Mackie *et al.*, 1999).

Further experimental work would be of great interest to improve our knowledge on the establishment of microorganisms in the foal digestive tract and on the different factors affecting microbial succession. This would help optimize this process through technical approaches, such as nutritional manipulations.

Roles of the GIT in gut health

The gut microbiota provides different roles for its host in terms of gut health. Work on humans and rodents have shown that commensal bacteria are able to modulate expression of genes involved in several important intestinal functions including angiogenesis, postnatal intestinal maturation, mucosal barrier fortification, and nutrient absorption (Hooper *et al.*, 2001). Even if these investigations have studied rodents, according to the authors, these findings apply to all mammals, and therefore to equines.

A major role of the equine microbiota which is now well known is its implication in the host nutrition.

Alleged roles of the GIT microbiota

Angiogenesis and GIT microbiota

Angiogenesis is a physiological process involving the growth of new blood vessels from pre-existing vessels. Work in rodents has suggested that gut bacteria were required for full intestinal blood vessel development (Stappenbeck *et al.*, 2002; Hooper, 2004). Authors have also shown that colonization by Bacteroidetes thetaiotaomicron alone stimulated villus capillaries network development, revealing that this single gut bacteria is sufficient to initiate the development program (Stappenbeck *et al.*, 2002).

Postnatal intestinal maturation

The intestinal microbiota, by its fermentative products (Volatile fatty Acids, VFA), has a trophic effect on the intestinal epithelium. Some work in the colon of rats bred in germ-free environments and rats colonised by conventional flora have suggested that intraluminal bacteria affect cell proliferation of the crypt (Alam *et al.*, 1994). Differentiation of epithelial cells is greatly affected by interactions with resident microorganisms (Gordon *et al.*, 1997; Hooper *et al.*, 2001). All three major short-chain fatty acids stimulate epithelial cell proliferation and differentiation in the large and small bowel *in vivo* (Frankel *et al.*, 1994).

Immunity and GIT microbiota

Immune system development and maturation

Studies in germ free mice have indicated that gut bacteria influence the maturation and function of several components of the intestinal mucosal immune system: microbiota drive induction of mucosal immunoglobulin A (Shroff *et al.*, 1995; Umesaki and Setoyama, 2000); the microflora also contribute to the development of intraepithelial lymphocytes (ILEs), as evidenced by the fact that numbers of ab T-cell receptor (TCR)-bearing intestinal IELs are reduced in germ free mice compared with conventional mice (Umesaki *et al.*, 1993; Umesaki *et al.*, 1999); some authors have shown that some specific bacteria such as Segmented Filamentous Bacteria (close to *Clostridium*) have a predominant role in immune system stimulation (Umesaki and Setoyama, 2000).

Barrier effect

The intestinal microbiota is a crucial line of resistance to colonization by exogenous microbes and, therefore, is highly relevant in prevention of invasion of tissues by pathogens. This is commonly referred to as the 'barrier effect' (Berg, 1996). Several mechanisms have been implicated in the barrier effect. *In vitro*, bacteria compete for attachment sites in the brush border of intestinal epithelial cells (Bernet *et al.*, 1994). Thus commensal microbes can prevent attachment and potential entry of pathogenic bacteria into the epithelial cells (Bernet *et al.*, 1994). Moreover, commensal bacteria compete for nutrients in ecological niches and maintain their collective habitat by administering and consuming all resources. Finally, intestinal bacteria can inhibit the growth of their competitors (pathogens) by producing antimicrobial substances: bacteriocins (Guarner and Malagelada, 2003; Fortun-Lamothe and Boullier, 2007).

A proven role of the GI tract microbiota in nutrition

The horse as a herbivore feeds on plants, essentially forages and grains. Plant cell-walls are mainly constituted of parietal carbohydrates such as cellulose, hemicelluloses and pectins that are not hydrolysable by the host endogenous enzymes. The primary role of the intestinal microbial ecosystem is to digest these large amounts of plant fibrous material in the hindgut. The processes, driven by hydrolytic and fermentative microorganisms, produce volatile fatty acids (VFA) which are then absorbed across the intestinal wall and provide energy for the animal (Argenzio *et al.*, 1974; Argenzio, 1975; Mac Bee, 1977; Prins, 1977). Every disturbance that breaks down the balance of the ecosystem can lead to incomplete fibre utilization and even illness of the animal. In addition to the fibre degrading function, microorganisms are also implicated from the gastric segment in the hydrolysis of rapidly digestible carbohydrates, sugars and starch, and probably in that of fat, glycolipids, phospholipids, triglycerides, but this latter has been less documented. Finally microorganisms are responsible for protein utilization in the hindgut.

Feed component utilization results from complex microorganism interactions organized in a trophic chain: this starts with the hydrolysis of large molecules to smaller molecules that are then fermented into end-products utilizable by the horse.

Microbial hydrolysis of the feed components

Cell-walls hydrolysis

The adhesion of fibrolytic microorganismes to plant particles is the first and essential step for the feed degrading process. This allows microorganisms to increase their retention time in the intestinal segments, which is vital for maintaining protozoa in the ecosystem as their division time is on average superior to that of small particles and liquids. Microscopic observations showed that ciliates in the caecum rapidly colonize plant tissues (Bonhomme-Florentin, 1985) in particular if partially degraded (Bonhomme-Florentin, 1969). Identically, electron microscope observations showed that in the equine caecum certain cocci are fixed by their 'capsule', others are attached by a diffuse and fibrous material, extracellular and electron dense and bacilli are adherent to the cell-walls via a thin material and modify their conformation to closely follow that of the wall (Bonhomme, 1986). The adherent proportion of microorganisms to caecal particles is high and probably close to 70-80% of the total microbial biomass (Julliand, 1996). Adhesion also ensures microorganisms increase the efficiency of their action by concentrating their hydrolytic enzymes on targeted tissues. In the intestinal contents, cellulolytic and hemicellulolytic activities are mainly related to the microbial population attached to particles in the solid fraction whereas they are lower for the free microbial population in the liquid fraction (Julliand, 1996; Jouany *et al.*, 2009).

Bacteria, fungi and protozoa are implicated in fibre breakdown. Both bacteria and protozoa (*Cycloposthium spp.* but also *Blepharocorys spp.*) were shown to have *in vitro* enzymatic activity involved in the degradation of hemicellulosic substrates; the xylan endo-1,3-p-xylosidase activity was significant, but the p-mannosidase activity was weak (Bonhomme-Florentin, 1988). The exact evaluation of the protozoal contribution to cell wall hydrolysis is impossible because it is difficult to be certain of the microbial origin of the cellulolytic activity. Ciliates exert a high predation on bacteria and fungi and ingest plant particles colonised by these microorganisms.

Julliand (1996) evaluated the enzymatic activity for total, solid-adherent and liquid-associated bacteria in the caecal contents of ponies. Carboxymethylcellulasic and xylanasic activities were essentially related to the solid-adherent bacteria and negligible in the liquid-associated bacteria. Carboxymethylcellulasic activities were lower than xylanasic activities which has recently been confirmed in both caecal and colonic contents (Jouany *et al.*, 2009). Both carboxymethylcellulasic and xylanasic activities were higher in the colon than in the caecum (Jouany *et al.*, 2009). Lower β-D-cellobiosidic and β-D-xylosidasic activities were reported for the liquid-associated bacteria compared to the solid-adherent bacteria (Julliand, 1996). The activity of the β-D-glucosidase were higher than that of a-L-arabinosidase (Jouany *et al.*, 2009).

Hydrolysis of cellulose and hemicelluloses by microbial enzymes leads to different hexose and pentose products of polysaccharide degradation.

Starch hydrolysis

Starch is degraded by starch-utilizing bacteria along the different segments of the GIT and by protozoa in the hindgut. It is not known if equine strains of fungi are capable of contributing to starch degradation. α-amylases which can hydrolyze α-(1,4) links anywhere in the starch molecule are expressed by certain strains of *Lactobacillus plantarum*, identified in the gastric content of horses (Giraud *et al.*, 1994). The α-amylolysis starts with the penetration of enzyme forming well or fissures in the starch grain which is then hydrolysed from the centre (amorphous) to the periphery (Kienzle *et al.*, 1997; Lynn and Cochrane, 1997).

Similarly to fibre degrading activity, recent work has evaluated the enzymatic activity of both solid-adherent and liquid-associated bacteria towards starch degradation in the hindgut. Amylolytic activity was essentially associated with the solid-adherent bacteria and was lower in the liquid-phase associated bacteria. The amylolytic activity appeared to be the same between the caecum and the right ventral colon (Jouany *et al.*, 2009).

Fat hydrolysis

It is not known in horses whether there are lipases produced by digestive micro-organisms. *In vitro*, bacteria and protozoa demonstrated a lipolytic activity and were responsible for the hydrolysis of triglycerides (Bonhomme-Florentin, 1976).

Protein hydrolysis

Some caecal bacteria isolated from pony caecal contents such as Bacteroides sp. (Baruc *et al.*, 1983; Maczulak *et al.*, 1985) have been reported to exhibit a proteolytic activity in the presence of peptones and amino acids or ammonia as the sole nitrogen source depending on the strains. Urea can be degraded by a very few bacteria possessing the enzyme urease (Baruc *et al.*, 1983; Maczulak *et al.*, 1985). The proteolytic activity of fungi has not been studied in horses. As for the ciliates, they would essentially play a role on proteins that are present in the solid or insoluble particles.

Fermentation end-products

Hexose and pentose products are fermented within the microbial cell cytoplasm into pyruvate using largely the Embden-Meyerhof-Parnas and pentose-phosphate pathways. Finally, pyruvate is transformed into products directly utilisable for the host maintenance and production requirements.

Organic acids

Pyruvate can be metabolised in different ways to various end-products including formate, acetate (C_2), propionate (C_3), butyrate (C_4), lactate, succinate, methanol, ethanol, carbon dioxide (CO_2) and dihydrogen (H_2). In the equine GIT ecosystem, the major compounds are acetate, propionate, butyrate as well as L – and D-lactate – lactates having comparatively low pKa (3.83 and 3.79, respectively). Parietal carbohydrates (cellulose and hemicellulosis) fermentation mainly generates acetate whereas that of highly fermentable carbohydrates tend to produce propionate and lactate. Lactate is known for its capability in decreasing the biotope pH (Van der Wielen, 2002) which in turn limits the development and activity of the microbial communities. Without glucose, lactate can be transformed into acetate in the presence of oxygen, by the pyruvate oxydase pathway (Divies *et al.*, 1994).

Methanogenic archaebacteria utilise either H_2 and CO_2 or formate, acetate, methylamine and methanol for the production of methane (CH_4).

The conversion of protein in ramified VFA (IsoAGV; isovalerate, isobutyrate and isocaproate) and ammonia (NH_3) is not well documented in equines.

Peptides, AA and ammonia

Microbial protein hydrolysis leads to the production of peptides which in turn are degraded into tri- or di-peptides and free amino-acids. Bacteria can also convert dietary and endogenous nitrogenous compounds into ammonia which is partly utilized by the microorganisms for growth together with amino acids produced by the activity of the bacterial proteases.

Long chain fatty acids

In theory, if microbial fat hydrolysis occurs in the horse GIT, it should produce LCFA but this has not been documented yet.

Gas

Microbial fermentations produce gas: methane (CH_4) or carbon dioxide (CO_2) which contributes to the anaerobic environment of the GIT ecosystem. Gas composition in the equine hindgut is not known. Intense fermentations can lead to excessive gas production in the stomach of horses, leading to gastric dilatation and even perforation (Murray, 2002).

Conclusion

Despite recent advances that have brought new insights into the diversity of the autochthonous microbial communities present in the different anatomic segments of the horse GIT, there is still a profound lack of knowledge about these microorganisms. We need to better understand the establishment of the bacterial communities in foals, measure correlations with the establishment of immunity and the impact on the future adult digestion. We need to highlight the potential role of the GI microflora in the health and wellbeing of the horse. As for the nutritional role of the microbial communities, it is probably even more essential than first thought and from the stomach to the colon the contribution of feed hydrolysis and fermentation has to be precisely evaluated to enable better recommendations in feeding practise.

References

Adam, K., 1951. The quantity and distribution of the ciliate protozoa in the large intestine of the horse. Parasitology 41: 301-311.

Adam, K., 1953. *In vivo* observations on the ciliate protozoa inhabiting the large intestine of the horse. Journal of general microbiology 9: 876-884.

Al Jassim, R.A.M., P.T. Scott, A.L. Trebbin, D. Trott and C.C. Pollitt, 2005. The genetic diversity of lactic acid producing bacteria in the equine gastrointestinal tract. FEMS Microbiology Letters 248(1): 75-81.

Alam, M., T. Midtvedt and A. Uribe, 1994. Differential Cell Kinetics in the Ileum and Colon of Germfree Rats. Scandinavian Journal of Gastroenterology 29(5): 445-451.

Alexander, F. and E. Davies, 1963. Production and fermentation of lactate by bacteria in the alimentary canal of the horse and pig. J. Comp. Path. 73: 1-8.

Alexander, F., M. Davies and A. Muir, 1970. Bacteriophage-like particles in the large intestine of the horse. Research in veterinary science 11(6): 592-593.

Argenzio, R., 1975. Functions of the equine large intestine and their interrelationship in disease. Cornell Vet 65: 303-327.

Argenzio, R., M. Southworth and C. Stevens, 1974. Sites of organic acid production and absorption in the equine gastrointestinal tract. Am J Physiol 226(5): 1043-1050.

Bailey, S.R., M.-L. Baillon, A.N. Rycroft, P.A. Harris and J. Elliott, 2003. Identification of Equine Cecal Bacteria Producing Amines in an *in vitro* Model of Carbohydrate Overload. Appl. Environ. Microbiol. 69(4): 2087-2093.

Baruc, C., K. Dawson and J. Baker, 1983. The caracterization and nitrogen metabolism of equine caecal bacteria. 8th ENPS.

Baruc, C.J., K.A. Dawson and J.P. Baker, 1983. The caracterization and nitrogen metabolism of equine caecal bacteria. 8th ENPS, University of Kentucky.

Batista, A., U. Chaves, Cte de Vasconcelos, O. Fishman and J.O. da Sylva, 1961. Flora micoteca intestinal de equinos e asininos no recife. Inst. Micol. Univ. Recife 326: 116.

Baverud, V., A. Gustafsson, A. Franklin, A. Aspan and A. Gunnarson, 2003. *Clostridium difficile*: prevalence in horses and environment and antimicrobial susceptibility. Equine Veterinary Journal 35(5): 465-471.

Berg, R.D., 1996. The indigenous gastrointestinal microflora. Trends in Microbiology 4: 430-435.

Bernet, M.F., D. Brassart, J.R. Neeser and A.L. Servin, 1994. Lactobacillus acidophilus LA 1 binds to cultured human intestinal cell lines and inhibits cell attachment and cell invasion by enterovirulent bacteria. Gut 35(4): 483-489.

Bonhomme, A., 1986. Attachment of horse cecal bacteria to forage cell walls. J. Vet. Sci. 48(2): 313-322.

Bonhomme-Florentin, A., 1969. Essais de culture *in vitro* des Cycloposthiidae, ciliés commensaux de l'intestin du cheval. Rôle de ces ciliés dans la dégradation de la cellulose. Protistologica 5(4): 519-522.

Bonhomme-Florentin, A., 1976. Lipolytic activity of ciliates and bacteria of horse's caecum. Animal physiology and biochemistry 282(17): 1605-1608.

Bonhomme-Florentin, A., 1985. Attachement des Ciliés du caecum de cheval aux fragments végétaux – Dégradation des chloroplastes – Attachement des bactéries aux Ciliés du caecum. Reprod. Nutri. Dévelop. 25(1A): 127-139.

Bonhomme-Florentin, A., 1988. Degradation of hemicellulose and pectin by horse caecum contents. British Journal of Nutrition 60: 185-192.

Breton, A., M. Dusser, B. Gailliard-Martinie, J. Guillot, L. Millet and G. Prensier, 1991. *Piromyces rhizinflata* nov. sp., a stritly anaerobic fungus from faeces of the saharian ass: a morphological, metabolic and ultrastructural study. FEMS Microbiol. lett. 82: 1-8.

Cann, A., S. Fandrich and S. Heaphy, 2004. Analysis of the Virus Population Present in Equine Faeces Indicates the Presence of Hundreds of Uncharacterized Virus Genomes. Virus Genes 30(2): 151-156.

Cotta, M.A., T.R. Whitehead and R.L. Zeltwanger, 2003. Isolation, characterization and comparison of bacteria from swine faeces and manure storage pits. Environ Microbiol 5(9): 737-745.

Da Veiga, L., F. Chaucheyras-Durand and V. Julliand, 2005. Comparative study of colon and faeces microbial communities and activities in horses fed a high starch diet. 3rd European Conference on Horse Nutrition, Hannover, Germany.

Daly, K. and S.P. Shirazi-Beechey, 2003. Design and evaluation of group-specific oligonucleotide probes for quantitative analysis of intestinal ecosystems: their application to assessment of equine colonic microflora. FEMS Microbiology Ecology 44(2): 243-252.

Daly, K., C.S. Stewart, H.J. Flint and S.P. Shirazi-Beechey, 2001. Bacterial diversity within the equine large intestine as revealed by molecular analysis of cloned 16S rRNA genes. FEMS Microbiology Ecology 38(2-3): 141-151.

De Fombelle, A., V. Julliand, C. Drogoul and E. Jacotot, 2001. Feeding and microbial disorders in horses : 1-effects of an abrupt incorporation of two levels of barley in a hay diet on microbial profile and activities. Journal of Equine Veterinary Science 21(9): 439-445.

De Fombelle, A., M. Varloud, A. Goachet, E. Jacotot, C. Philippeau, C. Drogoul and V. Julliand, 2003. Characterization of the microbial and biochemical profile of the different segments of the digestive tract in horses given two distinct diets. Animal Science 77: 293-304.

De Fombelle, A., M. Varloud, A.-G. Goachet, E. Jacotot, C. Philippeau, C. Drogoul and V. Julliand, 2003. Characterisation of the microbial and biochemical profile of the different segments of the digestive tract in horses fed two distinct diets. Animal Science 77: 293-304.

Divies, C., L. Frey, J. C. Hubert and H. de Roissart, 1994. Métabolisme d'autres substrats carbonés par les bactéries lactiques.' Les bactéries lactiques 1: 291-307.

Drogoul, C., 2000. Contribution à l'étude du fonctionnement de l'écosystème intestinal des équidés: influence de la granulométrie du fourrage sur la digestion des parois végétales chez le poney en relation avec les modifications du transit et de l'activité fibrolytique des écosystèmes caecal et coliques. Dijon, Université de Bourgogne: 169.

Dunkel, B. and P.A. Wilkins, 2004. Infectious foal diarrhoea: pathophysiology, prevalence and diagnosis. Equine Veterinary Education 16(2): 94-101.

Edwards, J.E., N.R. McEwan, A.J. Travis and R. John Wallace, 2004. 16S rDNA library-based analysis of ruminal bacterial diversity. Antonie van Leeuwenhoek 86(3): 263-281.

Elton, C., 1958. The Ecology of Invasions by Animals and Plants. London.

Endo, A., Y. Futagawa-Endo and L. Dicks, 2009. Lactobacillus and Bifidobacterium Diversity in Horse Feces, Revealed by PCR-DGGE. Current Microbiology 59(6): 651-655.

Endo, A., S. Okada and H. Morita, 2007. Molecular profiling of Lactobacillus, Streptococcus, and Bifidobacterium species in feces of active racehorses. The Journal of General and Applied Microbiology 53(3): 191-200.

Endo, A., S. Roos, E. Satoh, H. Morita and S. Okada, 2008. Lactobacillus equigenerosi sp. nov., a coccoid species isolated from faeces of thoroughbred racehorses. Int J Syst Evol Microbiol 58(4): 914-918.

Fonty, G., J.P. Jouany, E. Forano and P. Gouet, 1995. L'écosystème microbien du réticulo-rumen. In: Jarrige, R., Y. Ruckebusch, C. Demarquilly, M.H. Farce and M. Journet (eds.), Nutrition des ruminants domestiques, ingestion et digestion. INRA Editions, Paris, France, p. 299-338.

Fortun-Lamothe, L. and S. Boullier, 2007. A review on the interactions between gut microflora and digestive mucosal immunity. Possible ways to improve the health of rabbits. Livestock Science 107(1): 1-18.

Frankel, W., W. Zhang, A. Singh, D. Klurfeld, S. Don, T. Sakata, I. Modlin and J. Rombeau, 1994. Mediation of the trophic effects of short-chain fatty acids on the rat jejunum and colon. Gastroenterology 106(2): 375-380.

Frape, D., 2004. Equine nutrition and feeding. Blackwell Publishing, Oxford, United Kingdom.

Gaillard-Martinie, B., A. Breton, M. Dusser and J. Guillot, 1992. Contribution to the morphological, cytological and ultrastructural characterization of *Piromyces mae*, a strictly anaerobic rumen fungus. Curr. Microbiol. 24: 159-164.

Gaillard-Martinie, B., A. Breton, M. Dusser and V. Julliand, 1995. *Piromyces citronii* sp. nov., a stricly anaerobic fungus from the equine caecum: a morphological, metabolic and ultrastructural study. FEMS Microbiol. Lett. 130: 321-326.

Garner, H.E., D.P. Hutcheson, J.R. Coffman, A.W. Hahn and C. Salem, 1977. Lactic Acidosis: a Factor Associated with Equine Laminitis. J. Anim Sci. 45(5): 1037-1041.

Gerhards, H., S. Radicke and K. Hipp, 1991. Long term small intestinal fistulisation in ponies: surgical technique, postoperative care, and experimental effectiveness. Pferdeheilkunde 7: 243-248.

Geyer, H. and K. Drepper, 1973. Mikrobielle Einwirkungen auf die Verdauung beim Pferd. In: Giesecke, D. and H.K.H. Henderickx (eds.), Biologie und Biochemie der Verdauung. BLV Verlagsgesellschaft, München, Germany.

Giraud, E., A. Champailler and M. Raimbault, 1994. Degradatin of raw starch by a wild amylolytic strain of *Lactobacillus plantarum*. Applied and Environmental Microbiology 60: 4319-4323.

Gold, J.J., I.B. Heath and T. Bauchop, 1988. Ultrastructural description of a new chytrid genus of caecum anaerobe *Caecomyces equi* gen. nov., sp. nov., assigned to the Neocallimasticaecceae. BioSystems 21: 403-415.

Golomidova, A., E. Kulikov, A. Isaeva, A. Manykin and A. Letarov, 2007. The Diversity of Coliphages and Coliforms in Horse Feces Reveals a Complex Pattern of Ecological Interactions. Appl. Environ. Microbiol. 73(19): 5975-5981.

Goncalves, S., V. Julliand and A. Leblonde, 2002. Risk factors associated with colic in horses. Veterinary Research 33: 641-652.

Goodson, J., W.J. Tyznik, J.H. Cline and B.A. Dehority, 1988. Effects of an abrupt diet change from hay to concentrate on microbial numbers and physical environment in the cecum of the pony. Applied and Environmental Microbiology 54(8): 1946-1950.

Gordon, J.I., L.V. Hooper, M.S. McNevin, M. Wong and L. Bry, 1997. Epithelial cell growth and differentiation. III. Promoting diversity in the intestine: conversations between the microflora, epithelium, and diffuse GALT. Am J Physiol Gastrointest Liver Physiol 273(3): G565-570.

Grubb, S.A. and B.A. Dehority, 1976. Variations in colony counts of total anaerobic rumen bacteria influenced by media and cultural methods. Applied microbiology 31: 262-267.

Gruby, D. and O. Delafond, 1843. Recherches sur des animalcules se développant en grand nombre dans l'estomac et dans les intestins, pendant la digestion des animaux herbivores et carnivores. Compt. Rend. Acad. Sci 17: 1304-1308.

Guarner, F. and J.-R. Malagelada, 2003. Gut flora in health and disease. The Lancet 361(9356): 512-519.

Gürelli, G. and B. Göçmen, 2009. Intestinal ciliate composition found in the feces of the Cypriot wild donkey, Equus asinus Linnaeus, 1758. European Journal of Protistology In Press, Corrected Proof.

Hastie, P.M., K. Mitchell and J.M.D. Murray, 2008. Semi-quantitative analysis of Ruminococcus flavefaciens, Fibrobacter succinogenes and Streptococcus bovis in the equine large intestine using real-time polymerase chain reaction. British Journal of Nutrition 100: 561-568.

Hooper, L.V., 2004. Bacterial contributions to mammalian gut development. Trends in Microbiology 12(3): 129-134.

Hooper, L.V., M.H. Wong, A. Thelin, L. Hansson, P.G. Falk and J.I. Gordon, 2001. Molecular Analysis of Commensal Host-Microbial Relationships in the Intestine. Science 291(5505): 881-884.

Hungate, R.E., 1969. A roll tube method for cultivation of strict anaerobes. In: Norris, J.R. and D.W. Ribbons(eds.), Methods in Microbiology. Academic Press, London, United Kingdom, p. 117-132.

Ike, K., S. Mai and T. Ishii, 1985. Establishment of intestinal ciliates in new-born horses. Japanese Journal of Veterinary Science 47(1): 39-43.

Ike, K., R. Nuruki, S. mai and T. Ishii, 1983. Compositon of intestinal ciliates and bacteria excreted in feces of the race-horse. The Japanese journal of veterinary science 40(2): 157-163.

Imai, S., K. Inami, T. Morita, K. Ike and A. Ito, 1999. Intestinal Ciliate Composition Found in the Feces of Japanese Native Kiso Horse. Bulletin of Nippon Veterinary and Animal Science University 48: 33-38.

Jouany, J.P., B. Medina, G. Bertin and V. Julliand, 2009. Effect of live yeast culture supplementation on hindgut microbial communities and their polysaccharidase and glycoside hydrolase activities in horses fed a high-fiber or high-starch diet. Journal of animal science 87(9): 2844-2852.

Julliand, V., 1996. Etude de l'écosystème cæcal des équidés : aptitude à dégrader les polyolosides pariétaux, Caractérisation quantitative et qualitative des flores cellulolytiques bactériennes et fongiques dominantes. ENSBANA. Dijon, Université de bourgogne: 121.

Julliand, V., A. de Fombelle, C. Drogoul and E. Jacotot, 2001. Feeding and microbial disorders in horses: 3-Effects of three hay:grain ratios on microbial profile and activities. Journal of Equine Veterinary Science 21(11): 543-546.

Julliand, V., A. de Fombelle, C. Drogoul and E. Jacotot, 2001. Feeding and microbial disorders in horses: Part 3--Effects of three hay:grain ratios on microbial profile and activities. Journal of Equine Veterinary Science 21(11): 543-546.

Julliand, V., A. de Vaux, L. Millet and G. Fonty, 1999. Identification of Ruminococcus flavefaciens as the Predominant Cellulolytic Bacterial Species of the Equine Cecum. Appl. Environ. Microbiol. 65(8): 3738-3741.

Julliand, V., A. de Vaux, L. Villard and Y. Richard, 1996. Preliminary studies on the bacterial flora of the faeces taken from foals, from birth to twelve weeks. Effect of the oral administration of a commercial colostrum replacer. Pferdeheilkunde. 12: 209-212.

Julliand, V. and A.G. Goachet, 2005. Fecal microflora as a marker of cecal or colonic microflora in horses ? 19th Symposium, Tucson, Arizona, USA.

Julliand, V. and A.-G. Goachet, 2005. Fecal microflora as a marker of cecal or colonic microflora in horses ? 19th Symposium, Tucson, Arizona, USA.

Julliand, V., C. Riondet, A. De Vaux, G. Alcarz and G. Fonty, 1998. Comparison of metabolic activities between *Pyromyces citronii*, an equine fungal species, and *Pyromyces communis*, a ruminal species. Animal feed science and technology 70: 161-168.

Kern, D.L., L.L. Slyter, E.C. Leffel, J.M. Weaver and R.R. Oltjen, 1974. Ponies vs. Steers: Microbial and Chemical Characteristics of Intestinal Ingesta. J. Anim Sci. 38(3): 559-564.

Kern, D.L., L.L. Slyter, J.M. Weaver, E.C. Leffel and G. Samuelsons, 1973. Pony Cecum vs. Steer Rumen: The Effect of Oats and Hay on the Microbial Ecosystem. J. Anim Sci. 37(2): 463-469.

Kienzle, E., J. Pohlenz and S. Radicke, 1997. Morphology of starch digestion in the horse. J. Anim. Physiol. Anim. Nutr. 80: 213-216.

Klieve, A.V. and R.A. Swain, 1993. Estimation of ruminal bacteriophage numbers by pulsed-field gel electrophoresis and laser densitometry. Appl. Environ. Microbiol. 59(7): 2299-2303.

Kobayashi, Y., S. Koike, M. Miyaji, H. Hata and K. Tanaka, 2006. Hindgut microbes, fermentation and their seasonal variations in Hokkaido native horses compared to light horses. Ecological Research 21(2): 285-291.

Kollarczik, B., C. Enders, M. Friedrich and B. Gedek, 1992. Auswirkungen der rationszusammensetzung auf das keimsperktrum im jejunum von pferden. 1. Europäische Konferenz über die Ernährung des pferdes. Physiologie und Pathologie des Verdauungskanals – Konsequenzen für die Ernährung, Hannover, Germany.

Kulikov, E., A. Isaeva, A. Rotkina, A. Manykin and A. Letarov, 2007. Diversity and dynamics of bacteriophages in horse feces. Microbiology 76(2): 236-242.

Leao, M.I., F.Q. Almeida, A.U. Carvalho, S.C. Valadares Filho, A.A.M.A. Oliveira and R.C. Alvarenga, 1999. Technique for longterm ileal cannulation in horses. 16th Equine Nutrition and Physiology Symposium, Raleigh, NC, USA.

Li, J. and I.B. Heath, 1993. Chytridiomycetous gut fungi, oft overlooked contributors to herbivore digestion. Can. J. Microbiol. 39: 1003-1013.

Li, J., I.B. Heath and T. Bauchop, 1989. *Piromyces mae* and *Piromyces dumbonica*, two new species of uniflagellate anaerobic chytridiomycete fungi from the hindgut of the horse and elephant. Canadian Journal of Botanic 68: 1021-1033.

Liebetanz, E., 1910. Die parasitischen protozoen des wiederkäuermagen vorkommenden protozoen. Archiv für Protistenkunde 32: 111-170.

Lin, C. and D. Stahl, 1995. Taxon-specific probes for the cellulolytic genus Fibrobacter reveal abundant and novel equine-associated populations. Appl. Environ. Microbiol. 61(4): 1348-1351.

Loreau, M. and N. Behera, 1999. Phenotypic Diversity and Stability of Ecosystem Processes. Theoretical Population Biology 56(1): 29-47.

Lynn, A. and M.P. Cochrane, 1997. An evaluation of confocal microscopy for the study of strach granule enzymic digestion. Starch 49: 106-111.

Mac Bee, R.H., 1977. Fermentation in the hindgut. Microbial Ecology of the Gut. R. T. J. a. B. T. Clarke. London, Academic Press: 185-222.

Mackie, R.I., A. Sghir and H.R. Gaskins, 1999. Developmental microbial ecology of the neonatal gastrointestinal tract. Am J Clin Nutr 69: 1035-1045.

Mackie, R.I. and C.A. Wilkins, 1988. Enumeration of anaerobic bacterial microflora of the equine gastrointestinal tract. Appl. Environ. Microbiol. 54(9): 2155-2160.

Maczulak, A.E., K.A. Dawson and J.P. Baker, 1985. Nitrogen utilization in bacterial isolates from the equine cecum. Appl. Environ. Microbiol. 50(6): 1439-1443.

Marteau, P., P. Pochart, J. Dore, C. Bera-Maillet, A. Bernalier and G. Corthier, 2001. Comparative Study of Bacterial Groups within the Human Cecal and Fecal Microbiota. Appl. Environ. Microbiol. 67(10): 4939-4942.

Masri, M.D., A.M. Merritt, R. Gronwall and C.F. Burrows, 1986. Fecal composition in foal heat diarrhoea. Equine Veterinary Journal 18(4): 301-306.

McCreery, S., R.S. Fulghum and J.P. Baker, 1971. Microflora in the equine cecum. Journal of Animal Science 33((Suppl. 1)): 234.

Medina, B., I.D. Girard, E. Jacotot and V. Julliand, 2002. Effect of a preparation of Saccharomyces cerevisiae on microbial profiles and fermentation patterns in the large intestine of horses fed a high fiber or a high starch diet. J. Anim Sci. 80(10): 2600-2609.

Mentula, S., J. Harmoinen, M. Heikkila, E. Westermarck, M. Rautio, P. Huovinen and E. Kononen, 2005. Comparison between Cultured Small-Intestinal and Fecal Microbiotas in Beagle Dogs. Appl. Environ. Microbiol. 71(8): 4169-4175.

Merritt, A.M. and M.L. Campbell-Thompson, 1987. Effect of ranitidine on gastric acid secretion in young male horses. Am J Vet Res 48: 1511-1515.

Milinovich, G.J., P.C. Burrell, C.C. Pollitt, A.V. Klieve, L.L. Blackall, D. Ouwerkerk, E. Woodland and D.J. Trott, 2008. Microbial ecology of the equine hindgut during oligofructose-induced laminitis. 2(11): 1089-1100.

Milinovich, G.J., D.J. Trott, P.C. Burrell, E.L. Croser, R.A.M.A. Jassim, J.M. Morton, A.W. van Eps and C.C. Pollitt, 2007. Fluorescence *in situ* hybridization analysis of hindgut bacteria associated with the development of equine laminitis. Environmental Microbiology 9(8): 2090-2100.

Milinovich, G.J., D.J. Trott, P.C. Burrell, A.W. van Eps, M.B. Thoefner, L.L. Blackall, R.A.M.A. Jassim, J.M. Morton and C.C. Pollitt, 2006. Changes in equine hindgut bacterial populations during oligofructose-induced laminitis. Environmental Microbiology 8(5): 885-898.

Moore, B.E. and B.A. Dehority, 1993. Effects of diet and hindgut defaunation on diet digestibility and microbial concentrations in the cecum and colon of the horse. Journal of Animal Science 71: 3350-3358.

Morita, H., A. Nakano, M. Shimazu, H. Toh, F. Nakajima, M. Nagayama, S. Hisamatsu, Y. Kato, M. Takagi, H. Takami, H. Akita, M. Matsumoto, T. Masaoka and M. Murakami, 2009. *Lactobacillus hayakitensis, L. equigenerosi* and *L. equi*, predominant lactobacilli in the intestinal flora of healthy thoroughbreds. Animal Science Journal 80(3): 339-346.

Morita, H., C. Shiratori, M. Murakami, H. Takami, Y. Kato, A. Endo, F. Nakajima, M. Takagi, H. Akita, S. Okada and T. Masaoka, 2007. *Lactobacillus hayakitensis sp. nov.*, isolated from intestines of healthy thoroughbreds. Int J Syst Evol Microbiol 57(12): 2836-2839.

Morotomi, M., N. Yuki, Y. Kado, A. Kushiro, T. Shimazaki, K. Watanabe and T. Yuyama, 2002. *Lactobacillus equi sp. nov.*, a predominant intestinal *Lactobacillus* species of the horse isolated from faeces of healthy horses. Int J Syst Evol Microbiol 52(1): 211-214.

Murray, M.J., 2002. Diseases of the stomach. Manual of Equine Gastroenterology. Chatham, UK, Harcourt: 241-248.

Orban, J.I., J.A. Patterson, O. Adeola, A.L. Sutton and G.N. Richards, 1997. Growth performance and intestinal microbial populations of growing pigs fed diets containing sucrose thermal oligosaccharide caramel. J. Anim Sci. 75(1): 170-175.

Orpin, C., 1981. Isolation of Cellulolytic Phycomycete Fungi from the Caecum of the Horse. J Gen Microbiol 123(2): 287-296.

Orpin, C.G., 1981. Isolation of cellulolytic phycomycete fungi from the cæcum of the horse. Journal of Generic Microbiology 123: 287-296.

Ozeki, K., S. Imai and M. Katsuno, 1973. On the distribution of the ciliated protozoa in the large intestine of horse. Tohoku Journal of Agricultural Research 24(2): 86-101.

Pace, N.R., 1997. A Molecular View of Microbial Diversity and the Biosphere. Science 276(5313): 734-740.

Peloso, J.C., J. Schumacher, S.R. McClure, D.V. Hanselka, D.S. Householder and G.D. Potter, 1994. Technique for long-term ileal cannulation in ponies. Canadian Journal of Veterinary Research 58: 184-189.

Prins, R.A., 1977. Biochemical activities of gut microorganisms. In: Clarke, R.T.J. and T. Beauchop (eds.), Microbial Ecology of the Gut. Academic press, London, United Kingdom, p. 73-183.

Pulse, R.E., J.P. Baker and G.D. Potter, 1973. Effects of cecal fistilization upon nutrients digestion and retention indicator in horses. Journal of animal science 37(2): 488-492.

Rada, V., E. Vlková, J. Nevoral and I. Trojanová, 2006. Comparison of bacterial flora and enzymatic activity in faeces of infants and calves. FEMS Microbiology Letters 258(1): 25-28.

Ralston, S.L., D.E. Freeman and C.A. Baile, 1983. Volatile fatty acids and the role of the large intestine in the control of feed intake in ponies. Journal of animal science 57(4): 815-825.

Roger, T., C.H. Malbert and Y. Ruckebusch, 1990. Functional role of the equine ileocaecal sphincter. Journal of Gastrointestinal Motility 2(3): 1-6.

Sakaitani, Y., Y. Norikatsu, F. Nakajima, s. Nakanishi, H. Tanaka, R. Tanaka and M. Morotomi, 1999. Colonization of intestinal microflora in newborn foals. Journal of Intestinal Microbiology 13: 9-14.

Selim, A.S.M., P. Boonkumklao, T. Sone, A. Assavanig, M. Wada and A. Yokota, 2005. Development and Assessment of a Real-Time PCR Assay for Rapid and Sensitive Detection of a Novel Thermotolerant Bacterium, Lactobacillus thermotolerans, in Chicken Feces. Appl. Environ. Microbiol. 71(8): 4214-4219.

Shroff, K., K. Meslin and J. Cebra, 1995. Commensal enteric bacteria engender a self-limiting humoral mucosal immune response while permanently colonizing the gut. Infect. Immun. 63(10): 3904-3913.

Stahl, D.A., B. Flesher, H.R. Mansfield and L. Montgomery, 1988. Use of phylogenetically based hybridization probes for studies of ruminal microbial ecology. Appl.Environ.Microbiol. 54: 1079-1084.

Stappenbeck, T.S., L.V. Hooper and J.I. Gordon, 2002. Developmental regulation of intestinal angiogenesis by indigenous microbes via Paneth cells. Proceedings of the National Academy of Sciences of the United States of America 99(24): 15451-15455.

Suau, A., R. Bonnet, M. Sutren, J.J. Godon, G.R. Gibson, M.D. Collins and J. Dore, 1999. Direct analysis of genes encoding 16S rRNA from complex communities reveals many novel molecular species within the human gut. Appl.Environ.Microbiol. 65: 4799-4807.

Tajima, K., S. Arai, K. Ogata, T. Nagamine, H. Matsui, M. Nakamura, R. I. Aminov and Y. Benno, 2000. Rumen Bacterial Community Transition During Adaptation to High-grain Diet. Anaerobe 6(5): 273-284.

Teunissen, M.J., H.J.M. Op den Camp, C.G. Orpin, J.H.J. Huis in ,t Veld and G.D. Vogels, 1991a. Comparison of growth characteristics of anaerobic fungi from ruminant and non-ruminant herbivores during cultivation in a semi-defined medium. J. Gen. Microbiol. 137: 1401-1048.

Teunissen, M.J., A.A.M. Smits, H.J.M. Op den Camp, J.H.J. Huis in ,t Veld and G.D. Vogels, 1991b. Fermentation of cellulose and production of cellulolytic and xylanolytic enzymes by anaerobic fungi from ruminant and non-ruminant herbivores. Arch. Microbiol. 156: 290-296.

Tillotson, K., J.L. Traub-Dargatz, C.E. Dickinson, R.P. Ellis, P.S. Morley, D.R. Hyatt, R.J. Magnuson, W.T. Riddle, D. Bolte and M.D. Salman, 2002. Population-based study of fecal shedding of *Clostridium perfringens* in broodmares and foals. J. Am. Vet. Med. Ass. 220: 342-348.

Umesaki, Y. and H. Setoyama, 2000. Structure of the intestinal flora responsible for development of the gut immune systemin a rodent model. Microbes and Infection 2(11): 1343-1351.

Umesaki, Y., H. Setoyama, S. Matsumoto, A. Imaoka and K. Itoh, 1999. Differential Roles of Segmented Filamentous Bacteria and Clostridia in Development of the Intestinal Immune System. Infect. Immun. 67(7): 3504-3511.

Umesaki, Y., H. Setoyama, S. Matsumoto and Y. Okada, 1993. Expansion of aβ T-cell receptor-bearing intestinal intraepithelial lymphocytes after microbial colonization in germ-free mice and its independence from thymus. Immunology 79(1): 32-37.

Van der Wielen, P.W.J.J., 2002. Role of volatile fatty acids in competitive exclusion of *Salmonella enteritis*. Utrecht, The Nederlands, Faculty of Veterinary Medicine.

Varloud, M., G. Fonty, A. Roussel, A. Guyonvarch and V. Julliand, 2007. Postprandial kinetics of some biotic and abiotic characteristics of the gastric ecosystem of horses fed a pelleted concentrate meal. J. Anim Sci. 85(10): 2508-2516.

Varloud, M., E. Jacotot, G. Fonty, A. Guyonvarch and V. Julliand, 2004. Postprandial evolution of the microbial community and biochemical composition of stomach contents in equines. Reproduction nutrition development 44(Suppl. 1): S75.

Varloud, M., A. Roussel, P. Boisot and V. Julliand, 2006. A technique for the collection and the study of biochemical and microbial characteristics of postprandial gastric contents from conscious horses. Animal Feed Science and Technology 133(3-4): 259-274.

Vavra, J. and L. Joyon, 1966. Etude sur la morphologie, le cycle évolutif et la composition systématique de *Callimastix cyclopis*. Protistologica 2: 5-16.

Whitford, M.F., R.J. Forster, C.E. Beard, J. Gong and R.M. Teather, 1998. Phylogenetic Analysis of Rumen Bacteria by Comparative Sequence Analysis of Cloned 16S rRNA Genes. Anaerobe 4: 153-163.

Willing, B., A. Vörös, S. Roos, C. Jones, A. Jansson and J.E. Lindberg, 2009. Changes in faecal bacteria associated with concentrate and forage-only diets fed to horses in training. Equine Veterinary Journal 41: 908-914.

Wilson, K. and R. Blitchington, 1996. Human colonic biota studied by ribosomal DNA sequence analysis. Appl. Environ. Microbiol. 62(7): 2273-2278.

Wolter, R., J.C. Letourneau and M. Carcelem, 1979. Evaluation chez le poney de la digestibilité du mais fourrage déshydraté, des pulpes sèches de betteraves, de la luzerne déshydratée, du son de blé, de la paille de blé, et des pulpes de raisin. Annales de zootechnies 29(3): 93-100.

Yamano, H., S. Koike, Y. Kobayashi and H. Hata, 2008. Phylogenetic analysis of hindgut microbiota in Hokkaido native horses compared to light horses. Animal Science Journal 79(2): 234-242.

Yu, Z., M. Yu and M. Morrison, 2006. Improved serial analysis of V1 ribosomal sequence tags (SARST-V1) provides a rapid, comprehensive, sequence-based characterization of bacterial diversity and community composition. Environmental Microbiology 8(4): 603-611.

Yuki, N., T. Shimazaki, A. Kushiro, K. Watanabe, K. Uchida, T. Yuyama and M. Morotomi, 2000. Colonization of the Stratified Squamous Epithelium of the Nonsecreting Area of Horse Stomach by Lactobacilli. Appl. Environ. Microbiol. 66(11): 5030-5034.

Yuyama, T., S. Takai, S. Tsubaki, Y. Kado and M. Morotomi, 2004. Evaluation of a Host-Specific Lactobacillus Probiotic in Training-Horses and Neonatal Foals. Journal of Intestinal Microbiology 18(2): 101-106.

Feeding protected sodium bicarbonate attenuates hindgut acidosis in horses fed a high grain ration

J.D. Pagan[1], T.J. Lawrence[1], L.A. Lawrence[1] and C. Brown-Douglas[1]
Kentucky Equine Research, Versailles, Kentucky USA 40383

Abstract

Hindgut acidosis (HGA) is a potential problem in horses consuming large quantities of grain or fructan rich forages. Horses suffering from HGA may develop anorexia, colic, laminitis or display stereotypical behaviors such as wood chewing and stall weaving. Six exercised Thoroughbreds were used in a two period switch back design study to evaluate the effect of protected sodium bicarbonate (PSB) supplementation on hindgut acidosis in horses fed a high grain ration. Horses were fed a basal ration of unfortified sweet feed (9.32±0.76 g DMI/kg BW/day), timothy grass hay (9.11±0.26 g DMI/kg BW/day) and 50 g of loose salt per day. The treatments were 168 g/day (0.33±0.01 g/kg BW/day) of PSB added to the basal ration or the basal ration alone (control group). Feeding PSB attenuated a decrease in fecal pH and increase in fecal lactate concentration compared to the control group without producing a metabolic alkalosis. This suggests that feeding PBS may be effective in attenuating the HGA that can result from feeding high grain intakes to horses.

Keywords: protected sodium bicarbonate, hindgut acidosis, fecal pH

Introduction

Horses evolved as wandering herbivores with voluminous hindguts adapted to process large quantities of high fibre forage. Fibre fermenting bacteria that populate the hindgut depend on cellulose and hemicellulose as their primary energy substrates. Other populations of bacteria capable of rapidly fermenting soluble carbohydrates also inhabit the hindgut. When horses eat a diet high in fibre, the environment in the hindgut favors the fibre fermenting bacteria. When large grain meals are fed to horses, a portion of the starch may escape digestion in the small intestine and be rapidly fermented in the caecum and colon (Potter *et al.*, 1992). The accumulation of lactate, and the subsequent drop in pH, impairs the activity of lower-gut cellulolytic bacterial populations (Medina *et al.*, 2002). Furthermore, lactic acid accumulation increases the permeability of the large intestinal mucosa *in vitro* to toxins and larger molecules that have been implicated in the development of equine laminitis (Weiss *et al.*, 2000). Changes in the pH of the hindgut due to alterations in the microbial populations and acid profiles may result in hindgut acidosis (HGA). Horses suffering HGA may develop anorexia, colic or stereotypical behaviors such as wood chewing and stall weaving (Willard *et al.*, 1977; Johnson *et al.*, 1998).

Acute and chronic rumen acidosis are prominent production problems for ruminants fed diets rich in concentrate (Owens *et al.*, 1998). Dietary supplementation of sodium bicarbonate attenuates the decline in ruminal pH that is observed post feeding (Erdman, 1988), and may attenuate subacute rumen acidosis (SARA). Sodium bicarbonate has been shown to be effective in treating hindgut acidosis in horses when it is infused directly into the caecum via a caecal fistula (Willard *et al.*, 1977). Feeding raw sodium bicarbonate to horses is ineffective as a hindgut buffer since it is broken down in the stomach and small intestine resulting in a metabolic alkalosis characterized by increases in blood pH and bicarbonate concentrations (Schuback *et al.*, 2002). Kentucky Equine Research, Inc. (EquiShure™, Kentucky Equine Research, Inc., Versailles, KY, USA) in conjunction with Balchem Corporation has developed a protected sodium bicarbonate (PSB) that is encapsulated in vegetable oil. The following study was conducted to assess its effectiveness at treating HGA in horses fed a high grain ration.

Materials and methods

Six 5-year-old Thoroughbred horses (504±14.1 kg) were used in a study to evaluate the effect of PSB supplementation on hindgut acidosis in horses fed a high grain ration. The horses were in a regular training program for several months before beginning the study and were considered physically fit. The study utilized a switchback design with each period lasting 4 weeks with a 2 week washout period between periods. Before and during the study horses were exercised 3 times weekly on a high speed treadmill (5 min walk, 5 min trot, 5 min canter, 5 min walk) and walked for 1 hour on a mechanical walker 3 days per week. Each day the horses were turned out daily for 4 to 6 hours with muzzles to prevent grazing and housed overnight in 3.5x3.5 m box stalls. The horses were on a regular deworming and vaccination program and were in good health throughout the study.

Horses were fed a basal ration of unfortified sweet feed, timothy grass hay and 50 g of loose salt per day. Grain intakes ranged from 4 kg (2 horses) to 6 kg (4 horses) per day (9.32±0.76 g DMI/kg BW/day). Hay was fed at a rate of 5 kg (9.11±0.26 g DMI/kg BW/day) for the first 2 weeks of each period. Hay fed was decreased to 4 kg (7.29±0.20 g DMI/kg BW/day) for the latter half of each period. Horses were split into 2 groups and assigned to 1 of 2 treatments. The treatments were 168 g/day (0.33±0.01 g/kg BW/day) of protected sodium bicarbonate (PSB[a]) added to the basal ration or the basal ration alone (control group). Following a 2 week washout period the horses switched treatments for period 2. During the washout period the horses remained on the same exercise program and basal ration. Both the hay and grain portion of the diet were split into 2 equal feedings. The grain portion of the ration was fed at 07:00 h and 16:00 h and the hay portion of the ration was fed at 07:00 h and 22:00 h. Each grain meal contained 3.84±0.24 g/kg BW nonstructural carbohydrate (NSC). One half of the PSB (84 g) was added to each grain meal.

Venous blood and rectal fecal grab samples were taken at 2 hour intervals for an 8 hour period on day 15 of each period. The first samples (0 h) were taken immediately before the horses received their morning grain. Subsequent samples were taken at 2, 4, 6 and 8 h post feeding. pH, pCO_2, HCO_3^-, Na^+, K^+, Cl^-, and tCO_2 were measured in whole blood samples using an automated electrolyte and blood gas analyzer (Idexx Laboratories, Westbrook, USA). Volatile fatty acids (VFA), pH, and L- and D-lactate concentration were measured in fecal grab samples taken from the rectum. pH was measured in 50 g of faeces diluted in a 3:1 ratio with distilled water using a pH electrode (VWR International, San Dimas, CA 91773, USA). Fecal total VFAs were measured using gas chromatography and lactates were measured colorimetrically using a commercially available kit. Data were analyzed for time and treatment differences using paired t-tests with a significance level set at $P<0.05$. Values are expressed as means ± SEM.

During week 4 of each period, horses were fitted with collection harnesses and a 5 day complete fecal and urine collection was conducted. A 2 day harness adaptation preceded the 5 day collection. Fecal and feed samples were analyzed for dry matter (DM), crude protein (CP), acid detergent fibre (ADF), neutral detergent fibre (NDF), fat, ash, calcium (Ca), phosphorus (P), magnesium (Mg), potassium (K), sodium (Na), chloride (Cl), iron (Fe), zinc (Zn), copper (Cu), and manganese (Mn). Urine samples were analyzed for mineral contents using an Intrepid inductively coupled plasma (ICP) radial spectrometer after microwave digestion.

Results

Fecal pH in the control group decreased significantly from 0 h by 6 h post-feeding (Table 1). Fecal pH in the PSB group did not exhibit any significant fluctuations during the 8 hour sampling period. Fecal L-lactate and D-lactate were significantly higher ($P<0.05$) in the control group at 0 h. L-lactate was higher in the control 2 h and 6 h post feeding ($P<0.05$) and D-lactate was higher in the control 6 h post feeding. Fecal VFAs were significantly higher ($P<0.05$) in the PSB group than the control group at 0 h. VFA increased in the control group after feeding and was significantly higher than 0 h at 2, 4, 6 and 8 h post feeding ($P<0.05$). Fecal VFA was significantly higher than 0 h in the PSB group 6 h post feeding ($P<0.05$).

No significant differences were recorded in blood sodium and potassium (Table 2). Blood chloride concentration was significantly higher compared to 0 h within both groups 2 h post feeding. Chloride values remained elevated in the control group up to 8 h post feeding.

Blood pH was significantly lower compared to 0 h in both groups 8 h post feeding. There was no significant difference in tCO_2, PCO_2, or HCO_3 between the two groups at any time pre- or post feeding (Table 3).

There was a trend towards increased digestibility of NDF and hemicellulose in the PSB group ($P<0.10$) (Table 4).

The horses consumed 28 g of Na^+ and 64.5 g of fat/day from the PSB. Of this extra intake, 29% of the Na+ was recovered in the faeces, 29% was voided in the urine and 12% was retained (Table 5) while 29.1 g of additional fat (45% of PSB fat content) was voided in the faeces.

Table 1. Fecal pH, L-lactate, D-lactate and total volatile fatty acids (VFA) before and 2, 4, 6 and 8 h post feeding.

Sample time post-feeding (h)	Fecal pH		L-lactate (mmol/l)		D-Lactate (mmol/l)		Total VFA mM	
	Control	PSB	Control	PSB	Control	PSB	Control	PSB
0	6.46±0.16	6.42±0.11	1.72±0.43[a]	0.86±0.16	0.60±0.29[a]	0.28±0.09	31.2±1.5[a]	38.6±3.7
2	6.31±0.06	6.47±0.11	1.90±0.46[a]	0.81±0.27	0.79±0.33	0.35±0.12	38.2±3.8*	38.4±3.9
4	6.22±0.04	6.37±0.12	1.32±0.28	0.71±0.14	0.73±0.38	0.45±0.21	42.0±4.1*	48.3±5.0
6	6.05±0.08*	6.32±0.12	1.91±0.42[a]	1.02±0.26	1.12±0.45[a]	0.28±0.11	57.9±12.7*	52.2±3.5*
8	6.17±0.07	6.33±0.16	1.71±0.33	1.24±0.31	0.59±0.26	0.66±0.18	50.5±3.5*	51.9±6.2

PSB: protected sodium bicarbonate; *significantly different from 0 h ($P<0.05$); [a] significant treatment effect ($P<0.05$).

Table 2. Blood Na+, K+ and Cl- before and 2, 4, 6 and 8 h post feeding.

Sample time postfeeding (h)	Blood Na+ (mmol/l)		Blood K+ (mmol/l)		Blood Cl- (mmol/l)	
	Control	PSB	Control	PSB	Control	PSB
0	149.0±0.8	148.9±0.4	4.2±0.1	4.1±0.1	104.5±0.7	104.3±0.5
2	150.7±1.0	150.6±0.4	4.0±0.1	3.8±0.2	106.2±0.8*	105.9±0.8*
4	150.2±1.0	150.4±0.5	4.1±0.1	4.0±0.1	106.3±0.5*	105.7±0.4
6	150.5±0.9	150.6±0.6	4.1±0.1	4.1±0.1	106.3±0.4*	105.6±0.4
8	149.3±0.7	149.4±0.3	4.1±0.2	3.9±0.2	106.3±0.6*	105.0±0.2

PSB: protected sodium bicarbonate; *Significantly different from 0 h ($P<0.05$)

Table 3. Blood pH, tCO_2, PCO_2, and HCO_3 before and 2, 4, 6 and 8 h post feeding.

Sample time post-feeding (h)	Blood pH		Blood tCO_2 (mmol/l)		PCO_2 (mm Hg)		HCO_3^- (mmol/l)	
	Control	PSB	Control	PSB	Control	PSB	Control	PSB
0	7.43±0.008	7.44±0.006	32.2±0.74	32.8±0.60	49.8±1.42	50.0±0.85	30.7±0.69	31.1±0.53
2	7.42±0.005	7.42±0.003	31.5±0.94	32.7±0.96	49.67±1.23	51.6±1.39	30.0±0.91	31.1±0.84
4	7.42±0.002	7.43±0.002	32.2±0.93	32.7±0.70	50.5±1.54	50.1±0.90	30.6±0.64	31.0±0.64
6	7.42±0.006	7.43±0.003	31.8±0.82	32.6±0.52	50.2±1.25	50.6±0.88	30.3±0.78	31.2±0.46
8	7.41±0.010*	7.42±0.004*	31.0±0.58	32.2±0.54	50.2±1.78	50.6±0.66	29.4±0.53	30.5±0.49

PSB: protected sodium bicarbonate; *Significantly different from 0 h (P<0.05)

Table 4. Apparent digestibility of nutrients during 5 day complete collection trial.

Nutrient	Control (%)	PSB (%)
Dry matter	65.5±0.9	69.3±3.3
Crude protein	60.1±2.6	60.4±3.9
Acid detergent fibre	26.9±1.7	30.7±7.3
Neutral detergent fibre	34.1±3.4	42.4±5.6§
Hemicellulose	43.8±8.6	57.3±4.0§
Fat	80.0±3.3	76.1±2.1
Ash	47.2±4.2	46.1±4.7
Nonstructural carbohydrates	93.2±1.4	92.3±1.0
Calcium	31.1±3.7	26.9±5.5
Phosphorus	18.1±3.6	18.9±5.5
Magnesium	47.3±3.9	51.8±3.9
Potassium	74.9±1.6	77.2±1.3
Sodium	56.8±4.6	64.5±3.2
Iron	-8.1±4.2	-7.7±9.1
Zinc	9.4±21.7	-12.7±13.4
Copper	53.7±3.4	49.3±3.1
Manganese	22.5±4.8	29.0±4.5

PSB: protected sodium bicarbonate; §trend towards treatment difference (P<0.10).

Table 5. Na+ and fat intake, fecal excretion, urinary excretion and retention during 5 day complete collection digestion trial.

Treatment		Intake (g/d)	Fecal excretion (g/d)	Urinary excretion (g/d)	Retention (g/d)
Na+	Control	23.6±0.4	10.2±1.2	7.1±1.0	6.3±0.7
	PSB	51.5±0.3	18.3±1.6	23.6±1.4	9.6±1.8
Fat	Control	290.4±30.7	55.6±16.4	0	234.8±15.7
	PSB	354.9±30.5	84.7±10.7	0	270.2±20.4

PSB: protected sodium bicarbonate.

Discussion

Feeding 3.84±0.24 g/kg BW of NSC in a single meal resulted in a significant drop in fecal pH 6 h post feeding. Addition of PSB to the diet attenuated the drop in fecal pH. Horses receiving PSB had higher baseline (0 h) VFA and lower D-lactate and L-lactate concentrations. Milinovich *et al.* (2007) reported that microbial fermentation of oligofructose resulted in a marked decrease in caecal VFA concentrations, an increase in caecal lactate concentrations and a decrease in caecal pH. These conditions are known to have numerous detrimental effects in the horse including decreased cellulolytic and hemicellulolytic bacterial cell numbers, decreased fibre digestion rates and inhibition of VFA production and absorption (Hussein *et al.*, 2004). A trend towards improved hemicellulose and NDF digestibility suggests that there may have been an improvement in fibre fermentation resulting from a stabilized hindgut environment.

The encapsulation agent used to protect the sodium bicarbonate in the PSB is hydrogenated vegetable oil. Increased apparent absorption (intake - fecal excretion) of fat and sodium in the supplemented group indicates that some of the PSB was digested and absorbed although the site of absorption is unclear. That 29% of the extra Na+ and 45% of the extra fat contributed by the PSB was voided in the faeces suggests that a significant proportion of the PSB escaped digestion and absorption in the small intestine.

Sodium bicarbonate is sometimes administered as an alkalizing agent to racehorses in an attempt to enhance performance (Lawrence *et al.*, 1990) Levels of sodium bicarbonate used to produce a metabolic alkalosis generally range from 0.3-0.6 g/kg BW (Hodgson and Rose, 1994). The dose rate of PSB used in the current study contained 0.1 g $NaHCO_3$ /kg BW per feeding and this did not produce a metabolic alkalosis in the horses as evidenced by the lack of a significant treatment effect on blood pH, PCO_2 or HCO_3^- or tCO_2. This was probably due to a combination of the low dose rate and the protective coating which reduces the breakdown of sodium bicarbonate in the stomach and small intestine. Because of the purported performance enhancing effect of sodium bicarbonate, dose rates of sodium bicarbonate which result in blood tCO_2 levels of 37 mmol/l or higher are considered prohibited in most racing jurisdictions. The dose rate of PSB used in the present study did not significantly elevate tCO_2 and a dose rate of PSB containing 0.3 g HCO_3/kg BW/feeding did not produce prohibited levels of blood tCO_2 (Pagan, 2007: unpublished results).

Feeding PSB attenuated a decrease in fecal pH and increase in fecal lactate concentration compared to the control group. This suggests that feeding PBS may be effective in attenuating the HGA that can result from feeding high grain intakes to horses. More research is needed to evaluate how PSB supplementation affects intestinal epithelial health and integrity.

Acknowledgments

The authors would like to thank Delia Nash, Johanna Nicole, Dr. Jaye McCracken, Jade Clifford, Sharon Warner and Katrina Park for their assistance in conducting this study. Funding for this research was provided by Kentucky Equine Research, Inc. and Balchem Corporation.

References

Erdman, R.A., 1988. Dietary buffering requirements of the lactating dairy cow: A review. J. Dairy Sci. 71: 3246-3266.

Hodgson, D.R. and R.J. Rose, 1994. The Athletic Horse. W.B. Saunders Company, Philadelphia, PA, USA.

Hussein, H.S., L.A. Vogedes, G.C.J. Fernandez and R.L. Frankeny, 2004. Effects of cereal grain supplementation on apparent digestibility of nutrients and concentrations of fermentation end-products in the faeces and serum of horses consuming alfalfa cubes. J. Anim. Sci. 82: 1986-1996.

Johnson, K.G., J. Tyrell, J. Rowe and D.W. Pethick, 1998. Behavioural changes in stabled horses given nontherapeutic levels of virginiamycin. Equine vet J. 30(2): 139-43.

Lawrence, L., K. Kline, P. Miller-Graber, A. Siegel, E. Kurcz, M. Fisher and K. Bump, 1990. Effect of sodium bicarbonate on racing Standardbreds. J Anim Sci 1990. 68: 673-677.

Medina, M., I.D. Girard, E. Jacotot and V. Julliand, 2002. Effect of a preparation of *Saccharomyces cerevisiae* on microbial profiles and fermentation patterns in the large intestine of horses fed a high fibre or a high starch diet. J. Anim. Sci. 80: 2600-2609.

Milinovich, G.J., D.J. Trott, P.C. Burrell, E.L. Croser, R.A. Al Jassim, J.M. Morton, A.W. van Epps and C.C. Politt, 2007. Fluorescence *in situ* hybridization analysis of hindgut bacteria associated with the development of equine laminitis. Environ Microbiol 9: 2090-2100.

Owens, F.N., D.S. Secrist, W.J. Hill and D.R. Gill, 1998. Acidosis in cattle: a review. J. Anim. Sci., 76: 275-286.

Potter, G.D., F.F. Arnold, D.D. Householder, D.H. Hansen and K.M. Brown, 1992. Digestion of starch in the small or large intestine of the equine. Pferdeheilkunde 1: 107-111.

Schuback, K., B. Essén-Gustavsson and S.G. Persson, 2002. Effect of sodium bicarbonate administration on metabolic responses to maximal exercise. Equine Vet J Suppl. 2002 Sep;(34): 539-44.

Weiss, D.J., O.A. Evanson, B.T. Green and D.R. Brown, 2000. *In vitro* evaluation of intraluminal factors that may alter intestinal permeability in ponies with carbohydrate-induced laminitis. American Journal of Veterinary Research. 61: 858-61

Willard, J.G., J.C. Willard, S.A. Wolfram and J.P. Baker, 1977. Effect of diet on cecal pH and feeding behavior of horses. J. Anim. Sci. 45: 87.

Effect of collection time on the fermentative activity of equine faeces in the gas production technique

G. Desrousseaux[1,2], A.S. Santos[1], W.F. Pellikaan[2], J.W. Cone[2], L.M.M. Ferreira[1], C.M.V. Guedes[1] and M.A.M. Rodrigues[1]
[1]CECAV-UTAD, Animal Production Group, P.O. Box 1013, 5001-901 Vila Real, Portugal
[2]Animal Nutrition Group, Department of Animal Science, Wageningen University, Marijkeweg 40, 6709 PG, Wageningen, the Netherlands

Abstract

The present study intended to evaluate the influence of faecal collection time on the microbial activity of equine faecal inocula used in the gas production technique. Three adult horses, fed hay and concentrates 3 times a day, were used as faecal donors. Fermentation was carried out for 5 feedstuffs with faecal inocula obtained at four times: before the morning feed (t0) and 2 h (t2), 5 h (t5) and 8 h (t8) after the morning feed. Gas production curve fitting parameters and fermentation end-products were analyzed using the mixed procedure according to a 5×4 factorial design with feedstuff and collection time as main factors, whereas bacteriological counts were analyzed as a completely randomized design experiment using one-way ANOVA. Although there were no significant differences in pH and bacteriological counts, the proportions of propionate and butyrate and the extent of fermentation were the highest for t2 ($P \leq 0.05$). Moreover, there was an interaction ($P=0.0244$) for the asymptotic amount of gas evolved between the collection time and the type of feed. These results show that the collection time based on the feeding schedule of the donor horses had an influence on the fermentative activity of the faecal microflora.

Keywords: gas production technique, faeces, collection time

Introduction

The development of *in vitro* methodologies that are able to characterize the digestibility of feedstuffs and the availability of nutrients are important tools in animal nutrition studies. Originally developed for ruminants (Menke *et al.*, 1979), the cumulative gas production (GP) technique (Cone *et al.*, 1996) has been adapted to determine hind gut fermentation in simple-stomached animals such as poultry (Lan *et al.*, 2007), pigs (Van Laar *et al.*, 2002; Williams *et al.*, 2003; Cone *et al.*, 2005) and horses (Lowman *et al.*, 1996; Bush *et al.*, 2001; Zeyner *et al.*, 2005; Jansen *et al.*, 2007). The use of equine faeces may be used as a source of inoculum for *in vitro* fermentation studies on feedstuffs for equids (Lowman *et al.*, 1996) as an alternative to the use of rumen or caecal contents. The main objective of the present study was to evaluate the influence of collection time of feaces to be used as inocula on the subsequent fermentation of five different feedstuffs.

Material and methods

Three mature horses, fed with a standard diet (1.5% DM of BW) composed of grass hay (60% on DM basis) and commercial concentrate (40% on DM basis) 3 times a day, were used as faeces donors. Feedstuff (F) fermentation was studied after four faecal collection times (C): before the morning feed (9:00 h, t0), 2 hours (11:00 h, t2), 5 hours (14.00 h, t5) and 8 hours (17:00 h, t8) after the morning feed. Approximately 400 g of a proportionally mixed sample from all horses was mixed with 1,200 ml of a freshly-prepared CO_2-saturated nutritive medium (Menke *et al.* (1979) as modified by Steingass (1983)) at 39 °C under a constant CO_2 flow. Counts of *Streptococcus spp.*, *Lactobacillus spp.*, total anaerobic, cellulolyic and lactate-utilizing bacteria of the fresh faeces were conducted at each faecal collection time. Feedstuff samples (450 mg DM of barley, oats, meadow hay, wheat straw and alfalfa) were accurately weighed into 250 ml serum bottles and incubated with 60 ml filtered buffered faecal inoculum. There were two replicates bottles for each feedstuff per batch and three repetitions were performed for each collection time. After the incubation, pH was measured and a sample of fermentation fluid was stored at -20 °C for volatile fatty acids (VFA) analysis. VFA were determined by gas-liquid chromatography using pivalic acid as an internal standard. Gas production was recorded for 96 h using a fully automated system (Cone *et al.*, 1996). Gas curves were fitted by iteration using the Gompertz mono-phasic model as described by Bidlack and Buxton (1992) and Lavrenčič *et al.* (1997) as:

$$Y_t = B\left(e^{(-C)e^{-Dt}}\right) \quad (1)$$

where Yt = gas produced (ml/g OM) at time t (h); B = estimated asymptotic amount of gas produced (ml/g OM); C = specific gas production rate as affected by t and is governed by a constant D; D = constant factor describing the decay in gas production rate caused by the diminishing growth rate of microorganisms and increasing substrate limitation (Beuvink and Kogut, 1993). According to the Gompertz model, the fractional rate of gas production varies as a function of time and an average value (i.e. a = constant rate of gas production in h-1) can be calculated as:

$$a = D/C \quad (2)$$

Curve fitting parameter estimates and quantified fermentation end-product (volatile fatty acids: VFA) contents were analyzed using the mixed procedure according to a 5×4 factorial design with feedstuff and collection time as main factors whereas pH values were analyzed as a randomized design experiment using one-way ANOVA.

Table 1. Fermentation end-products and pH in fermentation fluids after 96 h incubation.

Parameters	pH	TVFA	VFA molar proportions (% of TVFA)				A:P
			HAc	HP	HB	HVal+*i*HVal	
Collection time (C)							
t0	6.66	3.12	41.3^{b}	34.7^{b}	14.2^{b}	9.75^{b}	1.205^{b}
t2	6.63	3.06	37.9^{a}	35.9^{c}	14.2^{b}	12.0^{c}	1.071^{a}
t5	6.63	2.97	41.6^{b}	33.4^{a}	13.5^{b}	11.5^{c}	1.253^{b}
t8	6.66	2.84	44.2^{c}	36.6^{c}	11.9^{a}	7.36^{a}	1.213^{b}
Pooled SEM	0.02	0.08	0.45	0.36	0.28	0.27	0.02
P-value	0.3868	0.0702	<0.0001	<0.0001	<0.0001	<0.0001	<0.0001
Feedstuff (F)							
Barley	6.51^{a}	3.45^{c}	38.7^{a}	39.0^{d}	13.7^{bc}	8.66^{a}	1.001^{a}
Oats	6.57^{b}	3.08^{b}	41.5^{b}	35.4^{c}	13.7^{bc}	9.35^{ab}	1.185^{b}
Alfalfa	6.77^{d}	2.84^{a}	40.7^{b}	32.8^{a}	14.1^{c}	12.4^{d}	1.245^{bc}
Meadow hay	6.67^{c}	2.74^{a}	41.9^{b}	34.9^{bc}	13.0^{ab}	10.3^{c}	1.204^{b}
Wheat straw	6.72^{cd}	2.87^{ab}	43.5c	33.7^{ab}	12.7^{a}	10.06^{bc}	1.292^{c}
Pooled SEM	0.02	0.09	0.51	0.41	0.31	0.31	0.02
P-value	<0.0001	<0.0001	<0.0001	<0.0001	0.0073	<0.0001	<0.0001
Interaction (C×F)							
P-value	0.9481	0.9365	0.0078	0.0075	0.1440	0.9558	0.0007

TVFA: sum of Acetic acid (HAc), Propionic acid (HP), *i*Butyric acid (*i*HB), Butyric acid (HB), *i*Valeric acid (*i*HVal) and Valeric acid (HVal), values in mM/g OM; HVal+*i*HVal: sum of valeric acid and *iso*valeric acid; A:P: Acetate:Propionate ratio; different superscripts indicate significantly different values within columns (P<0.05).

Results and discussion

Although there were no significant differences in pH values and TVFA, the proportion of propionic and butyric acid were highest at t2 while acetic was lowest (Table 1). In addition the extent of fermentation was greatest for t2 ($P \leq 0.05$) as shown in Table 1 and Table 2. At t2 most parameters for all feedstuffs studied were higher to those at other collection times (Table 2). Although t2 and t5 correspond to a collection time of two hours after a meal, the fact that significant differences exist between gas production profiles indicate that diurnal variation in faeces might be an important issue. Moreover there was an interaction ($P=0.0244$) for *B* between collection time and type of feed (Table 2). These results indicate that collection time based on the feeding schedule of donor horses has an influence on the fermentative activity of faeces. This is most likely to be explained by differences in microbial activities within the inocula at the different collection times, which is consistently reported in *in vitro* trials (Cone *et al.*, 2002; Mould *et al.*, 2005; Rymer *et al.*, 2005). Feeding schedules with 'fasting' periods where only a little fibre is available and then sudden introduction of concentrate meals may have induced bacterial shifts to more proteolytic and glucolytic types of fermentation respectively (De Fombelle *et al.*, 2003). This effect is reflected by the significant differences in VFA production and estimated amount of gas produced (*B)* depending on time of collection (De Fombelle *et al.*, 2001; Lattimer *et al.*, 2007).

Table 2. Parameter estimates of gas production profiles after incubation in buffered faecal fluid as affected by collection time (t0, t2, t5, t8) and type of feedstuff.

Parameters	B	C	D	a
Collection time (C)				
t0	245.8^{a}	3.2	0.14^{a}	0.04^{a}
t2	267.8^{b}	3.0	0.16^{b}	0.05^{b}
t5	247.9^{a}	3.6	0.14^{a}	0.04^{a}
t8	256.8^{ab}	3.2	0.14^{a}	0.04^{a}
SEM	5.15	0.19	0.005	0.003
P-value	0.0176	0.1265	0.0137	0.0091
Feedstuff (F)				
Barley	361.9^{d}	3.6^{b}	0.26^{d}	0.07^{c}
Oats	259.8^{c}	3.7^{b}	0.23^{c}	0.06^{c}
Alfalfa	188.7^{a}	2.7^{a}	0.13^{b}	0.05^{b}
Meadow hay	237.1^{b}	2.4^{a}	0.06^{a}	0.02^{a}
Wheat straw	225.3^{b}	3.7^{b}	0.06^{a}	0.02^{a}
SEM	5.76	0.21	0.005	0.003
P-value	<0.0001	<0.0001	<0.0001	<0.0001
Interaction (C×F)				
P-value	0.0244	0.4409	0.0735	0.6007

B: estimated asymptotic gas production (ml/g OM); *C*: specific rate of gas production as affected by t and governed by a constant *D*; *D*: constant factor describing the decay in gas production rate caused by the diminishing growth rate of microorganisms and increasing substrate limitation; *a*: constant rate of gas production in h-1; different superscripts indicate significantly different values within columns ($P<0.05$); SEM: standard error of the mean.

Conclusion

The present study showed that collection time may have an effect on the fermentative activity of equine faeces used as an inoculum source in *in vitro* studies. Standardization of the collection time should be undertaken. It may avoid errors, such as overestimating and underestimating of amounts of gas produced, VFA and the rate of feedstuff degradation, and allow a better reproducibility of *in vitro* feed digestibility studies.

References

Beuvink, J.M.W. and J. Kogut, 1993. Modeling gas production kinetics of grass silages incubated with buffered ruminal fluid. J. Anim. Sci. 71, 1041-1046.

Bidlack, J.E. and D.R. Buxton, 1992. Content and deposition rates of cellulose, hemicellulose, and lignin during regrowth of forage grasses and legumes. Can. J. Plant Sci. 72, 809-818.

Bush, J.A., D.E. Freeman, K.H. Kline, N.R. Merchen and G.C. Fahey Jr, 2001. Dietary fat supplementation effects on *in vitro* nutrient disappearance and *in vivo* nutrient intake and total tract digestibility by horses. J. Anim. Sci. 79, 232-239.

Cone, J.W., A.H. Van Gelder, G.J.W. Visscher and L. Oudshoorn, 1996. Influence of rumen fluid and substrate concentration on fermentation kinetics measured with a fully automated time related gas production apparatus. Anim. Feed Sci. Technol. 61, 113-128.

Cone, J.W., A.H. Van Gelder and H. Bachmann, 2002 Influence of inoculum source on gas production profiles. Anim. Feed Sci. Technol. 99, 221-231.

Cone, J.W., A.W. Jongbloed, A.H. Van Gelder and L. de Lange, 2005. Estimation of protein fermentation in the large intestine of pigs using a gas production technique. Anim. Feed Sci. Technol. 123-124, 463-472.

De Fombelle, A., V. Julliand, C. Drogoul and E. Jacotot, 2001. Feeding and microbial disorders in horses. 1. Effects of an abrupt incorporation of two levels of barley in a hay diet on microbial profile and activities. J. Equine Vet. Sci. 21, 439-445.

De Fombelle, A., M. Varloud., A.G. Goachet, E. Jacotot, C. Philippeau, C. Drogoul and V. Julliand, 2003. Characterization of the microbial and biochemical profile of the different segments of the digestive tract in horses given two distinct diets. Anim. Sci. 77, 293-304.

Jansen, W.L., J.W. Cone, S.N.J. Geelen, M.M. Sloet van Oldruitenborgh-Oosterbaan, A.H. Van Gelder, S.J.W.H. Oude Elferink and A.C. Beynen, 2007. High fat intake by ponies reduces both apparent digestibility of dietary cellulose and cellulose fermentation by faeces and isolated caecal and colonic contents. Anim. Feed Sci. Technol. 133, 298-308.

Lan, Y., B.A. Williams, M.W.A. Verstegen, R. Patterson and S. Tamminga, 2007. Soy oligosaccharides *in vitro* fermentation characteristics and its effect on caecal microorganisms of young broiler chickens. Anim. Feed Sci. Technol. 133, 286-297.

Lattimer, J.M., S.R. Cooper, D.W. Freeman and D.L. Lalman, 2007. Effect of yeast culture on *in vitro* fermentation of a high-concentrate or high-fiber diet using equine fecal inoculum in a Daisy II incubator. J. Anim. Sci 85, 2484-2491.

Lavrenčič, A., B. Stefanon and P. Susmel, 1997. An evaluation of the Gompertz model in degradability studies of forage chemical components. Anim. Sci. 64, 423-431.

Lowman, R.S., M.K. Theodorou, A.C. Longland and D. Cuddeford, 1996. A comparison of equine faeces or caecal digesta as sources of inoculum for *in vitro* fermentation studies using the pressure transducer technique. Anim. Sci. 62, 683A.

Menke, K., L. Raab, A. Salewski, H. Steingass, D. Fritz and W. Schneider, 1979. The estimation of the digestibility and metabolizable energy content of ruminant feedingstuffs from the gas production when they are incubated with rumen liquor *in vitro*. J. Agric. Sci. (Camb.) 93, 217-222.

Mould, F.L., K.E. Kliem, R. Morgan and R.M. Mauricio, 2005. *In vitro* microbial inoculum: A review of its function and properties. Anim. Feed Sci. Technol. 123-124, 31-50.

Rymer, C., J.A. Huntington, B.A. Williams and D.I. Givens, 2005. *In vitro* cumulative gas production techniques: History, methodological considerations and challenges. Anim. Feed Sci. Technol., 123-124, 9-30.

Steingass, H., 1983. Bestimmung des energetischen Futterwertes von wirtschaftseigenen Futtermitteln aus der Gasbildung bei der Pansenfermentation *in vitro*. Ph.D. Thesis. University of Hohenheim, Germany.

Van Laar, H., S. Tamminga, B.A. Williams, M.W.A. Verstegen and H.A. Schols, 2002. Fermentation characteristics of polysaccharide fractions extracted from the cell walls of maize endosperm. J. Sci. Food. Agric. 82, 1369-1375.

Williams, B.A., A. Awati, M.W. Bosch, S.R. Konstantinov and A.D.L. Akkermans, 2003. The effect of dietary fermentable carbohydrates on lactic acid concentration in small intestinal digesta of piglets at ten days after weaning. In: Proceedings of the British Society of Animal Science Winter Meeting, York, UK, p. 17.

Zeyner, A., B. Bosch, K.D. Markuske, M. Fürll and M. Krüger, 2005. Fermentation of selected substrates *in vitro* using a batch culture with equine faeces as inoculum. Pferdeheilkunde 21, 67-68.

Investigation of the buffering action and fermentation activity of Aquacid, when incubated *in vitro* with fibre and concentrate diets

M.J.S. Moore-Colyer
Royal Agricultural College Cirencester, Glos GL7 6JS, United Kingdom

Abstract

The foregut and hind gut buffering capacity of the supplement Aquacid was measured in two *in vitro* experiments, using fibre and concentrate diets. Aquacid buffered acidic conditions in the forgut, showing a significant (P<0.002) increase in pH from 2.137 for hay alone to 2.435 for hay + Aquacid. In *in vitro* hind gut conditions, Aquacid stimulated fermentation and increased lactate production by 1.87 mmol/litre, however, pH did not increase, indicating buffering activity. The results of these two experiments suggest that Aquacid may help in reducing acidity in the foregut and maintain hindgut pH despite an increase in lactate production.

Keywords: aquacid, buffer, foregut, hindgut

Introduction

Stomach ulcers and hindgut acidosis are debilitating disorders that affect all categories of performance horses. Many ant-acid preparations tend to have only a short-term foregut effect and rapidly lose their buffering capacity when passing through the acidic conditions of the stomach. These 2 experiments sought to determine if (a) the supplement Aquacid was an effective foregut buffer and (b) if any alterations occurred in any hind gut fermentation parameters after the supplement had undergone an *in vitro* foregut digestion procedure.

Materials and methods

Experiment 1

Two treatments hay ± Aquacid and hay alone were incubated for 6 hours in pepsin HCl solution at 37 °C to simulate stomach digestion. pH measurements were taken every hour for 6 hours totalling 12 readings. Differences between treatments were determined using Man Whitney U test (Genstat 12, Laws Agricultural Trust, 2009).

Experiment 2

- 16 replicate 125 ml serum bottles were prepared for each of the following 4 diets:
- 100 Releve (fibre mix)
- 100 Race-mix (cereal coarse mix)
- 70:30 Releve : Race mix
- 30:70 Releve : Race mix

Half the bottles i.e. 32 bottles had 0.02 g of aquacid added. The aquacid had undergone the following pre-digest treatment: 2-hour incubation (37 °C) with acid-pepsin, neutralised with sodium acetate buffer and then was incubated with pancreatin for a further 2 hours in order to simulate both stomach and small intestine digestion.

All 64 bottles (32 feed alone, 32 feed + Aquacid), were then incubated with an equine faecal inoculum using the *in vitro* gas production system of Theodorou *et al.* (1994).

Gas production measurements and dry matter losses were determined from 2 replicate bottles per treatment during a 68-hour incubation at 37 °C. Two replicate bottles per treatment were removed at three time points, of 6 hours, 12 hours and 18 hours. These bottles were then tested for lactate and pH levels. Differences between feeds were determined using analysis of variance (Genstat 12, Laws Agricultural Trust, 2009).

Results and discussion

Experiment 1

Hay alone had a mean pH of 2.137 which was significantly lower ($P<0.002$) than hay + Aquacid which had a mean pH of 2.435. These results indicate that Aquacid can exert some buffering action under highly acidic conditions that resemble pyloric stomach digestion for at least 6 hours.

Experiment 2

The pre-digestion performed on Aquacid did not prevent this supplement from altering hind gut fermentation parameters as 3 out of the 4 diets (except the 100% concentrate), containing the Aquacid supplement (Figure 1) showed a small increase in total gas production during the incubation.

Table 1 shows that the addition of Aquacid to the feeds significantly ($P<0.05$) increased the production of lactate by an average of 1.87 mmol/l. It is well documented than lactate accumulation drops the pH of the hind gut. However, no significant differences were noted between pH readings, 6.61 for feeds without Aquacid and 6.59 for feeds with Aquacid as indicated in Table 2.

Lactate concentration was significantly higher after 12 hours of incubation, which was >18 hours >6 hours. Lactate concentrations were also significantly ($P<0.05$) higher according to concentrate proportion with Race mix >30:70 >70:30 >100% Releve.

pH level across all feeds with and without Aquacid showed a significant drop in pH as incubation time increased. The feeds that contained a higher proportion of concentrates i.e. 30:70 and Race

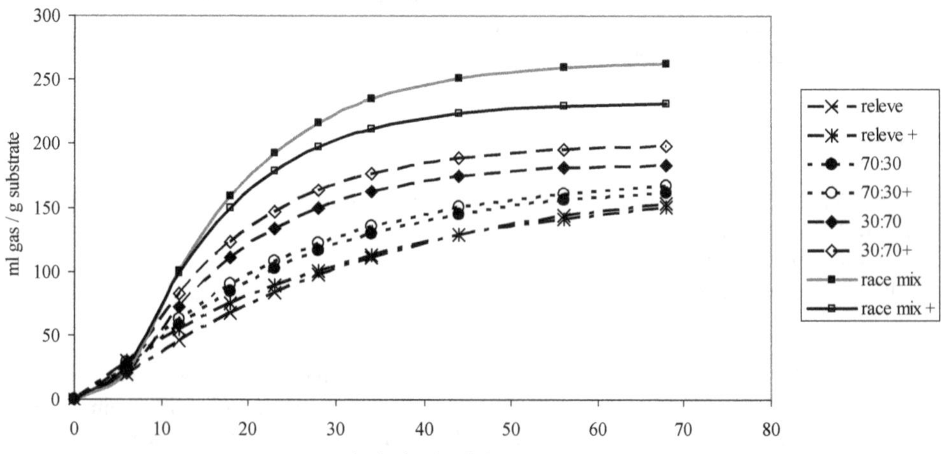

Figure 1. Mean cumulative gas production profiles for forage (100%) and forage:concentrate (70:30 and 30:70) and concentrate (100%) diets when incubated in vitro *with equine faecal inoculum with (+) or without Aquacid. Each value represents the mean of two bottles, while the line indicates the profile as described by the France* et al. *(1993) polynomial model.*

Table 1. Lactate concentrations (mmol/l) from 4 diets when incubated with an equine faecal inoculum for 6, 12 and 18 hours in the presence or absence of Aquacid.

Time	6 hours	12 hours	18 hours		s.e.d
All diets	9.60^{b}	10.24^{b}	7.61^{a}		0.917
Feed	100 Relev	70:30	30:70	100 Race mix	
	6.38^{a}	9.22^{b}	10.40^{b}	10.59^{b}	1.059
Treatment	No Aquacid	+ Aquacid			
	8.21^{a}	10.08^{b}			0.749

abcd Values in the same row with different superscripts are significantly (P<0.05) different; time represents lactate concentrations across all feeds and both treatments; feed shows lactate concentration for 4 feeds across the three time points and both treatments; treatment represents lactate concentrations with or without out Aquacid for four feeds across three time points.

Table 2. pH concentrations of 4 diets when incubated with an equine faecal inoculum for 6, 12 and 18 hours in the presence or absence of Aquacid.

Time	6 hours	12 hours	18 hours		s.e.d
all diets	6.772^{c}	6.555^{b}	6.491^{a}		0.0251
Feed	100 Relev	70:30	30:70	100 Race mix	
	6.693^{b}	6.635^{b}	6.570^{a}	6.527^{a}	0.0289
Treatment	No Aquacid	+ Aquacid			
	6.613	6.599			0.0205

abcd Values in the same row with different superscripts are significantly (P<0.05) different; time represents pH across all 4 feeds and both treatments; feed shows pH for 4 feeds across the three time points and both treatments; while treatment represents pH with or without out Aquacid for all four feeds across three time points.

mix had significantly lower pH than the more fibrous diets of 70:30 and 100 Relev. The addition of Aquacid did not significantly alter the pH of the four feeds across all three time points.

The results of these two experiments suggest that Aquacid may help in reducing acidity in the foregut. Moreover, despite the Aquacid stimulating an increase in lactate production the pH was maintained, indicating that Aquacid has an ability to maintain hind gut pH.

References

France, J., M.S. Dhanoa, M.K. Theodorou, S.J. Lister, D.R. Davies and D. Isac, 1993. A model to interpret gas accumulation profiles associated with *in vitro* degradation of ruminant foods. Journal of Theoretical Biology, 163: 99-111.

Genstat 12 Laws Agricultural Trust, 2009. Rothamstead Experimental Station, Harpendon, Hertfordshire, UK.

Theodorou, M.K., B.A. Williams, M.S. Dhanoa, A.B. Mc Allan and J. France, 1994. A simple gas production method using a pressure transducer to determine the fermentation kinetics of ruminant foods. Animal Food Science and technology, 48: 185-197.

Effect of three different forage-based diets on microbial flora, pH and viscosity of the equine hindgut

S. Muhonen[1], F.C. Wartena[2], A. Wesker[2] and V. Julliand[1]
[1]AgroSup Dijon, 26 Bd Dr Petitjean, BP87999, 21079 Dijon, France
[2]Wageningen University, P.O. Box 338, 6700 AH, Wageningen, the Netherlands

Abstract

The aim of this experiment was to investigate the effect of feeding various forage based diets differing in fibre composition on the microbial flora, digesta pH and digesta viscosity of the equine caecum, colon and faeces. Five geldings, fistulated in the caecum and right ventral colon, were randomised on three iso-energetic (13 MJ ME/100 kg BW and day) and iso-proteinaceous diets: (1) early harvested:late harvested grass haylage (80:20 ratio) (+small amount soybean meal) (diet G), (2) lucerne:late harvested grass haylage (80:20 ratio) (diet L) and 3) the more conventional diet of late harvested grass haylage:concentrate (oats, soybean meal) (65:35 ratio) (diet C) in a Latin-square design. Samples for microbial culturing, volatile fatty acids and pH measurements were taken from the caecum, colon and rectum after three weeks on each diet. Viscosity was measured on centrifuged caecal and colon fluid. Analysis of variance was performed. Total anaerobic and cellulolytic bacteria did not differ between diets. However, there was an effect of diet for pectinolytic (P=0.022) and xylanolytic (P=0.011) bacteria that had lower counts on diet G than diet C in both the caecum and colon. Lactate-utilizing bacteria were consistently higher (P=0.007) on diet C, which may be expected as this diet contained more starch, but amylolytic bacteria did not differ. The VFA ratio ((acetate+butyrate)/propionate) was lower on diet C compared to diet L (P=0.005) in the caecum and it was lower for diet C compared to diet L (P=0.013) and diet G (P=0.018) in the colon. Faecal pH was higher (P=0.012) on diet L. Viscosity did not differ between diets, but the colon data for one horse were very different from the other four horses which requires further investigation. In conclusion, the microbial flora in the equine hindgut was altered when feeding diets differing in fibre composition and maturity.

Keywords: haylage, lucerne, bacterial population, volatile fatty acids, caecum

Introduction

How different types of dietary fibre act in the large intestine depends on the extent to which they are digested (Stephen and Cummings, 1980). In horses it has been shown that forage maturity affects the extent of fibre digestion (Darlington and Hershberger, 1968). Fibre source and amount of fibre may affect the viscosity of intestinal contents (McRorie *et al.*, 2000). How differences in grass fibre composition between an early and late harvest or a legume forage affects the ecosystem of the equine hindgut is not well investigated. The aim of the study was to investigate the effect of feeding young grass haylage or lucerne haylage compared to the more conventional mature grass haylage:concentrate diet on the microbial flora and its activity, digesta pH and viscosity of the equine hindgut.

Material and methods

Five geldings, fistulated in caecum and right ventral colon, were randomised on three iso-energetic (13 MJ ME/100 kg BW and day) and iso-protein diets: (1) early harvested:late harvested grass haylage (80:20 ratio) (+small amount soybean meal) (diet G), (2) luzerne:late harvested grass haylage (80:20 ratio) (diet L) and 3) the more conventional diet of late harvested grass haylage:concentrate (oats, soybean meal) (65:35 ratio) (diet C) in a latin-square design. The horses were fed approximately 20% of the daily feed allowance at 08:00 h, 10:00 h and 16:00 h and 40% at 17:30 hours. The same ratios forage:forage or forage:concentrate were fed at all feeding occasions. Samples for microbial culturing, pH and viscosity measurements and volatile fatty acid (VFA) analysis were taken, at 12:00

hours, from caecum, colon and rectum after three weeks on each diet. Viscosity was measured on centrifuged caecum and colon fluid with a Micro-Ubbelohde Viscometer. The ANOVA was performed using the MIXED procedure (SAS Inst. Inc., Cary, NC). When significant main effects or significant treatment × segment interactions occurred, pair-wise t-tests were done to separate the main effect means. Differences were considered statistically significant at $P<0.05$.

Results and discussion

Total anaerobic and cellulolytic bacteria did not differ between diets (Table 1). However, there were effects of diet for pectinolytic ($P=0.022$) and xylanolytic ($P=0.011$) bacteria that had lower counts on diet G than diet C in the caecum and colon. On diet G the colon values for total anaerobic, xylanolytic and lactate-utilzing bacteria were lower than for the caecal values. This might be explained by the fact that diet G consisted of a young grass, which was probably highly digestible and to a large extent fermented early in the digestive tract. Lactate-utilizing bacteria were consistently higher ($P=0.007$) on diet C, which may be expected as this diet contained more starch (Medina *et al.*, 2002), but amylolytic bacteria did not differ. The microbial counts showed no interactions between treatment and segment of the large intestine.

Table 1. Microbial counts (log colony forming units/ml) of caecum and colon (right ventral) contents and faeces after 3 weeks of adaptation to forage based diets differing in fibre composition and maturity.

	Diet G[1]	Diet L[1]	Diet C[1]	SEM	*P*-values	
					Treatment	Segment
Total anaerobic bacteria						
Caecum	8.3^{A}	8.0	8.4	0.37	0.182	0.042
Colon	6.3^{B}	7.5	8.0			
Faeces	7.8^{A}	7.6	8.0			
Pectinolytic bacteria						
Caecum	7.2^{a}	7.3^{ab}	8.1^{b}	0.28	0.022	0.057
Colon	6.4^{a}	7.2^{ab}	7.2^{b}			
Faeces	7.4	7.2	7.8			
Xylanolytic bacteria						
Caecum	7.0^{aA}	7.2^{ab}	7.9^{b}	0.30	0.011	0.027
Colon	5.9^{aB}	6.8^{b}	7.0^{b}			
Faeces	7.2^{A}	7.1	7.7			
Cellulolytic bacteria						
Caecum	5.6	6.4	5.8	0.43	0.846	0.274
Colon	5.1	6.0	5.7			
Faeces	5.8	4.6	5.6			
Amylolytic bacteria						
Caecum	5.8	5.4	6.0	0.51	0.202	0.363
Colon	5.2	4.7	6.0			
Faeces	6.0	5.6	5.9			
Lactate-utilizing bacteria						
Caecum	6.6^{aA}	6.6^{a}	7.7^{bAB}	0.30	0.007	0.020
Colon	5.7^{aB}	6.4^{ab}	6.9^{bA}			
Faeces	6.8^{aA}	6.9^{a}	7.8^{bB}			

[1] Diet G: early harvested:late harvested grass haylage (80:20 ratio) (+small amount soybean meal), diet L: luzerne:late harvested grass haylage (80:20 ratio), diet C: late harvested grass haylage:concentrate (oats, soybean meal) (65:35 ratio); ^{ab} diet means within a row are different if superscripts differ ($P<0.05$); ^{AB} segment means within a parameter and column are different if superscripts differ ($P<0.05$).

The VFA's measured in the caecum, colon and faeces did not differ between diets (total VFA: caecum: diet G 60.9, diet L 62.1, diet C 57.3 mmol/l, colon: diet G 62.0, diet L 62.7, diet C 68.9 mmol/l, faeces: diet G 38.9, diet L 36.6, diet C 39,1 mmol/l, SEM = 5.95, *P*-value effect of feed = 0.959). However, the VFA ratio ((acetate+butyrate)/propionate) was lower on diet C compared to diet L (P=0.005) in the caecum and it was lower for diet C compared to diet L (P=0.013) and diet G (P=0.018) in the colon. A shift in VFA ratio can be expected when comparing a diet containing concentrate to all forage diets (De Fombelle *et al.*, 2003). Faecal pH was higher (P=0.012) on diet L and there was an interaction (P=0.002) between diet and digestive segment (Table 2). The higher faecal pH on diet L might be due to a higher buffering capacity of lucerne. Viscosity did not differ between diets (Table 2), but the colon data of one horse were very different from the other four horses which need further investigation. We have found no earlier references on viscosity of equine digesta.

In conclusion, the microbial flora in the equine hindgut was altered when feeding diets based on forages differing in fibre composition and maturity. How this might affect the water-holding capacity, fluid balance and the implications for the athletic horse have to be evaluated further. The importance of viscosity as a parameter for measuring changes in the hindgut of the horse needs further investigation.

Table 2. Caecal, colon (right ventral) and faecal pH and caecal and colon viscosity after 3 weeks of adaptation to forage based diets differing in fibre composition and maturity.

	Diet G[1]	Diet L[1]	Diet C[1]	SEM	*P*-values		
					Treatment	Segment	Treatment × segment
pH							
Caecum	7.0^{A}	7.0^{A}	7.0^{A}	0.08	0.012	0.027	0.002
Colon	7.0^{A}	7.1^{AB}	6.9^{AB}				
Faeces	6.5^{aB}	7.2^{bB}	6.7^{aB}				
Viscosity (mm^2/s)							
Caecum	1.1	1.1	1.1	0.10	0.771	0.092	0.823
Colon	1.2	1.2	1.3				

[1] Diet G: early harvested:late harvested grass haylage (80:20 ratio) (+small amount soybean meal), diet L: luzerne:late harvested grass haylage (80:20 ratio), diet C: late harvested grass haylage:concentrate (oats, soybean meal) (65:35 ratio); [ab] diet means within a row are different if superscripts differ (P<0.05); [AB] segment means within a parameter and column are different if superscripts differ (P<0.05).

References

Darlington, J.M. and T.V. Hershberger, 1968. Effect of forage maturity on digestibility, intake and nutritive value of alfalfa, timothy and orchardgrass by equine. J. Anim. Sci. 27, 1572-1576.

De Fombelle, A. M. Varloud, A.G. Goachet, E. Jacotot, C. Philippeau, C. Drogoul and V. Julliand, 2003. Characterization of the microbial and biochemical profile of the different segments of the digestive tract in horses given two distinct diets. Anim. Sci. 77, 293-304.

McRorie, J. S. Brown, R. Cooper, S. Givaruangsawat, D. Scruggs and G. Boring, 2000. Effects of dietary fibre and olestra on regional apparent viscosity and water content of digesta residue in porcine large intestine. Aliment. Pharmacol. Ther. 14, 471-477.

Medina, B. I.D. Girard, E. Jacotot and V. Julliand, 2002. Effect of a preparation of Saccharomyces cerevisiae on microbial profiles and fermentation patterns in the large intestine of horses fed a high fiber or a high starch diet. J. Anim. Sci. 80, 2600-2609.

Stephen, A.M. and J.H. Cummings, 1980. Mechanism of action of dietary fibre in the human colon. Nature 284, 283-284.

In vitro equine caecal fermentation of different casein levels

A.S. Santos[1], L.M.F. Ferreira[1], M. Cotovio[1], F. Silva[1], C.M.V. Guedes[1], J.W. Cone[2], R.J.B. Bessa[3] and M.A.M. Rodrigues[1]
[1]CECAV-UTAD, Animal Production Group, P.O. Box 1013, 5001-901 Vila Real, Portugal
[2]Animal Nutrition Group, Department of Animal Science, Wageningen University, Marijkeweg 40, 6709 PG, Wageningen, the Netherlands
[3]Interdisciplinary Centre of Research in Animal Health, Faculdade de Medicina Veterinária, Technical University of Lisbon, Lisboa, Portugal

Abstract

The effect of increasing protein nitrogen level from casein on fermentation parameters of caecal contents was evaluated using caecal contents from 3 cannulated horses. Four levels of protein nitrogen (3.7, 6.3, 12.5 or 25 mg of N in the form of casein) were studied using *in vitro* batch incubations, and the end-products of fermentation were evaluated. Results showed that the microbial caecal population might be capable of utilizing protein nitrogen. Furthermore different protein nitrogen concentrations may lead to different fermentation patterns.

Keywords: caecal fermentation, protein, in vitro *technique*

Introduction

The importance of equine hindgut fermentation has not yet been fully recognized. The extent to which hindgut bacteria utilize protein and non-protein nitrogen (NPN) is not well understood. Caecal bacteria show proteolytic activity and although caecal isolates were demonstrated to use ammonia and urea as nitrogen sources for microbial growth, many caecal bacteria require nitrogen sources other than ammonia or urea for growth (Maczulack *et al.*, 1985). This study aimed to evaluate the effects of increasing protein nitrogen (casein) levels to an N deficient incubation medium on fermentation parameters of caecal contents.

Material and methods

Caecal contents were collected from three horses (average BW 350 kg). Horses were fed to maintenance level according to INRA recommendations (INRA, 1990) with grass hay and concentrate (80:20). Caecal fluid was withdrawn 2 h after the morning meal into pre-warmed flasks, previously filled with CO_2, strained through four layers of cheesecloth and kept at 39 °C under CO_2. Incubations were performed with a modification of the procedure described by Cone *et al.*, (1996). The N-free buffer/mineral solution used contained (/l) 10.03 g $NaHCO_3$, 1.43 g Na_2HPO_4, 1.55 g KH_2PO_4, 0.15 g $MgSO_4{\cdot}7H_2O$, 0.52 g Na_2S, 0.017 g $CaCl_2.2H_2O$, 0.015 g $MnCl_2.4H_2O$, 0.002 g $CoCl_3.6H_2O$, 0.012 g $FeCl_3.6H_2O$ and 0.125 mg resazurin. To avoid a too high input of N, caecal fluid was diluted 1:10 with the buffer/mineral solution. To bind all N from the caecal fluid and to make N the limiting factor to fermentation, 10 g/l rapidly fermentable carbohydrates (glucose, 3.33 g/l; xylose, 3.33 g/l and soluble starch, 3.33 g/l) were added to the buffered caecal fluid and incubated at 39°C with continuous flushing of CO_2. During this incubation all available N from the caecal fluid was incorporated into bacterial N components (Cone *et al.*, 2009). After 2 h of incubation, 20 ml of the buffered caecal fluid, with rapidly fermentable carbohydrates, were added to incubation tubes containing 3.7, 6.3, 12.5 or 25 mg of N in form of casein. Duplicate samples were collected at six incubation times (0, 2, 4, 8, 12 and 24 h) and a duplicate repetition was performed. Samples were analyzed for pH, volatile fatty acids (VFA) and ammonia nitrogen ($N\text{-}NH_3$). Data were statistically analyzed using the mixed procedure of SAS according to a 4×6 factorial design with level (C) and sampling time (T) as main factors.

Results and discussion

All substrates and components were available in excess, with the exception of N, making N the limiting factor for microbial growth. Fermentation parameters are presented in Tables 1 and 2. Higher casein N levels increased accumulation of total VFA and decreased pH ($P<0.05$ and $P<0.001$ respectively) indicating higher microbial activity. These results are in accordance with Cone *et al.* (2009), who used rumen fluid in the gas production technique in an N depleted medium, reporting that gas production for casein was higher with higher amounts of casein. Ac:Pr decreased with casein level ($P<0.05$) indicating a higher increase in the proportion of propionate. This can be explained by higher propionic fermentation since there was an excess of rapidly fermentable carbohydrates, favoring the amylolytic microbial population. There was an interaction ($P<0.001$) between sampling time and casein level for VFA concentrations and also for Ac:Pr. This is mainly due to significant differences within the 24 h sampling time between the 12.5 and 25 mg casein level. It should be noted that for early sampling times (0, 2, 8 and 12 h) there was a similar behavior for all casein levels.

Values for N-NH3 (Table 2) varied with casein level ($P<0.001$), increasing until the 12.5 mg casein level, which can be in accordance with a higher amylolytic microbial activity, since proteolytic microorganisms tend to be amylolytic rather than cellulolytic (Siddons and Paradine, 1981). Nevertheless, the highest level of casein (25 mg casein N) showed a decrease ($P<0.05$) in N-NH3 concentrations, and this might explain the interaction between the two factors ($P<0.05$). Casein is a soluble N source, and therefore immediately available in the environment. Microorganisms have to breakdown casein into ammonia in order to incorporate it into microbial protein. Slowing down peptide breakdown would decrease the conversion of protein nitrogen to ammonia. Broderick and Wallace (1988) observed that peptides accumulate in the rumen of sheep fed casein as a protein supplement, but the same did not occur when a more slowly degradable protein (egg albumin) was used, concluding that rapidly hydrolysable proteins could lead to peptide accumulation. Wallace *et al.*

Table 1. VFA (mol/100 ml) and Acetate: Propionate ratio values (average ± sd) for in vitro *incubations of different casein N levels.*

	Acetate	Propionate	Butyrate	Total	Ac:Pr[1]
	mol/100 ml				
Casein N level (C)					
3.7	$0.38{\pm}0.05^{a}$	$0.15{\pm}0.04^{ab}$	$0.03{\pm}0.003^{a}$	$0.57{\pm}0.1^{a}$	$2.84{\pm}0.03^{ab}$
6.3	$0.39{\pm}0.05^{a}$	$0.13{\pm}0.04^{a}$	$0.04{\pm}0.003^{a}$	$0.55{\pm}0.1^{a}$	$2.91{\pm}0.03^{a}$
12.5	$0.51{\pm}0.05^{b}$	$0.25{\pm}0.04^{b}$	$0.07{\pm}0.003^{b}$	$0.83{\pm}0.1^{b}$	$2.79{\pm}0.03^{ab}$
25	$0.69{\pm}0.05^{c}$	$0.46{\pm}0.04^{c}$	$0.11{\pm}0.003^{c}$	$1.27{\pm}0.1^{c}$	$2.67{\pm}0.03^{b}$
Sampling time (T)					
0	$0.28{\pm}0.06^{a}$	$0.11{\pm}0.05^{a}$	$0.01{\pm}0.004^{a}$	$0.40{\pm}0.12^{a}$	$2.7{\pm}0.04^{a}$
2	$0.33{\pm}0.06^{a}$	$0.11{\pm}0.05^{a}$	$0.006{\pm}0.004^{a}$	$0.45{\pm}0.12^{a}$	$2.94{\pm}0.04^{abc}$
4	$0.33{\pm}0.06^{a}$	$0.13{\pm}0.05^{a}$	$0.005{\pm}0.004^{a}$	$0.46{\pm}0.12^{a}$	$2.78{\pm}0.04^{ab}$
8	$0.37{\pm}0.06^{b}$	$0.12{\pm}0.05^{a}$	$0.02{\pm}0.004^{ab}$	$0.51{\pm}0.12^{a}$	$3.02{\pm}0.04^{bc}$
12	$0.42{\pm}0.06^{c}$	$0.13{\pm}0.05^{a}$	$0.04{\pm}0.004^{b}$	$0.59{\pm}0.12^{a}$	$3.26{\pm}0.04^{c}$
24	$1.23{\pm}0.06^{d}$	$0.88{\pm}0.05^{b}$	$0.31{\pm}0.004^{c}$	$2.42{\pm}0.12^{b}$	$2.13{\pm}0.04^{d}$
Effects					
C	<0.05	<0.001	<0.001	<0.05	<0.05
T	<0.001	<0.001	<0.001	<0.001	<0.001
C*T	<0.001	<0.001	<0.001	<0.001	<0.001

[1] Ac:Pr, Acetate:Propionate ratio; values within columns with different superscripts are significantly different ($P<0.05$).

Table 2. Nitrogen ammonia (N-NH_3) concentration values (mg/100 ml) and pH values (average ± sd) for in vitro *incubations of different casein N levels.*

	N-NH_3 (mg/100 ml)	pH
Casein N level (C)		
3.7	$1.15±0.28^{a}$	$6.43±0.07^{b}$
6.3	$1.26±0.30^{b}$	$6.30±0.07^{ab}$
12.5	$1.42±0.27^{c}$	$6.22±0.07^{ab}$
25	$1.29±0.26^{b}$	$6.11±0.07^{a}$
Sampling time (T)		
0	$1.32±0.33^{ab}$	$6.31±0.09^{bc}$
2	$1.30±0.34^{abc}$	$6.56±0.09^{c}$
4	$1.16±0.34^{c}$	$6.55±0.09^{c}$
8	$1.20±0.33^{bc}$	$6.33±0.09^{bc}$
12	$1.38±0.33^{a}$	$6.10±0.09^{ab}$
24	$1.32±0.34^{ab}$	$5.73±0.09^{a}$
Effects		
C	<0.001	<0.05
T	<0.001	<0.001
C*T	<0.05	-

Values within columns with different superscripts are significantly different ($P<0.05$).

(1997) suggested that the rumen microbial ecology of the conversion of protein into ammonia should be viewed as a whole, and that oligopeptide breakdown can slow this process since there are very few species capable of this catabolic activity. With regard to the evolution of N-NH3 concentrations with time, values found at 0 h were equal as those found at 24 h (1.32 mg/100 ml), with intermediate values at 2, 4, 8 and 12 h. At early stages of incubation (0 and 2 h), higher values can be explained by an excess of ammonia released from casein during the initial incubation. Clearly, the release of N-NH3 is faster than the incorporation of N into microbial protein. Simialr findings were reported by Cone *et al.* (2005) upon the incubation of urea in an N-depleted faecal environment. The fact that N-NH3 concentrations tended to decrease and then increase, reaching higher values at 24 h can be due to microbial growth with time, and therefore a higher proteolytic activity.

Conclusion

Our results showed differences in VFA, N-NH3 and pH with casein level, and with sampling time, indicating that the microbial caecal population might be capable of utilizing protein nitrogen. Furthermore different protein nitrogen concentrations may lead to different fermentation patterns.

References

Broderick, G.A. and R.J. Wallace, 1988. Effects of dietary nitrogen source on concentrations of ammonia, free amino acids and fluroescamine-reactive peptides in the sheep rumen. J. Anim. Sci. 66, 2233-2238.

Cone, J.W., A.H. Van Gelder, G.J.W. Visscher and L. Oudshoorn, 1996. Influence of rumen fluid and substrate concentration on fermentation kinetics measured with a fully automated time related gas production apparatus. Anim. Feed Sci. Technol. 61, 113-128.

Cone, J.W., A.W. Jongbloed, A.H. Van Gelder and L. De Lange, 2005. Estimation of protein fermentation in the large intestine of pigs using a gas production technique. Anim. Feed Sci. Technol. 123-124, 463-472.

Cone, J.W., M.A.M. Rodrigues, C.M. Guedes and M.C. Blok, 2009. Comparison of protein fermentation characteristics in rumen fluid determined with the gas production technique and the nylon bag technique. Anim. Feed. Sci. Technol. 153, 28-38.

Martin-Rosset, W. (ed.), 1990. Alimentation des chevaux. INRA Editions, Paris, France.

Maczulack, A.E., K.A. Dawson and J.P. Baker, 1985. Nitrogen utilization in bacterial isolates from the equine cecum. Applied and Environmental Microbiology 50,1439-1443.

SAS, 1999. SAS/STAT User's Guide, Version 8. SAS Institute, Cary, NC, USA.

Siddons, R.C. and J. Paradine, 1981. Effect of diet on protein degrading activity in the sheep rumen. J. Sci. Food Agric., 32, 973-981.

Wallace, R.J., R. Onodera and M.A. Cotta, 1997. Metabolism of nitrogen-containing compounds. In: Hobson, P.N. and C.S. Stewart (eds.), The rumen microbial ecosystem, 2nd edition. Blackie Academic and Professional, London, United Kingdom, p. 283-328.

Individual effects on adaptation of equine intestinal microflora to fructans: indirectly measured by hydrogen exhalation test

A. Mößeler[1], I. Vervuert[1,2] and M. Coenen[1,2]
[1]Institute of Animal Nutrition, University of Veterinary Medicine Hannover Foundation, Bünteweg 2, 230559 Hannover, Germany
[2]Institute of Animal Nutrition, Nutrition Diseases and Dietetics, Faculty of Veterinary Medicine, University of Leipzig, 04159 Leipzig, Germany

Abstract

Laminitis can be experimentally induced in horses by administration of a single high dose of fructans but the question about individual reaction of horses consuming fructans with grass remains unclear. Six healthy trotter geldings were fed with Jerusalem artichoke (as a source of inulin-type fructan) at a dose of 1.5 g inulin/kg bodyweight per day in a single meal. Additionally the horses received hay (3 meals per day; about 1.5 kg/100 kg bw/day). To compare the uptake of fructans with other carbohydrates, grass meal was fed at the same dosage of 1.5 g hydrolysable carbohydrates/kg BW to all horses. Breath hydrogen and methane were measured on days 1, 3, 8 and 10 after first introduction of the fructan. Administration of Jerusalem artichoke resulted in a distinct rise of hydrogen exhalation (max.: 150±71 ppm, time max.: 339±85 min, area under curve: 52,434±24,460) in all horses, indicating a rapid fermentation in both the small and large intestine. Some horses (3 of 6) showed changes in hydrogen exhalation within the ten day period. The adaptation seemed to differ between the horses, as reduced (by nearly factor 6) as well as rising breath hydrogen concentrations and changes in methane exhalation were noticed in the course of time. On the other hand the exhalation pattern of hydrogen did not diffe; therefore there seems to be no shift in the localization of fructan fermentation in the course of the ten days lasting feeding trial. The variation of intestinal microfloral activity after inulin administration might be an interesting piece in the puzzle of understanding intestinal disorders and equine laminitis under the aspect of individual risk factors. Especially the finding of different modes of adaptation points out the complexity of the intestinal microflora.

Keywords: fructan, intestinal microflora, adaptation, exhalation

Introduction

Laminitis can be experimentally induced in horses by administration of a single high dose of fructans (Pollitt *et al.*, 2002). However the question about individual reaction of horses consuming fructans with grass remains unclear. In spring slow adaptation to grass is recommended to prevent laminitis (Meyer and Coenen, 2002) but there have been only few studies on adaptation of the equine intestinal microflora to fructans. Changes in the gut microflora can occur within a few hours when changes in substrate/feed are suddenly introduced (De Fombelle *et al.*, 2001), which may suggest that there may be a rapid adaptation to dietary fructan. While fistulation of animals is an invasive and expensive technique the hydrogen exhalation test can be performed easily, it is non-invasive and gives indirect information about the activity of the gastrointestinal microflora. A great advantage of this technique is that there are no influences of oxygen influx into the gut (as occurs in fistulated animals) on the intestinal microflora and offers the possibility of repeated measurements; therefore the activity of the intestinal microflora can be investigated after application of a test meal (kinetic within on day) and can also be useful for the monitoring of long-term adaptation processes (Coenen *et al.*, 2006).

Material and methods

Six healthy trotter geldings were fed with Jerusalem artichoke (as a source of inulin-type fructan) at a dose of 1.5 g inulin/kg bodyweight per day in a single meal. Additionally the horses received hay (3 meals per day; about 1.5 kg/100 kg bw/day). To compare the uptake of fructans with other carbohydrates, grass meal was fed at the same dosage of 1.5 g hydrolysable carbohydrates/kg BW to all horses. Before starting the exhalation sampling period the horses had no access to feed for 12 hours and during the test (lasting 10 hours) no hay was fed. Breath hydrogen and methane were measured on days 1, 3, 8 and 10 after first introduction of the fructan. Breath hydrogen and methane were measured every 0.5 hours over a 10 hour period on each of these days. Feed was withdrawn from horses 12 hours prior to starting breath hydrogen and methane measurements, and no hay was given during the ten hour experimental period. The breath samples were collected by using a tight fitting face mask at the end of exhalation and analyzed by gas chromatography. Breath hydrogen and methane were regarded as indirect indicators of gastrointestinal microbial activity. On day 3 of each trial the animals were given 500 g of a 50% glucose-mixture orally to use glucose as a marker of small intestine transit time. On that day blood samples were taken at the same time as breath samples to determine gastric emptying indirectly using plasma glucose levels as an indicator.

Results

Administration of Jerusalem artichoke resulted in a distinct rise of hydrogen exhalation (max.: 150±71 ppm, time max.: 339±85 min, AUC: 52,434±24,460) in all horses, indicating a rapid fermentation in both the small and large intestine. Small intestinal transit time (determined by glucose application) was 140±41 min (application of Jersualem Artichoke) and 150±67 min (grass meal). Comparing the exhalation pattern observed after feeding Jerusalem artichoke with those measured after feeding grass meal the much higher hydrogen and lower methane concentration in exhaled breath becomes obvious.

Some horses (3 of 6) showed changes in hydrogen exhalation within the ten day period. The adaptation seemed to differ between the horses, as reduced (by nearly factor 6) as well as rising breath hydrogen concentrations and changes in methane exhalation were noticed in the course of time (see Figures 1 to 5). On the other hand the exhalation pattern of hydrogen did not differ; therefore there seems to be no shift in the localization of fructan fermentation in the course of the ten days lasting feeding trial.

Discussion

The variation of intestinal microfloral activity after inulin administration might be an interesting piece in the puzzle of understanding intestinal disorders and equine laminitis under the aspect of individual risk factors. Especially the finding of different modes of adaptation (decrease of hydrogen exhalation with parallel increase of methane indicating an increase in methanogenic and eventually of hydrogen producing microorganisms, while in one horse rising hydrogen levels were found in

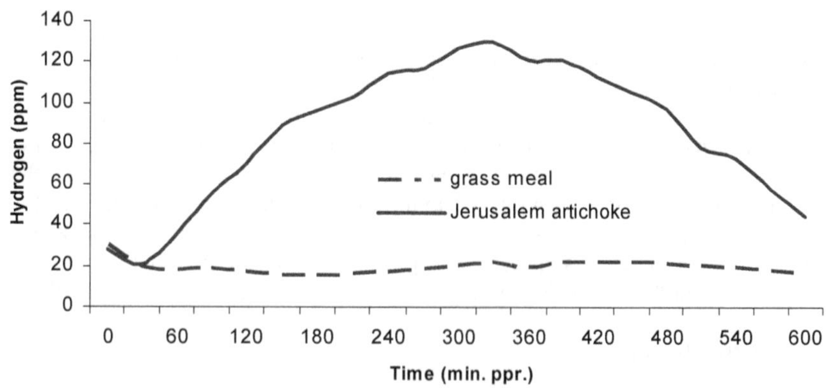

Figure 1. Mean hydrogen concentration in exhaled breath of six horses (mean of day 1, 3, 8 and 10) after intake of grass meal or Jerusalem artichoke.

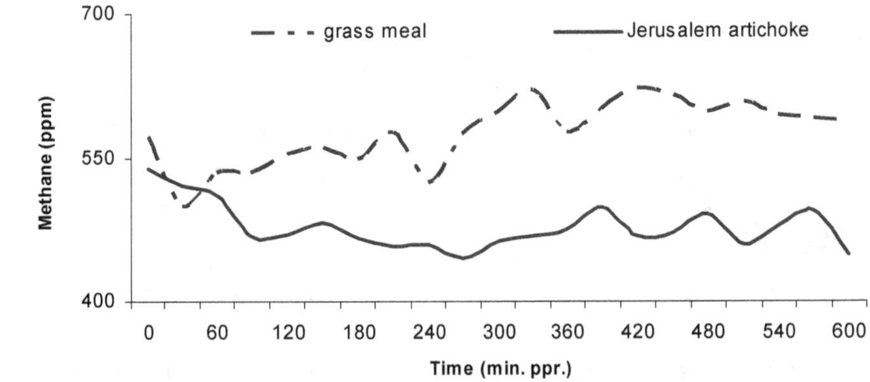

Figure 2. Mean methane concentration in exhaled breath of six horses (mean of day 1, 3, 8 and 10) after intake of grass meal or Jerusalem artichoke.

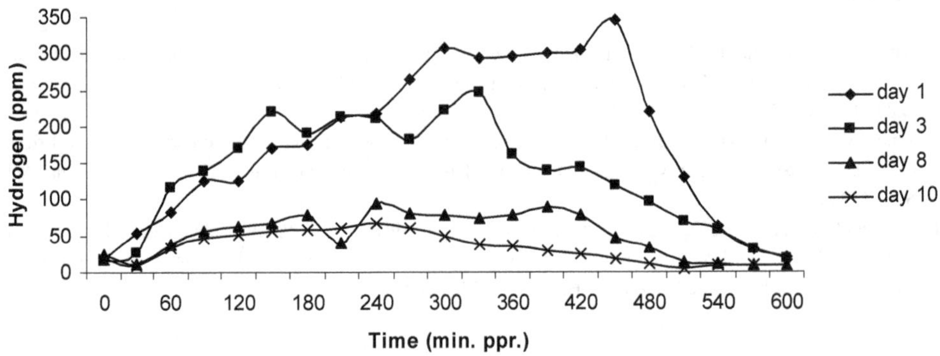

Figure 3. Hydrogen concentration (ppm) in exhaled breath of horse 3 after oral intake of Jerusalem artichoke during course of trial (on day 3 glucose was given additionally).

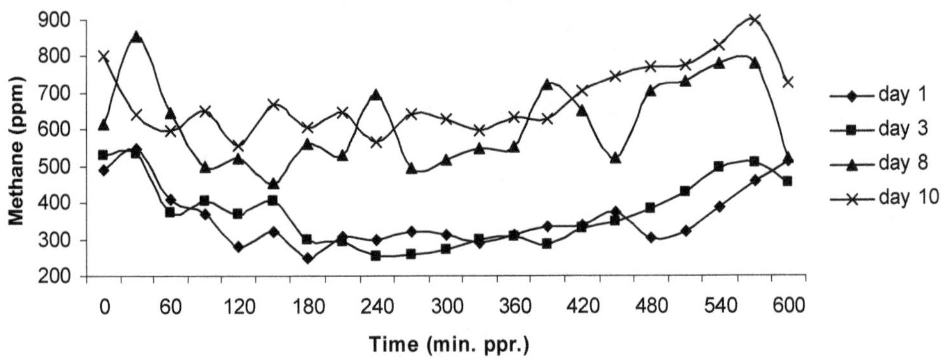

Figure 4. Methane concentration (ppm) in exhaled breath of horse 3 after oral intake of Jerusalem artichoke during course of trial (on day 3 glucose was given additionally).

the course of time with no appreciable effects on methane exhalation) points out the complexity of the intestinal microflora. But the meaning of these findings and possible consequences still remain unclear. In further studies horses or ponies that has been prone to laminitis should be used.

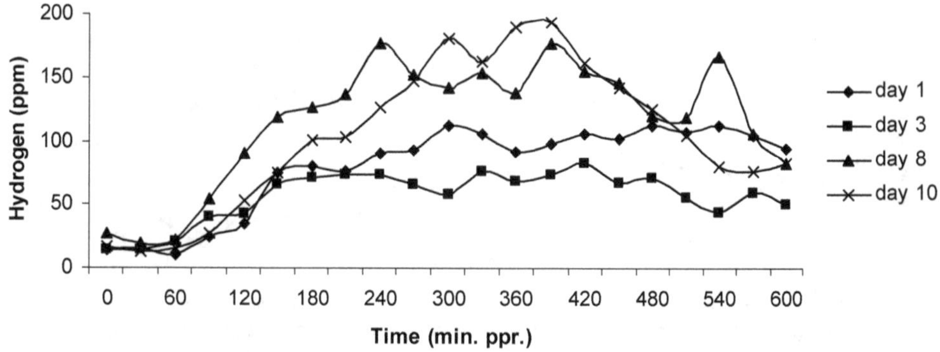

Figure 5. Hydrogen concentration (ppm) in exhaled breath of horse 6 after oral intake of Jerusalem artichoke during course of trial (on day 3 glucose was given additionally).

References

Coenen, M., A. Moesseler and I. Vervuert, 2006. Fermentative Gases in Breath Indicate that Inulin and Starch Start to Be Degraded by Microbial Fermentation in the Stomach and Small Intestine of the Horse in Contrast to Pectin and Cellulose. J. Nutr. 136: 2108-2110.

De Fombelle, A., E.V. Julliand, C. Drogoul and E. Jacotot, 2001. Feeding and microbial disorders in horses: 1: Effects of an abrupt incorporation of two levels of barley in a hay diet on microbial profile and activities. J. Equine Vet. Sci. 21: 439-445.

Meyer, H. and M. Coenen, 2002. Pferdefütterung. 4. erw. und aktualisierte Aufl. Parey Buchverlag, Berlin, Germany.

Pollitt, C.C. and A.W. van Eps, 2002. Equine laminitis; a new induction model based on alimentary overload with fructan. Proc. Austr. Equine Vet. Assoc., Bain-Fallon Memorial Lectures.

A comparative study of the apparent total tract digestibility of fibre in Icelandic and Danish Warmblood horses

R.B. Jensen[1], C. Brøkner[1], K.E. Bach Knudsen[2] and A.H. Tauson[1]
[1]Department of Basic Animal and Veterinary Sciences, Faculty of Life Sciences, University of Copenhagen, Bülowsvej 17, 1870 Frederiksberg, Copenhagen, Denmark
[2]Department of Animal Health and Bioscience, Faculty of Agricultural Sciences, University of Aarhus, Blichers Allé, P.O. Box 50 8830 Tjele, Denmark

Abstract

Four Icelandic (ICE) and four Danish Warmblood (DW) horses were used in a crossover study with two treatments to investigate the effect of breed and the effect of stage of maturity of haylage on the apparent total tract digestibility (ATTD) of a diet consisting of sugar beet pulp, black oats and early or late cut haylage. Fibre was analyzed as crude fibre (CF), acid detergent fibre (ADF), neutral detergent fibre (NDF) and dietary fibre (DF: non-starch polysaccharides (NSP) plus lignin). In haylage all analyzed fibre fractions increased with advancing stage of maturity, with the cell wall components cellulose, non-cellulose polysaccharides (NCP) and lignin causing this increase. Feeding early cut haylage resulted in a significantly ($P<0.05$) higher ATTD of NDF, cellulose, NCP, total NSP and DF compared to feeding late cut haylage. There was a significantly ($P<0.05$) higher ATTD of CF and DF in ICE than in DW. The DF analysis method gave the most appropriate description of the fibre fraction. The results suggested that ICE had higher ATTD of DF than DW, and this was caused by a tendency for a higher ATTD of cellulose.

Keywords: Icelandic horses, haylage, dietary fibre, digestibility

Introduction

Icelandic horses and ponies are commonly referred to as 'easy keepers', i.e. easy to keep in a good body condition. A tendency for a higher crude fibre (CF) (Slade and Hintz, 1969) and neutral detergent fibre (NDF) (Udén and Van Soest, 1982) digestibility has been found in ponies compared to other breeds and this might be a characteristic for the 'easy keeper'. However, others have not been able to detect any differences in the NDF digestibility between ponies and horses (Cuddeford *et al.*, 1995; Vermorel *et al.*, 1997). The dietary fibre analysis method gives a more complete description of the fibre components; hence, enables a more detailed description of the fibre fraction of feedstuffs and their digestibility (Bach Knudsen, 2001). The aim of the present study was to measure the apparent total tract digestibility (ATTD) of fibre in Icelandic and Danish Warmblood horses where traditional fibre analysis were extended with dietary fibre analysis.

Material and methods

Four Icelandic (ICE) and four Danish Warmblood (DW) horses were used to investigate the effect of breed and the effect of stage of maturity of haylage on the ATTD of a diet consisting of sugar beet pulp (SBP) black oats and early or late cut haylage. The experiment was designed as a crossover study consisting of two experimental periods. Each 21 day period consisted of 17 days of adaption to a diet and four consecutive days of total collection of faeces. The horses were housed in stalls with rubber mats and faecal collection was obtained from frequent collection of faeces from the rubber mats.

The feed rations were balanced to provide the same amount (14.5 g) of dry matter (DM)/kg body weight (BW) on each diet, with a 80:20 forage to concentrate ratio. The daily ration of haylage was divided into 5 meals and the concentrate, consisting of SBP and black oats, was fed twice daily. Two different 1st cut haylages from 2 different fields sown with the same grass mix (Crop-mix number 50,

Hunsballe Frø A/S, Holstebro, Denmark) consisting of 65% rye grass, 20% Meadow-grass and 15% timothy were cut on 25th of May and on 14th of June. The crops were cut with a mower conditioner and left for wilting until they were baled two days later. Cut crops were tedded twice during wilting, and the bales were wrapped immediately after baling with 8 layers of stretch film with 50% overlap. The DM % of the early and late cut haylages were 65 and 52% DM, respectively.

Detailed information on all of the chemical analysis can be found elsewhere (Jensen, 2009), and only the fibre analyses are presented here. Fibre in feedstuffs and faeces were analyzed as CF, acid detergent (ADF), NDF and dietary fibre (DF). The DF analysis determines cellulose, non-cellulose polysaccharides (NCP), total NSP (cellulose + NCP) and DF (total NSP + lignin).

All statistical analyses were performed using the general linear models procedure (GLM) in SAS® (Version 9.1, SAS Institute Inc. Cary, North Carolina, USA) according to the following model, where Y denotes the response variable analyzed: $Y_{ijk} = \mu$ + fixed effect of diet_i + fixed effect of breed_j + random effect of horse within breed_k + residual error_{ijk}. Results are presented as least square means (LS means) with root mean square error (RMSE) as measure of variance. Effects were considered significant if $P<0.05$ and a tendency if $P<0.10$.

Results and discussion

In haylage all analyzed fibre fractions increased with advancing stage of maturity in agreement with Ragnarsson (2009), the cell wall components cellulose, NCP and lignin causing this increase (Table 1). Feeding early cut haylage resulted in a significantly ($P<0.05$) higher ATTD of NDF, cellulose, NCP, total NSP and DF, than feeding late cut haylage (Table 2).

There was a significantly ($P<0.05$) higher ATTD of CF and DF in ICE than in DW, and a tendency for a higher ATTD of ADF, cellulose and total NSP in ICE than in DW (Table 2). As cellulose contributes to a major part of the fibre components in CF and ADF, and is a part of total NSP, cellulose may well have caused the differences in this study. No breed difference was found in the ATTD of NCP and NDF. A major part of the NCP fraction in grass consists of hemicelluloses, and hemicelluloses form a substantial proportion of NDF. This might explain the lack of any effect of the ATTD of NDF. When the DF analysis is applied to digestibility studies, detailed information on the fermentation of the different fibre fractions can be obtained, rendering this method of fibre analysis superior to traditional, methods of fibre analysis.

Table 1. Fibre composition (g/kg DM) of black oats, sugar beet pulp (SBP), early and late cut haylage.

	Oats		SBP		Early cut haylage		Late cut haylage	
Fibre components								
Cellulose	70		175		184		229	
NCP	172	(42)[a]	422	(250)[a]	199	(26)[a]	252	(24)[a]
Total NSP	243		597		383		480	
Klason lignin	57		34		67		106	
DF	299		631		450		587	
NDF	259		365		455		608	
ADF	103		232		254		342	
CF	105		170		240		321	

[a] Soluble non-cellulose polysaccharides (g/kg DM).

Table 2. The apparent total tract digestibility (%) of fibre of the experimental diets consisting of black oats, sugar beet pulp and either early (diet 1) or late (diet 2) cut haylage fed to Icelandic (ICE) and Danish Warmblood horses (DW).

	Diet		Breed		RMSE[1]	*P* values		
	Diet 1	Diet 2	DW	ICE		Diet	Breed	Horse
Cellulose	60	50	53	57	3.98	0.002	0.06	0.02
NCP	71	61	65	67	2.48	<0.001	0.15	0.01
Total-NSP	66	56	59	62	3.05	<0.001	0.08	0.01
DF	55	46	48	53	3.38	<0.001	0.03	0.01
NDF	60	51	55	56	3.11	0.001	0.29	0.01
ADF	51	46	46	51	4.31	0.06	0.08	0.01
CF	52	47	47	52	4.28	0.07	0.05	0.01

[1] Root mean square error; ADF: acid detergent; NCP: non-cellulose polysaccharides; NDF: neutral ditergent fibre; CF: crude fibre; DF: dietary fibre; NSP: non-starch polysaccharides.

The results suggest that ICE had higher ATTD of DF than DW, and this was probably due to a tendency for a higher ATTD of cellulose. Ragnarsson (2009) found no effect of breed on the ATTD of NDF, when comparing Standardbreds (STB) to ICE. However, STB had a significantly higher ATTD of organic matter than ICE. More research is needed to clarify breed differences in ATTD between ICE and other breeds.

References

Bach Knudsen, K.E, 2001. The nutritional significance of 'dietary fibre' analysis. Anim. Feed. Sci. Technol. 90(1-2), 3-20.

Cuddeford, D., R.A. Pearson, R.F. Archibald and R.H. Muirhead, 1995. Digestibility and gastro-intestinal transit time of diets containing different proportions of alfalfa and oat straw given to Thoroughbreds, Shetland ponies, Highland ponies and donkeys. Anim. Sci. 61, 407-417.

Jensen, R.B, 2009. Carbohydrates in equine nutrition. MSc Thesis. Copenhagen. University of Copenhagen, Faculty of Life Sciences.

Ragnarsson, S., 2009. Digestibility and metabolism in Icelandic horses fed forage-only diets. Doctoral Thesis. Uppsala, Swedish University of Agricultural Sciences.

Slade, L.M. and H.F. Hintz, 1969. Comparison of Digestion in Horses, Ponies, Rabbits and Guinea Pigs. J. Anim. Sci. 28(6), 842-843.

Udén, P. and P.J. Van Soest, 1982. Comparative digestion of timothy (*Phleum paratense*) fibre by ruminants, equines and rabbits. Br. J. Nutr. 47, 267-272.

Vermorel M., J. Vernet and W. Martin-Rosset, 1997. Digestive and energy utilisation of two diets by ponies and horses. Livest Prod Sci. 51(1-3), 13-19.

The effect of sugar beet pulp on caecal pH in Norwegian cold-blooded trotter horses

C. Brøkner[1], D. Austbø[2], J.A. Næsset[2], K.E. Bach Knudsen[3] and A.H. Tauson[1]
[1]University of Copenhagen, Faculty of Life Sciences, Department of Basic Animal and Veterinary Science, Grønnegaardsvej 3, 1870 Frederiksberg C, Denmark
[2]Department of Animal and Aquacultural Sciences, Norwegian University of Life Sciences, Box 5003, 1432 Ås, Norway
[3]Department of Animal Health and Bioscience, Faculty of Agricultural Sciences, University of Århus, Blichers Allé 20, 8830 Tjele, Denmark

Abstract

It was hypothesised that whole barley would lower caecal pH more than whole barley fed in combination with Betfor®. Four geldings were fed two concentrate diets: 2.06 kg whole barley + 300 g DM Betfor® (BB) and 2.06 kg whole barley (B). Timothy hay was fed as roughages in both diets. A pH electrode was inserted through a caecal cannula and pH was recorded. There was an overall tendency (P=0.07) to a dietary effect of Betfor® in terms of maintaining a more favourable caecal pH. Therefore, in conclusion, adding Betfor® to a whole barley meal prevented a drop in post-prandial caecal pH.

Keywords: caecal, pH, beet fibre

Introduction

Feeding 2 g barley starch per kg body weight (BW) in one meal has been shown to lower the caecal pH from around 6.9-7 to 6.1 (Austbø, 2005), a potentially harmful level. However, feeding whole barley in combination with molassed sugar beet pulp (Betfor®) did not result in decreased caecal pH. This experiment aimed at quantifying the effect of Betfor® on caecal environment. It was hypothesised that whole barley would lower caecal pH more than whole barley fed in combination with Betfor®.

Material and methods

Experimental design

A study with four horses and two diets was conducted in order to determine the effect of Betfor® on caecal pH. A change over design with a sequence of 17 days adaptation to hay and whole barley mixed with Betfor® was followed by 2 consecutive sampling days, where after Betfor® was removed from the diet. After one washout day, sampling was repeated for 2 consecutive days.

Animals

Four geldings of Norwegian Cold-blooded Trotter (age 5-16 years) with an average BW of 544 kg and a mean body condition score of 6 on a scale from 1 to 9 were used. The horses were fitted with a permanent ceacal cannula close to the ileo-caecal junction. The horses were housed under the same conditions in individual 3x3 m stalls on wood shavings during the whole experimental period.

Diets

The two diets consisted of 2.06 kg whole barley (equivalent to 2 g starch per kg BW) (B) or 2.06 kg whole barley mixed with 300 g dry Betfor® (BB). The Betfor® was soaked in 1 L water before mixing with barley. All horses were fed 1.7 kg timothy hay as roughage and had free access to water. The concentrate meals were fed at Time 0 and hay was fed 2 ½ h later. The remaining 5.1 kg hay, 600 g dry Betfor® and 150 g Multivitamins were divided into two meals and fed after ended sampling in the afternoon and before the night.

Sampling

A pH electrode, attached to a logger, was inserted into the caecum through the cannula and pH was recorded at 1 min intervals for 9 hours, starting at Time 0. The pH meters were cleaned and calibrated daily before the morning meal.

Statistical analyses

Area under the curve (AUC) was calculated for caecal pH and evaluated by analysis of variance using the PROC GLM procedure in SAS version 9.2 with feed, dietary treatment and horses as fixed effects in the model. A *P*-value less than 0.05 was considered significant. The model used was:

$$y_{ijk} = \mu + \alpha_i + \beta_j + \gamma_k + \varepsilon_{ijk} \quad (1)$$

where y is response variable (i.e. pH recordings), α is diet, β is day (within diet), γ is horses, ε is residuals; i = 1, 2; j = 1, 2; k = 1-4.

Results and discussion

Whole barley lowered caecal pH but the decrease was smaller when barley was fed in combination with Betfor® (Figure 1). The first 90 min. of sampling, the caecal pH oscillated around 7.1 when horses were fed whole barley in combination with Betfor®, whereas whole barley alone resulted in a caecal pH around 6.9. The lowest pH was recorded 6 hours postprandially, after which caecal pH began to rise again. The lowest pH was 6.2 for B and 6.5 for BB. There was an overall tendency (P=0.07) to a dietary effect on lowering caecal pH (Table 1).

In the present experiment, the starting point for caecal pH differed from Brøkner *et al.* (unpublished), who reported a starting caecal pH around 6.7 regardless of diet. However, Austbø (2005) reported a caecal pH around 6.9-7 in horses fed either timothy hay or pelleted barley, using the same horses

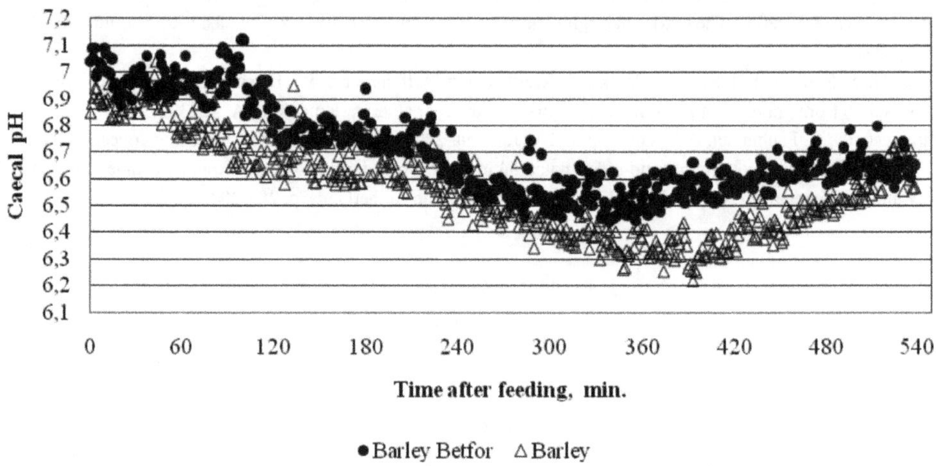

Figure 1. The effect of feeding whole barley and whole barley in combination with Betfor® on caecal pH (mean of 4 horses). The concentrate meals were fed at time 0 and hay at time 150.

Table 1. Dietary effect on caecal pH as evaluated by area under curve (AUC).

Diets	AUC
Barley/Betfor®	3,621
Barley	3,551
P-value	0.07

and method. Thirteen hours prior to the morning meal and the beginning of sampling, the horses were fed either timothy hay or timothy hay with Betfor® which either indicates a long term effect of Betfor® on caecal environment, or individual variations among horses. In agreement with the present experiment Austbø (2005) reported the lowest caecal pH (6.1) 5 hours postprandially. The 1 hour difference could be related to the delay in arrival of digesta in caecum perhaps due to eating rate, different meal size or larger surface area, which facilitated a more rapid starch fermentation of pelleted barley as compared to whole barley.

The effect of whole barley on the caecal environment was most likely due to low small intestinal digestibility of barley starch (Meyer *et al.*, 1995). Betfor® is high in dietary fibre (Brøkner *et al.*, 2009), which has been shown to have nutritional significance due to physiochemical properties such as cation exchange capacity and prolonged intestinal passage time due to the high water binding capacity of beet fibre (Bach Knudsen, 2001). Philippeau *et al.* (2009) reported that the redox potential was higher (more positive) for a fibre-based concentrate as compared to ground barley. Fibre therefore potentially can have a protecting effect on caecal pH by binding protons resulting from starch fermentation and thereby indirectly preventing a lowered caecal pH.

In conclusion, adding Betfor® to a whole barley meal appeared to reduce the drop in post-prandial caecal pH. Dietary fibre with properties like Betfor® may therefore potentially be used to contribute to maintaining equine gut health. However, the results are preliminary and the work needs to be repeated with more horses.

References

Austbø, D., 2005. pH-verdien i blindtarmen hos hest, effekt av rasjon og tid. Husdyrforsøksmøtet, Norge.

Bach Knudsen, K.E., 2001. The nutritional significance of 'dietary fibre' analysis. Animal feed science and technology. 90. 3-20.

Meyer, H., S. Radicke, E. Kienzle, S. Wilke, D. Kleffken and M. Illenseer, 1995. Investigations on preileal digestion of starch from grain, potato and manioc in horses. Journal of veterinary medicine A, 42, 371-381.

Philippeau, C., C. Faubladier, A.G. Goachet and V. Julliand, 2009. Is there an impact of feeding concentrate before or after forages on colonic pH and redox potential in horses. In: Lindner, A. (ed.), Applied equine nutrition and training. Wageningen Academic Publishers, Wageningen, the Netherlands, p. 203-208.

Part 5. Nutrition and metabolic disease

Recent advances in the understanding of laminitis and obesity

P.A. Harris[1] *and R. Geor*[2]
[1]*Equine Studies Group, WALTHAM Centre for Pet Nutrition, Leicester, Freeby Lane, Waltham-on-the-Wolds, Leicestershire, LE14 4RT, United Kingdom*
[2]*Department of Large Animal Clinical Sciences, College of Veterinary Medicine, Michigan State University, G-100 Veterinary Medical Center, East Lansing, MI 48824-1314, USA*

Abstract

Laminitis, insulin resistance and obesity are terms that are now frequently heard at equine scientific and lay conferences. This review provides an historical background to the various 'laminitis' theories and discusses some of the more recent research that has been published in this area.

Keywords: metabolic syndrome, insulin, pathophysiology, weight loss, fructose

Historical background

It is perhaps difficult to find a 'lay' horse magazine that does not discuss laminitis or obesity at some point during the year. Certainly general interest in these conditions has increased recently, due to the apparent 'obesity epidemic' in humans and horses plus the postulated link between equine obesity and an increased risk of laminitis. Laminitis, however, is not a 'modern' condition; it has been recognised for well over 2000 years. Aristotle referred to it around 350 BC as 'Barley Disease' presumably because it was associated even then with excessive consumption of grain. This cause was confirmed in more recent times by Garner and colleagues in the mid 1970s (Garner *et al.*, 1975) when they demonstrated that the administration of a large bolus of starch could trigger episodes of laminitis. The first full description or review was possibly commissioned by the Greek Emperor Constantine in the 4th century AD, although it was not published until 900 AD. This noted, as we do today, that the condition (referred to as 'gout as well as barley disease') could have multiple causes. In this early review they included as possible causes travelling on hard surfaces, overeating and drinking too much cold water when overheated. Interestingly, considering today's link between obesity and laminitis – this review apparently recommended, in addition to mild bleeding, exercise and diet restriction as part of the treatment/management regimen.

The term 'founder' was in use by at least the end of the 16th century and apparently arose from the term 'Morfounde' used by mariners when referring to vessels that had been driven under the sea by a succession of waves (Wagner and Heymering, 1999). The actual term 'laminiti' was first used in the 18th century, at which point a clear distinction between acute and chronic laminitis was starting to be made. Numerous theories as to the cause and means of treatment were presented. Around this time the use of cold-water soaking as a treatment protocol was advocated; this has gone in and out of 'fashion' as a treatment regimen, and is now back in vogue with the advent of new cryotherapeutic techniques (Van Eps and Pollitt, 2004).

The early 1900's led to more precise definitions as to the potential causes and nature of laminitis, which had by now become the most widely accepted term for the condition. The association with retained placenta was recorded at this time. Exponential growth in our knowledge and understanding of laminitis started in the mid 1900s when Obel gave a detailed description of the anatomy and histopathology of the horse's foot with laminitis and established the classification scheme for describing the clinical severity of the condition (Obel grades 1-IV : Obel, 1948), which is still in use today. In the 1970's Garner and colleagues (Garner *et al.*, 1975) established the carbohydrate overload model (CHO) for the experimental induction of laminitis, and started to investigate the potential role of the gastrointestinal flora, lactic acid and endotoxaemia in laminitis (Garner, 1975,

Garner *et al.*, 1977, Garner *et al.*, 1978); this leads to today's 'toxic' theory of laminitis. Adams began using more modern techniques such as angiography to determine effects of laminitis on the circulation of the foot (Adams, 1972), which ultimately led to today's 'vascular' theory of laminitis (Hood *et al.*, 1993; Elliott, 2007; Robertson *et al.*, 2009a). In the late 1970s/early 1980s (True *et al.*, 1978), clinical reports were published of a link between bedding horses on black walnut shavings and outbreaks of laminitis – this has lead to a plethora of published papers which have confirmed the role of leucocyte activation and cytokines in this condition and could be said to have led to the 'inflammatory' theory of laminitis (Black *et al.*, 2006; Riggs *et al.*, 2007).

It is around this time – the early 1980s – that a link with insulin resistance was recognised when Coffman and Colles (1983) suggested that laminitic ponies were significantly less sensitive to insulin than others. Other researchers subsequently proposed insulin resistance as a contributing factor in the apparent association between obesity and laminitis (Jeffcott *et al.*, 1986; Field and Jeffcott, 1989). This area was further developed by Johnson and colleagues, as well as the researchers at Virginia Tech., in the early 2000s, with the introduction of terms such as 'prelaminitic metabolic syndrome' and 'equine metabolic syndrome' to describe horses and ponies with apparent increased susceptibility to laminitis (Johnson, 2002; Hoffman *et al.*, 2003; Treiber *et al.*, 2005, 2007). More recently Australian researchers (Asplin *et al.*, 2007) have shown that laminitis can be induced, in healthy ponies, through maintaining a high plasma insulin concentration (with normal glucose). Collectively, this body of work has lead to today's 'endocrine' theory of laminitis.

In the late 1990s, investigations provided evidence to suggest basement membrane breakdown is a central mechanism in the development of laminitis, associated with up regulation of matrix metalloproteinases (MMP2 and MMP9: Pollitt and Daradka, 1998; Kyaw-Tanner and Pollitt, 2004). This work was the foundation for the 'enzymatic' theory of laminitis. Finally work presented in the early 2000s provided the first scientific support for a possible genetic component to this condition (Splan *et al.*, 2005; Trieber *et al.*, 2006) and it is likely that for many cases of laminitis one or more of the above proposed pathophysiological pathways are involved in the actual triggering and development of the condition (Harris *et al.*, 2006).

In the later part of the 20th century there was increased recognition of the importance of pasture-associated laminitis as farms diversified and ponies/horses were more often kept on 'improved' pastures established for high-producing dairy cows. In the late 1990s a potential link between fructans and laminitis was suggested (Longland *et al.,* 1998) and in the early 2000s the very active Australian research group led by Prof. Chris Pollitt showed that the administration of a bolus dose of a specific oligofructose (an inulin-like fructan) could reliably result in laminitis (Pollitt, 2002; Van Eps and Pollitt, 2006).

The above paragraphs provide a brief, non-exhaustive overview of some of the key developments in laminitis research from centuries BC to today. As we enter the 2nd decade of the 21st century we have a number of core experimental models of laminitis: carbohydrate overload (high starch low fibre meal :CHO); fructan overload (FR); black walnut extract administration (BWE), prolonged hyperinsulinaemia (INS) and a number of interlinking theories, including the vascular, enzymatic, toxic, inflammatory and endocrinopathic theories. There are exciting new investigative methods being employed such as proteomics (Carter *et al.,* 2009a) and hoof lamellar extracellular fluid sampling (Nourian *et al.,* 2010). However, many questions remain concerning the pathophysiology of naturally occurring laminitis. For example, how do these experimental models (and theories on pathogenesis derived from these models) relate to the naturally occurring disease? Additionally, why do some animals seem to be more predisposed to laminitis than others?

Currently there are several groups around the world working on various aspects of laminitis but despite the high demand for more knowledge and advice, funding is limited. Last year the American

Association of Equine Practitioners (AAEP) conducted a survey of its members with respect to defining and prioritizing equine health research needs. Of the 10 most important diseases or disease categories that respondents felt required further research, laminitis had the highest return at 63%. Laminitis was also ranked the highest in the previous survey carried out in 2003; highlighting the continued significance of this condition. This paper will try to provide insights into some the more recent advances in our understanding of laminitis and obesity, concentrating on papers and abstracts published in the last few years that deal with the pathophysiology or the dietary management. It will not attempt to cover work on treatment nor veterinary management which are outside the scope of this review. Some of the work cited has only been presented in abstract form at the time of writing this review; it is therefore possible that interpretations may alter when the full paper is published. Due to its very specific nature the emphasis will not be on BWE-induced laminitis, although some of this work is included for comparative or illustrative purposes.

Some recent advances with respect to the pathophysiology of laminitis

Firstly, recent work has confirmed our assumption that laminitis is a systemic condition that is manifest in the hoof (Riggs *et al.*, 2007, Wattle and Hansson, 2009) and that weight bearing might be the final trigger for the development of clinical signs (Leise *et al.*, 2009a). Secondly, whilst most researchers would agree that the pathogenesis of laminitis is multifactorial in nature and there may be many routes to the final end stage of the condition (Bailey *et al.*, 2004, Harris *et al.*, 2006) recent work has begun to clarify the differences and similarities between the experimental models and the naturally occurring disease. With respect to the naturally occurring disease, perhaps the most recent concept that may enable us to solve the question 'why this individual rather than that one' is that any vascular, inflammatory, haematological perturbations, that may occur as a result of the final triggering factor, may be exacerbated by pre-existing metabolic and/or endocrine imbalances within that individual animal. Insulin resistance may, for example, lower the threshold for the development of laminitis in the face of hindgut disturbances that initiate the cascade of events which ultimately lead to the clinical expression of laminitis (Harris and Geor, 2007). These themes will be explored below.

Is inflammation key?

Recent work has confirmed that dysregulation of the extracellular matrix may be an important component of the laminitis process in all forms of the condition and that it may occur in association with, or independent of, inflammatory processes evident during the development of laminitis (Loftus *et al.*, 2009, Coyne *et al.*, 2009). In at least 3 of the current models (BWE, CHO, FR), there is evidence of systemic inflammation, leuckocyte activation and infiltration of the lamellae, release of chemokines and subsequent monocyte /macrophage infiltration (Belknap *et al.*, 2007; Falerios *et al.*, 2009; Tadros *et al.*, 2009). However, there is some controversy over the time course of the events with some groups suggesting that any leucocyte activation may be a reaction to, rather than a cause of the lamellar pathology, at least in OF induced laminitis (Visser and Pollitt, 2009a). It does appear that the time course of lamellar inflammation may differ according to the model being evaluated with many of the inflammatory events apparently peaking around the onset of lameness with OF rather than in the early developmental stages as with BWE (Leise *et al.*, 2009b). This suggests that whilst inflammation is important it may not necessarily be the initiating cause of the lamellar pathology in all episodes of laminitis. This does raise the question of how much we can extrapolate from these experimental models to the field situation. However, as we currently cannot study the time course of events in the real time it is possible that earlier peaks in responses are being missed especially when samples are collected from horses with naturally acquired laminitis, when the prodromal phase is usually missed (Loftus *et al.*, 2009). This is where some of the new research techniques, that, for example, may enable interstitial fluid to be continuously evaluated, could be of immense value.

Importance of basement membrane breakdown

Initially it was thought that up-regulation of MMPs (Clutterbuck *et al.*, 2010) in particular of MMP-2 and MMP-9 was responsible for the disruption to the basement membrane (which then resulted in laminitis in conjunction with weight bearing). However, recently this has been disputed as work has suggested that the degradation of the lamellar basement membrane occurs prior to changes in MMP (2 and 9) expression and activation (Visser and Pollitt, 2009b). The possible role of MMP-14 and ADAMTS-4 in basement membrane disruption is currently under investigation (Kyaw-Tanner *et al.*, 2008; Coyne *et al.*, 2009). Once again there are differences in the models in that there is considerable compromise of the basement membrane zone with CHO and OF but little evidence of basement membrane breakdown with the BWE model (Faleiros *et al.*, 2009) and there was no global separation at the lamellar dermal/epidermal interface among ponies with laminitis following prolonged hyperinsulinaemia (Nourian *et al.*, 2009).

Role of endotoxin

Although endotoxaemia has been closely associated with the risk of laminitis (Parsons *et al.*, 2007), the role of endotoxin as a primary instigator of laminitis remains controversial (Bailey *et al.*, 2009); and it may be that other bacterial and non bacterial products that are absorbed from the gastrointestinal tract are more important in the pathogenesis of laminitis. However, although an acute bolus of endotoxin does not result in laminitis, this does not mean that it is not a contributory factor (Toth *et al.*, 2008). Endotoxin (LPS) does stimulate the production of certain vasoactive mediators (but not endothelin 1) by equine endothelial cells (Menzies Gow *et al.*, 2008) and work showed that following OF administration LPS increased to reach a peak of 2.4±1.0 pg/ml at 8 h (Bailey *et al.*, 2009). Furthermore recent work has shown that at 24 h after an IV injection of LPS (20 ng of *E. coli* O55:B5 LPS/kg) insulin sensitivity (as determined by the frequent sampled intravenous glucose tolerance test [fsivGTT]) was significantly reduced, whilst the acute insulin response to a glucose load was significantly increased (Toth *et al.*, 2008). The authors suggested that such disturbances in glucose and insulin dynamics could contribute to the development of laminitis in horses.

Does oxidative damage have a part to play?

Oxidative damage may play a role in laminitis secondary to increased free radical formation perhaps in response to one or more of the following: the development of insulin resistance, glucose auto-oxidation, reperfusion injury, neutrophil adherence to the damaged/activated endothelium and migration into the tissues (Black *et al.*, 2006; Bailey *et al.*, 2004; Kronfeld *et al.*, 2005). Higher concentrations of urinary TBARS (thiobarbituric acid reactive substances) were reported (Neville *et al.*, 2004) in chronic laminitic ponies when compared to ponies who did not suffer from laminitis. However, recently a study (Treiber *et al.*, 2009) looking at markers of inflammation and redox status in pastured ponies with and without a prior history of laminitis found no differences in markers of antioxidant function, although serum concentrations of TNFα were higher in previously laminitic ponies. Interestingly recent work has shown that laminar tissue from both control and BWE treated horses is devoid of superoxide dismutase (SOD) and this suggests that the equine digital laminae may be highly susceptible to damage by the superoxide anion (Loftus *et al.*, 2006). End products of lipid peroxidation (hydroxy- 2- nonenal – 4-HNE) were elevated in the laminar tissue of horses with BWE induced laminitis but not the control animals; there were no differences between the groups for lung, liver or the intestinal tissue (Yin *et al.*, 2009). Further investigation is needed to determine the precise role of free radicals in the pathogenesis of laminitis, which in turn will help to inform whether antioxidant supplementation (what and when and for how long) is likely to be of any value in the treatment or management of the condition.

Vasodilation vs. vasoconstriction

The role of changes in digital blood flow remains controversial with some authors stating that vasodilation is an early component of laminitis (Pollitt and Davies, 1998) and others that venous constriction plays a role (Hood *et al.*, 2001; Elliott, 2007; Robertson *et al.*, 2009b). The increase in hoof wall temperature tends to occur just before the onset of clinical signs, and may be preceded by a period of decreased temperature (Van Eps and Pollitt, 2004). The potential value of cryotherapy in the management of laminitis does not necessarily indicate that vasodilation is a central pathogenic mechanism. Cryotherapy will reduce enzymatic activity and pro-inflammatory cytokine expression (Van Eps *et al.*, 2010). Increased shunting of blood through ateriovenous anastomoses rather than through capillaries, could contribute to the increase in hoof wall temperature that has been reported.

Recent studies have provided evidence of vascular dysfunction in laminitis, with pathology confined to the small veins within the laminar dermis (Robertson *et al.*, 2009b). Digital veins are more sensitive to certain vasoconstrictor agents (including serotonin) than the arteries (Peroni *et al.*, 2006; Robertson *et al.*, 2009). It is possible that vasoactive substances may not directly influence regional blood flow but might result in an alteration in endothelial function (Eades *et al.*, 2007). Insulin resistance may add to this dysfunction through platelet and leucocyte activation, increased endothelin production as well as production of mediators of inflammation and oxidant stress (Geor and Frank, 2009, Robertson *et al.*, 2009). High concentrations of insulin in particular may have adverse effects on vascular function if the endothelial function is already compromised (Berhane *et al.*, 2009) and work has shown that equine endothelial cells under conditions of anoxia/reoxygenation do produce reactive oxygen species (Rebiere de Pouyade *et al.*, 2009). Interestingly, some of these vasoconstrictor agents may work synergistically and their action may be potentiated in the presence of corticosteroids. This may contribute to the postulated but as yet unproven increased risk of laminitis following systemic glucocorticoid administration (Bailey and Elliott, 2007).

It has been suggested that certain vasoactive amines absorbed from the intestinal tract could contribute to pasture associated laminitis (Elliott and Bailey, 2006). These amines are present in the caecal fluid of horses and ponies and levels increase when NSC rich pasture is consumed, with plasma levels being higher when ponies are on spring/summer pastures (Bailey *et al.*, 2003). Furthermore, *in vitro* models have demonstrated that the addition of starch or OF (inulin) to caecal fluid resulted in a concentration and time related increase in the concentration of vasoactive amines, and amines such as tyramine, tryptamine and phenylethylamine can cause constriction of isolated digital blood vessels directly or secondarily through their effects on 5-HT and thromboxane concentrations (Bailey *et al.*, 2004, Elliott, 2007). It is also worth noting that one amine present in the caecum, histamine, can cause vasodilation. Other amines, such as methylamine, may be converted to toxic metabolites in the blood stream or blood vessel wall (unpublished data). Any involvement of amines in pasture induced laminitis is, therefore, likely to be complex.

Relatively recent work has shown that faecal amine concentrations increase in response to the feeding of inulin (Crawford *et al.*, 2007). It is therefore possible that these vasoactive amines may form part of the 'laminitis trigger factors' that enter the peripheral circulation following gastrointestinal disturbance. In the carbohydrate overload model of acute laminitis, the vasoconstrictor amines tyramine, tryptamine and phenylethylamine were found to be markedly increased in the plasma of horses, six hours after starch was administered (Botteon *et al.*, 2008). Further work is needed in this area.

Role of fructan/oligofructose

'Fructan' is a collective term for oligo- and polyfructosyl sucrose. Depending on the number of fructose molecules, fructans can be described as oligosaccharides (<10 monosaccharide units) or polysaccharides (>10 units). Fructans can be divided broadly into three groups characterized by their glycosidic linkages (Longland and Byrd, 2005). Inulin is a fructose polysaccharide, which contains a mixture of linear fructose polymers and oligomers linked by β(2—1) bonds. A glucose molecule typically resides at the end of each fructose chain and is linked by an α(1—2) bond, as in sucrose. All dicotyledonous plants and some monocotyledonous plants contain such *linear inulin fructans*. Other monocotyledons such as timothy and wheat form *linear levan fructans* with β(2—6) bonds. Many temperate grasses contain typically *branched fructan molecules (graminans)* which have both β(2—1) and β(2—6) bonds. Oligofructose can be defined as a fructose oligosaccharide containing 2-10 monosaccharide residues. Oligofructose derived from chicory contains both fructose chains and fructose chains with terminal glucose units (GF_n). Synthesized oligofructose contains only fructose chains with glucose end units or GF_n molecules. Both types of oligofructose contain β(2—1) linkages between the fructose molecules. As we understand more about fructans and their effect on the horse it is likely that we will need to be more precise in our use of the terms OF, inulin, fructan etc.

Workers have been able to describe the changes in the equine hindgut bacterial populations (Milinovich *et al.*, 2006) and to show that certain Streptococci routinely become established as the dominant type of bacteria by as early as 2 h post oligofructose (OF: 'Raftilose' from chicory) administration (Milinovich *et al.*, 2007). The lead author also detected two novel bacterial species within the genus streptococcus in the hindgut of horses in which laminitis was experimentally induced (Milinovich *et al.*, 2008). The team in Australia also showed in 2007 (Nourian *et al.*, 2007) that laminitis is present ultra-structurally by 24 h (or earlier as the earliest samples were collected at 24 h when the first signs of lameness were seen) after OF induction at 10 g/kg bwt, and confirmed that the lamellar dermo-epidermal attachment at the basement membrane zone was a key target. Recent work has shown that the dose of oligofrucose needed to induce laminitis even in adult horses is perhaps lower than previously thought; in a recent study 3 out of 8 given 5.0 g of OF/kg BW developed laminitis as did 4 out of 4 given 7.5 g/kg BW (Kalck *et al.*, 2009). Interestingly there was no effect on insulin although there were changes in the glucose dynamics. This reduction in the dose required to initiate laminitis when provided as a bolus (post an adaptation phase) may be very important when relating this model to the situation in the field: ponies potentially can ingest a significant proportion (up to 40% of their daily dry matter intake [DMI]) within 3 h turnout (Ince *et al.*, 2005. Ince pers. comm.); grass can contain high concentrations of fructans (Longland and Byrd, 2005) and therefore it is possible that horses could ingest sufficient fructan from a high yielding fructan rich pasture in a relatively short space of time to initiate a laminitic episode (Longland and Byrd, 2005). Grazing a high NSC providing pasture may in turn promote the development of insulin resistance which may lower the threshold for laminitis to be triggered and various predisposing events and/or threshold lowering factors may be additive in nature (Harris *et al.*, 2006, Harris and Geor, 2007).

As mentioned above different fructan types are found in different plants, and these differences may influence the rate and extent of their digestion in the horse, as does the molecular size of the fructan. It is therefore important to note that the OF used in most experimental studies may not be representative of the more complex fructans present in many temperate grasses (Longland and Byrd, 2005) and data obtained with use of the oligofructose laminitis model may not be applicable to grass/forage fructans (Longland and Harris, 2009a).

What about glucose and insulin?

Glucose is important in maintaining lamellar integrity and has been shown to be essential for hoof explants in culture. Culture without glucose or inhibition of glycolysis causes basement membrane zone separation under tension (Pass *et al.*, 1998; French and Pollitt, 2004). Originally it was thought that the link with insulin resistance could be through a direct consequence of glucose deprivation especially as it has been suggested that the hoof utilises glucose at a high rate (Wattle and Pollitt, 2004). However, more recent studies have cast doubt on this theory by showing that lamellar tissue (assessed using explants) may not be dependent on insulin for glucose uptake (Asplin *et al.*, 2006). These workers went on to explore whether high concentrations of insulin (with a variable rate influsion of 50% glucose solution to prevent hypoglycaemia) could result in laminitis independent of any gastrointestinal trigger. Specifically, an intravenous infusion of insulin via a euglycaemic hyperinsulinaemic clamp technique for up to 72 h induced OBEL grade 2 laminitis (at 55.4±5.5 h), in ponies with a mean serum insulin concentration of 1,036±55 lU/mL vs. 14.6 lU/ ml in controls which did not develop laminitis (Asplin *et al.*, 2007). Subsequently laminitis has been induced in healthy standardbred racehorses within 48 h of the onset of hyperinsulinaemia (De Laat *et al.*, 2010). Recently it has been shown that IGF-1 receptors are present on lamellar epithelial cells and insulin may activate these receptors resulting in cellular proliferation (Bailey and Chockalingham, 2009). This could explain some of the histological changes observed in lamellar tissue from animals subjected to the insulin induced (INS) model (Nourian et al., 2009). Nourian et al., (2009) concluded that 'Prolonged hyperinsulinaemia causes unique lamellar lesions normally characteristic of acute and chronic laminitis. Lamellar proliferation may be an insulin effect through its mitogenic pathway. Aberrant lamellar mitosis may lengthen and weaken the lamellar, distal phalanx attachment apparatus and contribute to the clinical signs that developed.' Interestingly the ultrastructural changes characterised by Van Eps and Pollitt (2009) during the recovery phase – 7 days after induction by OF – suggest that the surface area of lamellar attachment apparatus was reduced and weakened even though all the epidermal compartments were enveloped in normal appearing basement membrane. Premature resumption of athletic exercise could, therefore, rupture the surviving lamellar attachments and result in reoccurrence of laminitis in an apparently recovered horse.

Adaptation to high carbohydrate diets in man may be associated with decreased insulin sensitivity as large fluctuations in glucose and insulin following meals high in sugar may supply inappropriate signals of energy availability to the glucose regulatory system, thereby altering insulin sensitivity of the tissue. This in turn may result in changes in metabolic signalling both within and between cells (Kronfeld *et al.*, 2005; Treiber *et al.*, 2006). It is important to note that grazing pastures at certain times of the year can result in marked fluctuations of blood glucose and insulin in a similar way to the feeding of large cereal based meals (Byrd *et al.*, 2006, Treiber *et al.*, 2008), and in an inbred herd of Welsh and Dartmoor ponies, an inulin-resistant phenotype was associated with a far higher risk for development of laminitis when grazing spring pasture when compared to non-insulin resistant ponies (Treiber *et al.*, 2006). In this context, it has been argued that laminitis may be triggered in a chronically insulin resistant horse or pony under conditions that exacerbate IR and/or hyperinsulinemia. Such conditions could include grazing pastures with a high non-structural carbohydrate content (e.g. during spring or when pastures are stressed by drought or frost), consumption of other feeds rich in starch and sugars (e.g. sweet feeds), overfeeding that induces or worsens obesity, the administration of corticosteroids, or an episode of endotoxaemia (Geor and Harris, 2009).

Is fructose important?

In man it has been recognised recently that associated with our change in lifestyle over the last 50 years or so 'has been the substantial increase in the amount of dietary fructose consumption due to the high intake of sucrose and high fructose corn syrup a common sweetener used in the food industry' and that this may be linked with the development of insulin resistance and the human metabolic syndrome (Basciano *et al.*, 2005). It has been suggested that diets enriched with sucrose or fructose can lead to insulin resistance independent from increased energy intake and whole body or visceral fat accumulation. Changes in insulin sensitivity consistent with this hypothesis were seen when obese/overweight men and women were provided with a diet containing 25% of total calories as fructose but not when glucose was used (Stanhope *et al.*, 2009). Our recent work has shown that 2 weeks of dietary fructose supplementation at 1 g/kg three times daily resulted in decreased insulin sensitivity (as measured by minimal model analysis of the intravenous glucose tolerance test: fsivGTT) and increased resting insulin concentrations in insulin resistant (overweight) but not control healthy mares. There was no effect of dietary glucose or inulin (fructan) (Geor *et al.,* 2009). Further work is needed to determine the role of dietary carbohydrates in the development of insulin resistance in horses, especially those already genetically or otherwise at greater risk.

Obesity and laminitis

Obesity in horses is increasingly being recognised as a globally important welfare issue. For example, in a study of 319 pleasure riding horses in Scotland, 32% were found to be obese (BCS 6 on a 6-point scale) and a further 35% were considered fat (BCS 5/6; Wyse *et al.*, 2008). The recent BEVA-EBM study on pasture associated laminitis concluded that '*overweight animals that develop laminitis tend to have more severe signs than those of optimal weight. When laminitis does occur, overweight animals are more likely to die of the disease than their thinner counterparts.' (*pers comm. BEVA Newsletter 2009).

In most cases, the reason for a horse being overweight is that they have stored excess energy/calories as fat i.e. they have been overfed relative to their activity level. Many horses that spend most of their time in stables with occasional hacks may not require any more than maintenance energy intakes, and yet many are fed much more than this. Similarly, animals turned out to pasture at certain times of the year might be getting several times their energy/calorie requirement (Geor *et al.,* 2007; Geor and Harris, 2009a,b). It is perhaps surprising that more animals do not become overweight, and this may reflect individual differences in activity levels and feed intake when out in the field, as well as ability to convert feed to fat. Certain breeds (e.g. Quarter Horses and Morgans) and types (ponies) are perhaps more prone to obesity and would benefit from close attention to diet and exercise.

However, many factors influence an individual's propensity for weight gain. Seasonality is perhaps an underestimated factor in that under feral conditions, horses and ponies tend to gain body mass (fat) during summer months when food is abundant and then lose body fat during the winter. This natural pattern tends to be disturbed with domestication and the provision of rugs/stabling and feed throughout the winter months. A recent study highlighted the differences in appetite in native ponies of varying starting BCS allowed to eat *ad libitum* on a forage chaff based complete feed either in the summer or the winter (Dugdale *et al.,* 2008). Appetite was greatest in summer ponies in thin and moderate BCS (summer, 159.11±9.09 gDM/kgBM$^{0.75}$/d, 3.96±0.17% of BM as DMI; winter, 114.58±6.92 gDM/kgBM$^{0.75}$/d, 2.98±0.16% of BM as DMI). Conversely, the appetite of fat ponies was similar between seasons (summer, 83.58±2.34 gDM/kgBM$^{0.75}$/d, 2.02±0.01% of BM as DMI; winter, 80.73±13.14 gDM/kgBM$^{0.75}$/d, 1.96±0.25% of BM as DMI). Fat ponies tended to gain very little body mass in summer (0.01±0.06 kg/d), but showed average daily gains of 0.33±0.2 kg/day in winter. BCS changed little throughout the studies in fat ponies, but increased significantly (c. 3 points on a 1-9 scale), in the thin animals in both summer and winter. Interestingly ultrasonic fat depth at the 12th intercostal space (range 0.65-30.67 mm), but not the rump (range 4.79-28.77 mm), correlated well with total body fat percentage (range 4.19-33.37%), derived from deuterium oxide dilution (12th intercostal space, r=0.82; gluteal region, r=0.59).

The 'syndrome' of obesity, insulin resistance and laminitis in mature horses has been referred to as either 'peripheral Cushing's syndrome' or the equine 'metabolic syndrome' (EMS; Johnson *et al.,* 2006) although this description has been disputed (Kronfeld *et al.,* 2005). A recently published consensus statement recommended retention of EMS to describe this phenotype (Frank *et al.*, 2010). The majority of equids with EMS exhibit the following characteristics:

- Generalized obesity or increased adiposity in specific locations, such as the nuchal ligament region (cresty neck).
- Insulin resistance characterized by hyperinsulinaemia or abnormal glycaemic and insulinaemic responses to oral or intravenous glucose and/or insulin challenges.

- A predisposition towards laminitis. Clinical or subclinical laminitis that has developed in the absence of recognized causes such as colic, colitis or retained placenta.

Obesity is observed in the majority of EMS cases but the condition also may occur in leaner individuals. Although obesity has been associated with laminitis in some studies, it has not been established whether this condition directly raises risk of laminitis or if the increased risk is due to other factors such as IR and inflammation, which are associated with the animal being obese as discussed below.

Studies in other species have demonstrated that chronic obesity induces a low-grade inflammatory state. Additionally, this inflammation is likely to be a key component in the pathogenesis of obesity-associated IR (Muoio and Newgard, 2008). In horses, associations between obesity, blood mRNA expression of TNFα and IL-1β and IR have been reported, suggesting that systemic inflammation may also play a role in the IR of equine obesity (Vick *et al.*, 2007). Recent data have shown that CD14, TLR2 PIK3R1 and INFKB were significantly up regulated in liver tissue from horses with laminitis and those with clinical signs of metabolic disease compared with controls (Stokes *et al.*, 2009). It is important to recognize that not all obese horses are insulin resistant and that IR can occur in non-obese animals and without raised basal plasma insulin concentrations (Bailey e*t al.*, 2008; Carter *et al*., 2009b).

Diet appears to play an important role in the triggering of laminitis in horses or ponies with this EMS phenotype, particularly the ingestion of pasture, forage or other feeds (e.g. cereal grains or sweet feeds) high in non-structural carbohydrates (NSC; simple sugars, starches and/or fructans) and the effect of dietary NSC on insulin sensitivity may be magnified in equids with pre-existing IR (Hoffman *et al.*, 2003; Hess *et al.*, 2005; Treiber *et al.*, 2006; Carter *et al.*, 2009b). In summer, but not in winter, laminitis-prone ponies had higher serum insulin concentrations (Bailey *et al*., 2008) when compared to age- and BCS-matched control ponies, suggesting that consumption of summer pasture (high in NSC including fructans) was important. A marked increase in the serum insulin concentrations of laminitis-prone ponies was recorded during the transition from winter to spring in association with an increase in the forage water soluble carbohydrate content (simple sugars and fructans: Treiber *et al.*, 2008a). The most marked increases were seen in ponies that subsequently developed laminitis.

Prediction of increased risk

In a group of inbred ponies with a high incidence of laminitis, generalized obesity, regional accumulation of neck crest adipose tissue ('cresty neck'), hyperinsulinaemia (>30 IU/ml) and hyperleptinemia (>7.5 ng/ml) were predictors of the development of laminitis when ponies were subsequently exposed to spring pasture (Carter *et al.*, 2009 b,c)

Laminitis-prone ponies, some of which did not have elevated basal plasma insulin concentrations but were shown to be IR using the fsivGTT, demonstrated exaggerated increases in serum insulin concentrations in response to the feeding of inulin, a type of fructan (Bailey *et al.*, 2007).

Prior work suggested that the insulin response to a low dose dexamethasone test could detect ponies with an abnormal insulin response, which in turn could be linked with increased insulin resistance and an increased risk of laminitis (Bailey *et al.*, 2007). More recent work has suggested that, similar to the seasonal effects on the cortisol response to this test, there is a seasonal effect on the insulin responses and the current recommendation would be to carry out the DST in spring to optimise both the diagnosis of PPID and laminitis susceptibility (Borer *et al.*, 2009). However, further work is needed to characterise the value of this test in the individual animal (Menzies-Gow *et al.*, 2009).

Is the location of fat deposition important?

In humans, visceral (abdominal) adiposity is more closely linked to risk for diabetes and cardiovascular disease than generalized obesity, and measurement of waist circumference is a better indicator of abdominal fat accumulation than is body mass index (Lee *et al.*, 2006). In horses and ponies, there may be a similar association between regional adiposity and disease risk (Carter *et al.*, 2009b).

A recent small study suggested that in horses fed on hay, regardless of their level of insulin sensitivity, the expression of IL-1b and IL-6 was significantly higher in the nuchal adipose tissue than in tissue from omental/retroperitoneal and mesocolonic AT (Burns *et al.*, 2009). However, as far as lipogenic activity is concerned recent work has shown that mesenteric adipose tissue in the horse has much greater activity compared with adipose tissue taken from the nuchal crest region or the liver (Suagee *et al.*, 2010). Interestingly this study suggested that despite previous work showing that adipose tissue has GLUT4 protein expression, acetate was found to be the primary carbon source for fatty acid synthesis rather than glucose. However, none of these animals were obese. Obesity as defined by a BCS of 8 or more out of 9 and in particular significant deposition of subcutaneous fat in the neck region (cresty neck: Carter *et al.*, 2009b) has been associated with insulin resistance. It would be interesting to repeat some of the adipose work with samples from obese animals and/or those being fed high sugar/starch diets.

Fat stores within the neck crest may be the most readily appreciated superficial index of adiposity in the pony (Carter *et al.*,. 2009b). A strong correlation has been found in horse cadavers between total carcase fat and the depth of the crest fat measured at the level of the 4th cervical vertebra (Znamirowska, 2005). It is possible therefore that mean neck circumference or neck crest scoring may provide indirect estimations of IR and whole body adiposity (Znamirowska, 2005; Carter *et al.*, 2009b). Certainly one study (Frank *et al.*, 2006) showed a clear association between the magnitude of IR and mean neck circumference. However, the results from a more recent trial (Dugdale *et al.*, 2010) suggests that, similar to BCS in general, crest scoring systems may be relatively insensitive indices of early weight loss and the accompanying beneficial changes on IR, at least in Native pony mares.

Management

The reader is referred elsewhere for more information with respect to pasture and general management (Longland and Byrd, 2006; Harris and Geor, 2007; Geor and Harris, 2009). Discussion of dietary supplements that might improve weight loss/level of IR are also beyond the scope of this review (see Tinworth *et al.*, 2010). However, we have highlighted below some recent work with practical relevance to the feeding management of horses and ponies with obesity and/or an increased risk of laminitis.

NSC content

Many of the more recent papers or reviews providing advice for managing the obese/laminitic/insulin resistant animal make reference to providing feeds and/or forages that do not provoke a marked glucose or insulin response. However, many factors can influence the glucose and insulin response to any particular ration (Harris and Geor, 2009; Vuervert *et al.*, 2009). Despite this our current recommendation would be to feed, those animals that have IR or a history of laminitis or are potentially at risk of developing IR or laminitis, diets that have been shown to result in low or low-moderate glucose/insulin responses in at least one group of animals. Analytical values do not guarantee such a response. Whilst this recommendation may be practical for commercially purchased feeds it does not deal with the forage component. It is typically recommended to feed hay with a NSC content of less than 10% (this is starch/sugar/fructan content). Without actually analysing the hay this can be difficult to determine as even mature looking hays if produced from stressed pastures can contain high NSc content. Soaking hay has been recommended as a means to reduce the WSC content, however recent work has suggested that the efficacy of this is both variable and unpredictable (Longland *et al.*, 2009). Soaking hay should therefore be used as an adjunct to choosing a low NSC containing hay. We also do not have sufficient scientific evidence to know if the 10% level is optimal but certainly based on other work it would seem, at the present time, to be a sensible goal.

Caution is required in using the ethanol soluble carbohydrate fraction as a measure of the simple sugar fraction of the feed in question (and based on this, the assumption that WSC-ESC estimates the fructan content) as some smaller molecular weight fructans and other carbohydrate oligomers may also be soluble in 80% ethanol (Longland and Harris, 2009b). Finally, it is important to note that the McLeary colorimetric method ('megazyme') for fructan determination, results in variable and substantially underestimated values for the fructan content of certain feeds and forages including timothy (Longland and Harris, 2009a).

Weight loss

Given the increasing interest in obesity it is perhaps not surprising that recently there has been increasing focus on ways to safely and effectively promote weight loss. The optimal rate of weight loss for equidae without compromise to overall health is unknown, although on the basis of data in other species, a target loss of 1% BM weekly may be considered appropriate (German *et al.*, 2007). A few of the more recent papers are outlined below.

Researchers in Belgium (Van Weyenberg *et al.*, 2008) evaluated the effects of a weekly weight loss, of around 1% of the 9 obese ponies estimated ideal body weight, on insulin sensitivity as determined through an oral glucose tolerance test (OGTT). Interestingly they found that the initial energy restriction programme (70% of maintenance energy requirements based on estimated ideal body weight) did not maintain the desired constant weight loss and they had to further restrict the animals to 50% (after week 6) and then to as low as 35% (after week 14). It is important to note that these ponies were fed based on their estimated ideal body weight which was stated as being on average around 78% of their initial body weight, but unfortunately how this was specifically determined was not clarified. This meant that effectively the ponies were receiving around 0.5% of their actual BW on an as fed basis during the last four weeks of the trial. Whilst no clinical problems were noted in this trial, fasting plasma triglycerides (TG) and non-esterified fatty acids (NEFAs) were significantly increased. By the end of the 18 week trial the ponies had lost around 18% of their initial body weight and the BCS had reduced from 8/9 (out of 9) to 4/5 (out of 9). The need to reduce the energy during the study to maintain BW loss is a common theme in some of the studies reported to date (Van Weyenberg *et al.*, 2008; Gordon *et al.*, 2009). It is possible that, like other species, as equidae reduce their body weight, as a result of restricted energy intake, they will reduce their energy expenditure, making it difficult to continue to maintain weight loss. This emphasises the need to individually tailor any weight loss programme and monitor it closely.

Another study (Dugdale *et al.*, 2010) looked at weight loss in five mature (10±2 y), overweight/obese pony mares (BM, 257±20 kg: body condition score (BCS) 6.8/9±0.5) over a 12 week period. Animals were individually housed and provided with a chaff-based, complete diet at 1% of actual BM as DMI daily (~65% of the initial energy requirements based on obese body weight). BM decreased by 4.3±1.1% during the first week and thereafter by 0.7±0.1% of BM at the end of this first week. Fat comprised 47±20% of BM loss. Fatter animals lost relatively more fat. Mean serum NEFA concentrations increased only very slightly and remained within the reference range and the plasma TG and total cholesterol concentrations remained unaltered. Although no clinical adverse signs were seen with the decreased feeding activity, the time spent in 'play' and rest increased by 36±11% and 438±95% respectively. Interestingly apparent digestibility of the ration did not alter significantly following the 12 weeks of dietary restriction.

From these two studies it would appear that clinically-useful rates of weight loss (~1.0% BM weekly) in obese, native pony breeds requires that energy intake is limited to less than 70% of maintenance requirements and potentially much less than this. However, it is not advised that such dietary programmes are undertaken without close veterinary supervision and means to enrich the environment of any dietary restricted animal are implemented. Much more work is needed to determine practical regimens to promote safe weight loss especially in obese ponies.

A recent study (Gordon *et al.*, 2009) examined weight loss in overweight QH and TB horses provided hay (1% BW) and a reduced calorie feed (0.5%BW) with and without exercise for a 12 week period compared to horses fed a control diet without exercise. The authors reported more weight loss with the diet (at an estimated ~60% of their energy requirements) and exercise regimen but they also reported reducing the low calorie feed to 0.3% BW after 6 weeks in this group and therefore a direct comparison with the non exercise group cannot be made. At the end of the trial the control group had also lost weight and were not significantly different from the diet group despite receiving an estimated ~140% more energy. The authors considered that this could have resulted from environmental differences, in that the control animals had not been stabled unlike the other groups and had experienced a harsh winter plus they were able to freely exercise unlike the others. Plasma NEFA concentrations decreased in restricted diet plus exercise treatment.

Effect of weight loss on insulin and glucose dynamics

In the Belgium study, the weight loss was associated with a reduced glucose peak, decreased baseline insulin values and a decreased area under the curve for glucose and insulin in response to the oral glucose tolerance test. Despite the comparatively modest weight loss in the Dugdale *et al.* (2010) study, a marked reduction or correction of the initial hyperinsulinaemia was observed suggesting potential benefit greater than the weight loss might suggest. However, a control stable BW group was not evaluated. In another study where horses were fed different types of hay to either to maintain (115% of calculated NRC - DE requirements), gain (150%) or lose weight (88%) over a 130 day period there were marked individual responses in the amount of weight gain and loss that was achieved. In this study insulin sensitivity (as determined by the Euglycaemic clamp) was apparently significantly higher in the gain group (BCS 6.5±0.5) than both the maintenance and lose weight group at day 130 (Owens *et al.*, 2009). However, those animals that did not change weight as expected were excluded from the analysis.

Certainly the importance of exercise with or without a weight loss regimen on body weight and/or level of insulin resistance has been discussed over the years (Freestone *et al.*, 1991; Treiber *et al.*, 2006; Treiber, 2008b; Gordon *et al.*, 2009). A very recent study looking at the effect of exercise in obese animals concluded that 'moderate exercise training without concurrent dietary restriction does not mitigate insulin resistance in overweight or obese horses. A more pronounced reduction in adiposity or higher volume or intensity of exercise may be necessary for improvement in insulin sensitivity in such horses.' (Carter *et al.*, 2010). Much more work is needed to determine the quantity of physical activity required to positively influence insulin sensitivity in obese animals and to evaluate the effect of weight loss with and without exercise on IR.

Monitoring weight loss

A research group in Portugal (Santos *et al.*, 2009) outlined the potential to discriminate small changes in fat tissue depth using real-time ultrasound and suggested that BCS might not be of value in such instances. Gordon *et al.*, (2009) found that changes in BW were associated with changes in BCS (average of -52 kg associated with an average BCS change of -2.5). In ponies (Dugdale *et al.*, 2010) there was no change in BCS despite a 10-12% decrease in body weight suggesting that BCS may be an insensitive measure of initial body weight change during a weight loss programme at least in overweight/obese ponies. In the longer Belgian study with a mean ± SD weight loss of 18.2±1.8%, over the 18 weeks, they reported a 4 point decrease in BCS from outset values using a comparable 9 point BCS system (Van Weyenberg *et al.*, 2008). The difference in the utility of BCS may depend on several factors including the breed/type of animal being evaluated, as the BCS scoring system used in both cases was originally developed in QH mares (Henneke *et al.*, 1983) and may not be suitable for ponies. Other potential confounding factors include the amount of weight loss and/or the starting BCS – in that for the more obese animals there may be more internal fat that is lost initially (resulting in a reduction in weight but no change in the externally palpated BCS).

In the Dugdale *et al.* (2010) study in addition to BCS, girth measurements and ultrasound-derived measures of subcutaneous fat depth overlying the gluteal region and 12th intercostal space (rib-eye) were recorded weekly. Body fat content was estimated at the beginning and end of the study by deuterium oxide dilution methods. This study highlighted that small changes in BCS at the higher end of the 1 - 9 point scale were associated with a significant increase in fat content and therefore such a linearly ordinal BCS system may in fact be a relatively insensitive tool for monitoring changes in body composition in overweight/obese animals. Heart and belly girths, rump width and subcutaneous fat depth at rib-eye decreased significantly with time and BM, but neck circumference and rump fat depths did not change consistently as the BM decreased. This is in contrast to the Gordon *et al.* (2009) study where rump fat thickness did apparently significantly reduce. Again this could reflect starting BCS and/or breed differences etc. This suggests that further work is needed in this area, but at the moment BCS should not be relied upon to monitor initial weight loss especially in obese/overweight native ponies.

Conclusion

Unfortunately the above can only provide a review of only a relatively small proportion of the more recent studies that have been undertaken into laminitis/obesity in the last few years. Just searching for publications including the term laminitis and equine/horse since the year 2004 in the scientific literature search engine 'Scopus' revealed 100 papers and in Pub Med over 200. When obesity was included over 400 records were found. This short review does, however, highlight the diversity of the work being undertaken and the advances that are being made into understanding these important conditions.

References

Adams, O.R., 1972. Vascular changes in experimental Laminitis. In Proceedings of the 18th Annual Meeting of the American Association of Equine Practitioners, San Francisco p 359.

Asplin, K.E., C.M. McGowan, C.C. Pollitt, J. Curlewis and M.N. Sillence, 2006. Gluocse uptake in the equine hoof Proceedings of the Australian Equine Science Symposium June 2006 Gold Coast Qld Vol1 p 26.

Asplin, K.E., M.N. Sillence, C.C Pollitt and, C.M. McGowan, 2007. Induction of laminitis by prolonged hyperinsulinaemia in clinically normal ponies. Veterinary Journal, 174, 530-535

Bailey, S.R., L.M. Katz, Y. Berhane, T. Samuels, N. De Brauvere, C.M. Marr and J. Elliott, 2003. Seasonal changes in plasma concentrations of cecum-derived amines in clinically normal ponies and ponies predisposed to laminitis. Am. J. Vet. Res. 64(9): 1132-1138.

Bailey, S.R., C.M. Marr and J. Elliott, 2004. Current research and theories on the pathogenesis of acute laminitis in the horse. Vet J. Apr;167: 129-142.

Bailey, S.R., N.J. Menzies-Gow, P.A. Harris, J.L. Habershon-Butcher, C. Crawford, Y. Berhane, R.C. Boston and J. Elliott, 2007. Effect of dietary fructans and dexamethasone administration on the insulin response of ponies predisposed to laminitis. J. Am Vet Med Assic 231, 1365-1373.

Bailey, S.R. and S. Chockalingham, 2009. Proliferative effects of insulin on equine lamellar epithelial cells mediated by the IGF-1 receptor. In proceedings of 2nd AAEP Foundation Equine Laminitis Research Worskhop p26.

Bailey, S.R. and J. Elliott, 2007. The Corticosteroid laminitis story : 2. Science of it, when and how. Equine Veterinary Journal 39, 7-11.

Bailey, S.R., J.L. Habsershon-Butcher, K.J. Ransom, J. Elliott and N. Menzies Gow, 2008. Hypertension and insulin resistance in a mixed-breed population of ponies predisposed to laminitis. Am J Vet Res; 69: 122-129.

Bailey, S.R., H.S. Adair, C.R. Reinemeyer, S.J. Morgan, A.C. Brooks, S.L. Longhofer and J. Elliott, 2009. Plasma concentration of endotoxin and platelet activation in the developmental stage of oligofructose –induced laminitis. Veterinary immunology and immunopathology 129, 167-173.

Basciano, H., L. Federico and K. Adeli, 2005. Fructose, insulin resistance and metabolic dyslipidemia. Nutr. Metabol. 2, 5.

Berhane, Y., H. Jordan, S.R. Bailey, H. Zerpa, K.E. Borer, N.J. Menzies-Gow, P.A. Harris and J. Elliott, 2009. Modulation of endothelium- dependent relaxation of equine digital arteries by insulin and hypoxia. In Proceedings of the 2nd AAEP Foundation Equine Laminitis Research Workshop p36.

Belknap, J.K., S. Giguere, A. Pettigrew, A.M. Cochran, A.W. Van Eps and C.C. Pollitt, 2007. Lamellar proinflammatory cytokine expression patterns in laminitis at the developmental stage and at the onset of lameness : innate vs adaptive immune response. Equine Vet J 39: 42-47.

Black, S.J., D.P. Lunn, C. Yin, M. Hwang, S.D. Lenz and J.K. Belknap, 2006. Leucocyte emigration in the early stages of laminitis Vet Immunol Immunopath 109: 161-166.

Borer, K.E., N.J. Menzies-Gow, Y. Berhane, S.R. Bailey, P.A. Harris and J. Elliott, 2009. Seasonal influence on insulin and cortisol results from overnight dexamethasone suppression tests (DST) in normal and previously laminitic ponies In Proceedings of the 2nd AAEP Foundation Equine Laminitis Research Workshop p38.

Botteon, P.T.L., V.P. Ferraz, M.B. Gloria, M.M. Melo, F.O. Paes Leme, R.R. Faleiros and G.E.S. Alves, 2008. Plasmatic concetrations of tryptamime, tryamine and phenluethylamine in horses under the effect of carbohydrate overload and non-steroid antiinflammatory compounds. Pesq.Vet.Bras. 28(6): 299-302.

Burns, T.A., R.J. Geor, M.C. Mudge, L.J. McCutcheon and J.K. Belknap, 2009. Proinflammatory cytokine and chemokine expression profiles of various adipose tissue depots of insulin-resistant and insulin-sensitive light breed horses. In proceedings of 2nd AAEP Foundation Equine Laminitis Research Worskhop p27.

Byrd, B., K. Treiber, D. Kronfeld, W. Staniar, R. Geor and P.A. Harris, 2006. Circadian variation of pasture NSC and insulin concentration in horses. Proc. Am. Soc. Anim. Sci. 84, 330-331.

Carter, R., M. De laat, C.C. Pollitt and H. Galantino-Homer, 2009a. differences in lamellar protein expression during laminitis induced by a 48 h prolonged euglycaemi-hyperinsulinemic clamp. In proceedings of 2nd AAEP Foundation Equine Laminitis Research Worskhop p41.

Carter, R.A., K.H. Treiber, R.J. Geor, L. Douglass and P. Harris, 2009b. Prediction of incipient pasture-associated laminitis from hyperinsulinemia, hyperleptinemia, and generalized and localized obesity in a cohort of ponies – Equine Vet Journal 41: 171-178.

Carter, R.A., W.B. Staniar, T.A. Cubitt, P. Harris and R.J. Geor, 2009c. Apparent adiposity assessed by standardised scoring systems and morphometric measurements in horses and ponies Veterinary Journal 179: 204-210.

Carter, R.A., J. McCutcheon, E. Valle, E.N. Meilahn and R.J. Geor, 2010. Effects of exercise training on adiposity, insulin sensitivity and plasma hormone and lipid concentrations in overweight or obese insulin-resistant horses Am J Vet Res 71(3): 314-321.

Clutterbuck, A.L., P. Harris, D. Allaway and A. Mobasheri, 2010. Matrix metalloproteinases in inflammatory pathologies of the horse. The Veterinary Journal 183, 27-38.

Coffman, J.R. and C.M. Colles, 1983. Insulin tolerance in laminitic ponies. Can J Comp Med; 47: 347-351.

Coyne, M.J., H. Cousin, J.P. Loftus, P.J. Johnson, J.K. Belknap, C.M. Gradil, S.J. Black and D. Afandari, 2009. Cloning and expression of ADAM- related metalloproteases in equine laminitis. Vet Immunol Immunopath 129: 331-341.

Crawford, C., M.F. Sepulveda, J. Elliott, P.A. Harris and S.R. Bailey, 2007. Dietary fructan carbohydrate increases amine production in the equine large intestine: Implications for pasture- associated laminitis. J Anim Sci 85: 2949-2958.

Dugdale, A.H.A., G.C. Curtis, D.C. Knottenbelt, P.A. Harris and C.M. Mc Argo, 2008. Changes in body condition and fat deposition in ponies offered an ad libitum chaff based diet. In Proceedings of 12th Congress of the ESVCN Vienna p 39.

Dugdale, A., G.C. CurtiS, P. Cripps, P.A. Harris and C.M. Mc Argo, 2010. Effect of dietary restriction on body condition, composition and welfare of overweight and obese pony mares. Equine Veterinary Journal (in press).

Eades, S.C., A.M. Stokes, P.J. Johnson, C.J. LeBlanc, V.K. Ganjam, P.R. Buff and R.M. Moore, 2007. Serial alterations in digital hemodynamics and endothelin-1- immunoreactivitym platelet- neutrophil aggregation, and concentrations of nitric oxide, insulin, and glucose in blood obtained from horses following carbohydrate overload. Am J Vet Res 68: 87-94.

Elliott, J. and S.R. Bailey, 2006. Gastrointestinal derived factors are potential triggers for the development of acute equine laminitis. Journal of Nutrition 136: 2103S-2107S.

Elliott, J., 2007. Pasture induced laminitis: new insights into the cause In Proceedings of the 1st International WALTHAM-RVC Laminitis Conference Equine Veterinary Journal p 15 -22.

Faleioros, R.R., B.S. Leise, M. Watts, P.J. Johnson, S.J. Black and J.K. Belknap, 2009. Laminar Chemokine mRNA concentrations in horses with carbohydrate overload- induced laminitis. In proceedings of 2nd AAEP Foundation Equine Laminitis Research Worskhop p21.

Field, J.R. and L.B. Jeffcott, 1989. Equine laminitis – another hypothesis for pathogenesis. Med Hypotheses; 30: 203-210.

Frank, N., S.B. Elliott, L.E. Brandt and D.H. Keisler, 2006. Physical characteristics, blood hormone concentrations and plasma lipid concentrations in obese horses with insulin resistance. J.A.V.M.A. 228: 1383-1390.

Frank, N., R.J. Geor, S.R. Bailey, A.E. Durham and P.J. Johnson, 2010. Equine metabolic syndrome: ACVIM Consensus Statement. J Vet Int Med 24: 467-475.

Freestone, J.F., K.J. Wolfsheimer, S.G. Kamerling, G. Church, J. Hamra and C. Bagwell, 1991. Exercise induced hormonal and metabolic changes in Thoroughbred horses: effects of conditioning and acepromazine Equine veterinary journal, 23: 219-223.

French, K.R. and C.C Pollitt, 2004. Equine Laminitis : glucose deprivation and MMP activation induce dermo-epidermal separation *in vitro* Equine Veterinary Journal 36: 261-266.

Garner, H.E., J.R. Coffman and A.W. Hahn, 1975. Equine Laminitis of alimentary origin : An experimental model. Am J Vet Res 36: 441-444.

Garner, H.E., J.N. Moore and J.H. Johnson, 1978. Changes in the caecal flora associated with the onset of laminitis. Equine Vet J 4: 249-252.

Garner, H.E., D.P. Hutcheson, J.R. Coffman, A.W. Hahn and C. Salem, 1977. Lactic acidosis: a factor associated with equine laminitis. J Anim Science 45: 1037-1041.

Geor, R.J., K.H. Treiber, R.A. Carter and P. Harris, 2007. Metabolic predispositions to laminitis: obesity and insulin resistance. Proceedings of the 1st International WALTHAM® – Royal Veterinary College Laminitis Conference. Harris, P.A., S.J. Hill, J. Elliott and S.R. Bailey (eds.), Equine Veterinary Journal Ltd, Newmarket. pp 23-28.

Geor, R.J. and N. Frank, 2009. Metabolic syndrome – from human organ disease to laminar failure in equids Vet Immunol Immunopath 129: 151-154.

Geor, R.J. and P.A. Harris, 2009a. Dietary management of obesity and insulin resistance: countering risk for laminitis In: Geor, R.J. (ed.), Veterinary Clinics of North America (equine Practice). Elsevier Saunders Philadelphia, USA, p. 51-66.

Geor, R.J. and P.A. Harris, 2009b. Dietary Management of the obese horse. In: Robinson, N.E. and K.A. Sprayberry (eds.), Current Therapy in Equine Medicine. Elsevier, St Louisp, USA, p. 59-65.

Geor, R.J., J. De Silva, K. Meyers, T. Smith and P. Harris, 2009. Effects of short-term adaptation to dietary carbohydrates on glucose and insulin dynamics in healthy and overweight/obese insulin resistant mares. In proceedings of 2nd AAEP Foundation Equine Laminitis Research Worskhop p 28.

German, A.J., S.L. Holden, T. Bissot, R.M. Hackett and V. Biourge, 2007. Dietary energy restriction and successful weight loss in obese client-owned dogs. J. Vet. Intern. Med. 21: 1174-1180.

Gordon, M.E., M.L. Jerina, R.H. Raub, K.A. Davison, J.K. Young and K.K. Williamson, 2009. The effects of dietary manipulation and exercise on weight loss and related indices of health in horses. Comparative Exercise Physiology 6(1): 33-42.

Harris, P.A., S.R. Bailey, J. Elliott and A. Longland, 2006. Countermeasures for Pasture- associated laminitis in Ponies and Horses J Nutr 136: 2114S-2121S.

Harris, P.A. and R. Geor, 2007. Nutritional Countermeasures to laminitis. In: Harris, P.A., s.j. Hill, J. Elliott and S.R. Bailey (eds.), The latest findings in laminitis research The 1st WALTHAM- Royal Veterinary college Laminitis Conference p 29-38.

Harris, P.A. and R.J. Geor, 2009. Primer on Dietary carbohydrates and utility of the glycemic index in equine nutrition. In: Geor, R.J. (ed.), Veterinary Clinics of North America (equine Practice). Elsevier Saunders. Philadelphia, USA, p. 23-38.

Hess, T.M., D.S. Kronfeld, K.H. Treiber, B.M. Byrd, W.B. Staniar and R.K. Splan, 2005. Laminitic metabolic profile in genetically predisposed ponies involves exaggerated compensated insulin resistance. J. Anim. Phys. Anim. Nutr. 89: 430-431.

Hoffman, R.M., R.C. Boston, D. Stefanovski, D.S. Kronfeld and P.A. Harris, 2003. Obesity and diet affect glucose dynamics and insulin sensitivity in Thoroughbred geldings. J Anim Sci 81: 2333-2342.

Hood, D.M., D.A. Grosenbaugh, M.B. Mostafa, S.J. Morgan and B.C. Thomas, 1993. The role of vascular mechanisms in the development of acute equine laminitis J Vet Intern Med 7: 228-234.

Hood, D.M., I.P. Wagner and G.W. Brumbaugh, 2001. Evaluation of hoof wall surface temperature as an index of digital vascular perfusion during the prodromal and acute phases of carbohydrate- induced laminitis in horses American Journal of Veterinary Rest 62: 1167-1172.

Ince, J.C., A.C. Longland, M. Moore-Colyer, J. Newbold, C. Drakley and P.A. Harris, 2005. a pilot study to estimate the intake of grass by ponies with restricted access to pasture Proceedings of the British Society of animal Science 109.(Abstr.).

Jeffcott, L.B., J.R. Field, J.G. McLean and K. Odea, 1986. Glucose tolerance and insulin sensitivity in ponies and standardbred horses EVJ 18: 97-101.

Johnson, P.J., 2002. Equine metabolic syndrome (peripheral Cushing's syndrome). Vet. Clin. N. Am.: Equine Pract. 18: 271-293.

Johnson, P.J., S.K. Ganjam, S.R. Turk and P.R. Buff, 2006. Obesity paradigm: An introduction to the emerging discipline of adipobiology. Proc. Am. Ass. equine Practnrs. 52: 41-50.

Kalck, K.A., N. Frank, S.B. Elliott and R.C. Boston, 2009. Effects of a low- dose oligofructose treatment administered via nasogastric intubation on induction of laminitis and associated alterations in glucose and insulin dynamics in hroses Am J Vet Research 70: 624-632.

Kronfeld, D., K. Treiber, T. Hess and R. Boston, 2005. Insulin resistance in the horse: definition, detection and dietetics. J Anim Sci. 83: E22-E33 77).

Kyaw-Tanner, M. and C.C. Pollitt, 2004. Equine Laminitis : increased transcription of matix metalloproteinase- 2 (MMP-2) occurs during the developmental phase Equine Vet J 36: 221-225.

Kyaw-Tanner, M.T., O. Walltle, A.W. van Eps and C.C. Pollitt, 2008. Equine laminitis : membrane type matrix metalloproteinase-14 (MMP14) is involved in acute phase onset. Equine Vet J 40: 482-487.

De Laat, M.A., M. McGowan, M.N. Sillence and C.C. Pollitt, 2010. Equine laminitis: induced by 48 h hyperinsulinaemia in standardbred horses. Equine Vet J 42: 129-135.

Lee, S., F. Bacha, N. Gungor and S.A. Arslannon, 2006. Waist circumference is an independent predictor of insulin resistance in black and white youths. J Pediatr 148: 188-194.

Leise, B.S. P.J. Johnson, R.R. Flaerios, S. Black, M.R. Watts and J. Belknap, 2009a. Hind laminar proinflammatory response is present after carbohydrate overload. In proceedings of 2nd AAEP Foundation Equine Laminitis Research Worskhop p.23.

Leise, B.S. P.J. Johnson, R.R. Flaerios, S. Black and J. Belknap, 200b9. Laminar inflammatory gene expression in the carbohydrate overload model of equine laminitis In proceedings of 2nd AAEP Foundation Equine Laminitis Research Worskhop p 48.

Loftus, J.P., J.K. Belknap, K.M. Stankiewicz and S.J. Black, 2006. Laminar xanthine oxidase, superoxide dismutase and catalase activities in the prodromal stage of black walnut induced equine laminitis Equine Vet J 39: 48-53.

Loftus, J.P., P.J. Johnson, J.K. Belknap, A. Pettigrew and S.J. Black, 2009. Leukocyte- derived and endogenous matrix metalloproteinases in the lamellae of horses with naturally acquired and experimentally induced laminitis Vet Immuno immunopath 129: 221-230.

Longland, A. and B. Byrd, 2005. Fructans and laminitis. J Nutr. 136: 2099S-2102S.

Longland, A., A. Cairns and M. Humphreys, 1999. Seasonal and diurnal changes in fructan concentration in Lolium perenne: implications for the grazing management of equines predisposed to laminitis. 16th Equine Nutritional Physiology Society Symposium; Raleigh, NC, USA, p. 258-259.

Longland, A.C., C. Barfoot and P.A. Harris, 2009. The loss of water- soluble carbohydrate and soluble protein from nine different hays soaked in water for up to 16 hours. J. Equine Vet Science 29: 383-384.

Longland, A.C. and P.A. Harris, 2009a. Comparison of a colorimetric and a high performance liquid chromatography method for the determination of fructan in pasture grasses for horses. J. Equine Vet Science 29: 376-377.

Longland, A.C. and P.A. Harris, 2009b. The ethanol-soluble carbohydrate content and composition of some common horse feeds. J. Equine Vet Science 29: 396-397.

Menzies-Gow, N.J., S.R. Bailey, P.A. Harris and J. Elliott, 2009. Use of insulin results from overnight dexamethasone suppression test to identify previously laminitic ponies. In Proceedings of the 2nd AAEP Foundation Equine Laminitis Research Workshop p 51.

Menzies-Gow, N.J., S.R. Bailey, Y. Berhane, A.C. Brooks and J. Elliott, 2008. Evaluation of the induction of vasoactive mediators from equine digital vein endothelia cells by endotoxin Am J Vet Res 69: 349-355.

Milinovich, G.J., D.J. Trott, P.C. Burrell, A.W. Van Eps, M.B. Thoefner, L.L. Blackall, R.A.M. Al Jassim, J.M. Morton and C.C. Pollitt, 2006. Changes in equine hindgut bacterial populations during oligoructose-induced laminitis. Enviromental microbiology 8: 885-898.

Milinovich, G.J., D.J. Trott, P.C. Burrell, E.L. Croser, R.A.M. Al Jassim, J.M. Morton, A.W. Van Eps and C.C. Pollitt, 2007. Fluorescence in situ hybridization analysis of hindgut bacteria associated with the development of equine laminitis. Environmental Microbiology, 9: 2090-2100.

Milinovich, G.J., P.C. Burrell, C.C. Pollitt, A. Bouvet and D.J. Trott, 2008. *Streptococcus henryi* sp. nov. and *Streptococcus caballi* sp. nov., isolated from the hindgut of horses with oligofructuose-induced laminitis. International Journal of Systematic and Evolutionary Microbiology 58: 262-266.

Muoio, D.M. and C.B. Newgard, 2008. Molecular and metabolic mechanisms of insulin resistance and β-cell failure in type 2 diabetes. Nature Reviews 9: 193-205.

Neville, R.F., T. Hollands, S.N. Collins and F.V. Keyte, 2004. Evaluation of urinary TBARS in normal and chronic laminitic ponies. Equine Vet J. 36: 292-294.

Nourian, A.R., G.I. Ba, A.W. Van Eps and C.C. Pollitt, 2007. Equine laminitis : ultrastructural lesions detected 24-30 hrs after induction with oligofructose. Equine Vet journal 39: 360-364.

Nourian, A.R. K.E. Asplin, C.M. McGowan, M.N. Sillence and C.C. Pollitt, 2009. Equine Laminitis : Ultrastructural lesions detected in ponies following hyperinsulinaemia. Equine Veterinary Journal 41: 671-677.

Nourian, A.R., P.C. Mills and C.C. Pollitt, 2010. Development of an intra-lamelaar microdialysis method for laminitis investigations in horses. The Veterinary Journal 183: 22-26.

Obel, N., 1948. Studies on the histopathology of acute laminitis (thesis) Uppsalla Almqvist and Wiksells Boktrychkeri AB, Sweden.

Owens, K., S. Pratt-Phillips, C. Kollar, R. Gray and J. Zimmerman, 2009. the effect of changes in body condition on insulin sensitivity in horses Je Equine Vet Sci. 29: 358-359.

Parsons, C.S., J.A. Orsini, R. Krafty, L. Capewell and R. Boston, 2007. risk factors for development of acute laminitis in horses during hsopitalizatin : 73 cases (1997 -2004) JAVMA 230: 885-889.

Pass, M.A., S. Pollitt and C.C. Pollitt, 1998. Decreased glucose metabolism causes separation of hoof lamellae *in vitro* : A trigger for laminitis? Equine Vet J 26: 133-138.

Peroni, J.F., J.N. Moore, M.E. Noschkam e Grafton, M. Aceves-Avila, S.J. Lewis and R.P. Robertson, 2006. Predisposition for venoconstriction in the equine laminar dermis:implications in equine laminitis. J Appl Phsyiol 100: 759-763.

Pollitt, C.C. and C.T. Davies, 1998. Equine laminitis: its development coincides with increased sublamellar blood flow. Equine Veterinary Journal Supplement 26: 125-132.

Pollitt, C.C., 2002. Pathogenesis and therapy of equine laminitis. In: Proceedings of the British Equine Veterinary Association Annual General Meeting and Sir Frederick Smith Lecture, London. pp 17-20.

Pollitt, C.C. and M. Daradka, 1998. Equine laminitis basement membrane pathology: loss of type IV collagen, type VII collagen and laminin immunostaining.Equine veterinary journal. Supplement, 26: 139-144.

De la Rebiere e Pouyade, G., A. Mouithys-Mickalad, A. Salcicia, J. Ceusters, G. Deby-Dupont and D. Serteyn, 2009. Anoxia/reoxygenation of equine endothelia celss produce reactive oxygen species: an oxymetric and elctron paramagnetic resonance investigation In proceedings of 2nd AAEP Foundation Equine Laminitis Research Worskhop p 53.

Riggs, L.M., T. Franck, J.N. Moore, T.M. Krunkosky, D.J. Hurley, J.F. Peroni, G. De la Rebiere and D.A. Serteyn, 2007. Netutrophil myeloperoxidase measurements in plasma laminar tissue and skin of horses given black walnut extract. Am J vet Res 68: 81-86.

Robertson, T.P., S.R. Bailey and J.F. Peroni, 2009. Equine Laminitis: a journey to the dark side of venous. Vet Immunol Immunopath 129: 164-166.

Robertson, T.P., E. Noschka, M. Lopes, S. Kwon, J.F. Peroni and J.N. Moore, 2009. Contractile responses of laminar blood vessels isolated from hrose given black walnut heartwood extract, carbohydrate overload or LPS infusion: implications for equine laminitis. In proceedings of 2nd AAEP Foundation Equine Laminitis Research Worskhop p 54.

Santos, A.S., E. Verhees, W.F. Pellikaan, A.F.B. Van der Poeal and R.S. Silva, 2009. Using real time ultrasound measurements to monitor small variations in body condition of horses In: EAAP book of abstracts. Wageningen Academic Publishers, Wageningen, the Netherlands, p. 423.

Splan, R., D. Kronfeld, K. Treiber, T. Hess and W. Staniar, 2005. Genetic predisposition for laminitis in ponies. Proceedings of the 19th Symposium of the Equine Science Society; 2005; Tuscon Arizona, USA, p. 219-220.

Stanhope, K.L., J.M. Schwarz, N.L. Keim, S.C. Griffen, A.A. Bremer, J.L. Graham, B. Hatcher, C.L. Cox, A. Dyachenko, W. Zhang, J.P. McGahan, A. Seibert, R.M. Krauss, S. Chiu, E.J. Schaefer, M. Ai, S. Otokozawa, K. Nakajima, T. Nakano, C. Beysen, M.K. Hellerstein, L. Berglund and P.J. Havel, 2009. Consuming fructose-sweetened, not glucose-sweetened, beverages increases visceral adiposity and lipids and decreases insulin sensitivity in overweight/obese humans. J Clin Invest. 119: 1322-1334.

Stokes, A.M., M.L. Keowen, M. McGreachy, K. Carlisel and F. Garza, 2009. Potential role of the Toll- like receptor signalling pathway in equine laminitis. In proceedings of 2nd AAEP Foundation Equine Laminitis Research Workshop p. 56.

Suagee, J.K., B.A. Col, M.V. Crisman, J.G. Wearn, L.J. McCutchoen and R.J. Geor, 2010. De novo fatty acid synthesis and NADPH generation in equine adipose and liver tissue. Comparative biochemistry and physiology Part B 155: 322-326.

Tadros, L., N. Frank and D.W. Horohov, 2009. Inflammatory cytokine blood messenger RNA expression during the development of oligfructose-induced laminitis in horses. In proceedings of 2nd AAEP Foundation Equine Laminitis Research Workshop p. 22.

Tinworth, K.D., P.A. Harris, M.N. Sillence and G.K. Noble, 2010. Potential treatments for insulin resistance in the horse: A comparative multi-species review. Vet J (in press).

Toth, F., N. Frank, S.B. Elliott, R.J. Geor and R.C. Boston, 2008. Effects of an intravenous endotoxin challenge on glucose and insulin dynamics in horses. Am J vet Res 69: 82-88.

Treiber, K.H., D.S. Kronfeld, T.M. Hess, R.C. Boston and P.A. Harris, 2005. Use of proxies and reference quintiles obtained from minimal model analysis for determination of insulin sensitivity and pancreatic beta-cell responsiveness in horses. Am J Vet Res 66: 2114-2121.

Treiber, K.H., D.S. Kronfeld, T.M. Hess, B.M. Byrd and R.K. Splan, 2006. Evaluation of genetic and metabolic predispositions and nutritional risk factors for pasture-associated laminitis in ponies. J Am Vet Med Assoc 228: 1538-1545.

Treiber, K.H., T.M. Hess, D.S. Kronfeld, R.C. Boston, R.J. Geor and P.A. Harris, 2007. Insulin resistance and compensation in laminitis-predisposed ponies characterised by the minimal model. Pferdeheilkunde 23: 237-240.

Treiber, K.H., R.A. Carter, P.A. Harris and R.J. Geor, 2008a. Seasonal changes in energy metabolism of ponies coincides with changes in pasture carbohydrates: Implications for laminitis. J Vet Int Med 22: 735-736 (abstract).

Treiber, K.H., R.J. Geor, R.C. Boston, T.M. Hess, P.A. Harris and D.S. Kronfeld, 2008b. Dietary energy source affects glucose kinetics in trained Arabian geldings at rest and during endurance exercise. J. Nutr. 138: 964-970.

Treiber, K., R. Carter, L. Gay, C. Williams and R. Geor, 2009. Inflammatory and redox status of ponies with a history of pasture- associated laminitis. Veterinary Immunology and Immunopathology 129: 216-220.

True, R.G., J.E. Lowe and J. Heissen, 1978. Black walnut shavings as a cause of acute laminitis. Proc. Am. Ass. equine Practnrs. 24: 511-516.

Van Eps, A.W. and C.C. Pollitt, 2004. Equine laminitis: Cryotherapy reduces the severity of the acute lesion. Equine Veterinary Journal 36: 255-260.

Van Eps, A.W. and C.C. Pollitt, 2006. Equine laminitis induced with oligofructose. Equine veterinary journal 38: 203-208.

Van Eps, A.W. and C.C Pollitt, 2009. Equine laminitis model: lamellar histopathology 7 days after induction with oligofructose. Equine Vet J 41: 735-740.

Van Eps, A., B. Leise, M. Watts, C. Pollitt and J. Belknap, 2010. Digital hypothermia inhibits early laminar inflammatory signalling in the oligofructose laminitis model. J Vet Int Med 24: 690 (abstract).

Van Weyenberg, S., M. Hesta, J. Buyse and G.P.J. Janssens, 2008. The effect of weight loss by energy restriction on metabolic profile and glucose tolerance in ponies. JAPAN 92: 538-545.

Vervuert, I., K. Voigt, T. Hollands, D. Cuddeford and M. Coenen, 2009. Effect of feeding increasing quantities of starch on glycaemic and insulinaemic responses in healthy horses. Veterinary Journal 1: 67-72.

Vick, M.M., A.A. Adams, B.A. Murphy, D.R. Sessions, D.W. Horohov, R.F. Cook, B.J. Shelton and B.P. Fitzgerald, 2007. Relationship among inflammatory cytokines, obesity and insulin sensitivity in the horse. J. Anim. Sci. 85: 1144-1155.

Visser, M.B. and C.C. Pollitt, 2009. Temporal changes in basement membrane proteolysis and protease expression during laminitis development. In proceedings of 2nd AAEP Foundation Equine Laminitis Research Worskhop p. 33.

Visser, M.B. and C.C. Pollitt, 2009b. Leucocyte inflitration and the role of IL-6 during the development of oligofructose induced laminitis In proceedings of 2nd AAEP Foundation Equine Laminitis Research Worskhop p. 59.

Wagner, I.P. and H. Heymering, 1999. Historical perspectives on laminitis Veterinary clinics of North America. Equine Practice 15: 295-309.

Wattle, O. and C.C Pollitt, 2004. Lamellar metabolism Clinical Techniques in equine Practice 13: 22-33.

Wattle, O. and E. Hansson, 2009. Histopathology of equine hoof wall, skin and chestnut in acute spontaneous laminitis. In proceedings of 2nd AAEP Foundation Equine Laminitis Research Worskhop p. 60.

Wyse, C.A., K.A. McNie, V.J. Tannahil, J.K. Murray and S. Love, 2008. Prevalence of obesity in riding horses in Scotland. Vet Rec 162: 590-591.

Yin, C. A. Pettigrew, J.P. Loftus, S.J. Black and J.K. Belknap, 2009. Tissue concentrations of 4-HNE in the balck walnut extract model of laminitis: indicatin of oxidant stress in affected laminae. Vet immunol Immunopath 129: 211-215.

Znamirowska, A., 2005. Prediction of horse carcass composition using linear measurements. Meat Sci. 69: 567-560.

Nutritional risk factors for the development of hyperlipaemia in a population of donkeys

F. Burden, E. Hazell-Smith and A. Trawford
The Donkey Sanctuary, Sidmouth, Devon, EX10 0NU, United Kingdom

Introduction

Hyperlipaemia is a metabolic disorder to which donkeys are particularly susceptible with mortality rates of 65% being reported. Feeding practices are commonly cited risk factors for the development of this disease; however, to date there has been no scientific data to support this contention. The aim of this study was to determine nutrition related risk factors for the development of hyperlipaemia in a population of donkeys.

Methods

Records of all donkeys resident on Donkey Sanctuary farms (n=3,619) between 1st January 2005 and 31st December 2008 were examined and cases of hyperlipaemia recorded. For each case two matched control cases were chosen at random and a retrospective case-control study was carried out. Potential risk factors that were examined included: body condition score (BCS), feeding of concentrate, concentrate type, concentrate feeding regime, changes in feeding and access to grazing. Odds ratio (OR) analyses were carried out to identify risk factors that would benefit from further study, statistical significance was held at $P<0.05$.

Results and discussion

459 cases of hyperlipaemia were recorded. Risk factors found to be significantly associated ($P<0.05$) with the development of hyperlipaemia included donkeys being given concentrate feeds (OR=2.37), changes in feeding practices either 2 (OR=3.23) or 4 weeks (OR=3.03) before disease onset and weight loss within the past month of more than 10 kg (OR=6.74). BCS was not significantly associated with the development of hyperlipaemia.

The results of this study show that careful management of feeding regimes for donkeys is extremely important in order to avoid hyperlipaemia. Ideally donkeys should be fed with fibre *ad lib* with concentrate feeds being avoided where possible. When concentrate feeds are administered care must be taken to ensure that changes in feed are made gradually. Donkeys have specific nutritional needs and it is essential that they are addressed to avoid the onset of this serious disease.

Validation of the deuterium oxide (D_2O) dilution method for the measurement of body fat content in living ponies

A. Dugdale[1], G. Curtis[1], P.A. Harris[2], E. Milne[3] and C.M.G. Argo[1]
[1]*University of Liverpool, School of Veterinary Science, Leahurst Campus, Chester High Road, Neston, Wirral, CH64 7TE, United Kingdom*
[2]*Equine Studies Group, Waltham Centre for Pet Nutrition, Freeby Lane, Waltham-on-the-Wolds, Leicestershire, LE14 4RT, United Kingdom*
[3]*Rowett Research Institute, Greenburn Rd, Bucksburn, Aberdeen, AB21 9SB, United Kingdom*

Introduction

Excessive or depleted body fat is associated with increased risk of morbidity and mortality in Equidae. Currently, body composition assessment is dependent on subjective body condition scores (BCS). This study aimed to develop an accurate method for the determination of actual body fat content (total body lipid (TBL) or white adipose tissue (WAT)), in living Equidae.

Materials and methods

Seven healthy, mature (13±2yr, 212±14 kg), Welsh section A pony mares, destined for euthanasia (for non-research purposes) were weighed (±1 kg) and BCS appraised (1 (emaciated) to 9 (obese); after Henneke *et al.*, 1983). Blood samples were collected by jugular venepuncture, before and 4 h after D_2O (99.8 atom percent excess) administration through a contralateral venous catheter (Fuller *et al.*, 2004). D_2O infusions for obese (BCS 7 to 9, n=2), moderate (BCS 4 to 6, n=3) and thin (BCS 1 to 3, n=2) ponies were 0.11, 0.12 and 0.13 g/kg respectively. Plasma was stored at -80 °C in air-tight tubes pending analysis by gas isotope ratio mass spectrometry, following zinc reduction of the stable hydrogen isotopes (Wong *et al.*, 1987). After euthanasia, total WAT mass (excluding intra-muscular WAT) was recorded (±0.1 kg) (developed from Webb and Weaver, 1979), before proximate chemical analyses of homogenised total body tissues (AOAC International, 2000).

Results and discussion

Coefficients of determination (R^2) following independent regression of D_2O-derived body fat estimates on TBL and WAT contents were 0.99 and 0.98 respectively. Bland-Altman plots confirmed that the D_2O dilution technique provided a useful alternative to either dissection or chemical analysis in the measurement of body fat content. D_2O dilution under-estimated TBL by 1.78±1.2% and WAT by 0.7±1.8%. Calibration factors were slightly increased with increased BCS. Importantly, body fat content increased exponentially as opposed to linearly with increased BCS.

This study provides the first validation of a minimally-invasive protocol for the accurate and objective measurement of body fat content in living Equidae.

Acknowledgements

The study was supported by World Horse Welfare and the BBSRC.

References

AOAC International, 2000. Official methods of analysis of AOAC International. 17th Edition. Ed: Horowitz W. AOAC International, Gaithersburg, MD, USA.

Fuller, Z., C.A. Maltin, E. Milne, G.S. Mollison, J.E. Cox and C. McG. Argo, 2004. Comparison of calorimetry and the doubly labelled water technique for the measurement of energy expenditure in Equidae. Anim. Sci. 78, 293-303.

Henneke, D.R., G.D. Potter, J.L. Kreider and B.F. Yeates, 1983. Relationship between condition score, physical measurements and body fat percentage in mares. Equine Vet. J. 15, 371-372.

Webb, A. and B.M.Q. Weaver, 1979. Body composition of the horse. Equine Vet. J. 11, 39-47.

Wong, W.W., L.S. Lee and P.D. Klein, 1987. Deuterium and oxygen-18 measurements on microliter samples of urine, plasma, saliva and human milk. Am. J. Clinical Nutr. 45, 905-913.

Comparison of two, practical weight loss protocols for the management of overweight and obese horses and ponies

G.C. Curtis[1], C.F. Barfoot[2], A.H.A. Dugdale[1], P.A. Harris[3], D. Grove-White[1] and C.M.G. Argo[1]
[1]University of Liverpool, Department of Veterinary Science, Leahurst Campus, Chester High Road, Neston, Wirral, CH64 7TE, United Kingdom
[2]Mars Horsecare, UK Ltd Old Wolverton, Bucks, MK12 5PZ, United Kingdom
[3]Equine Studies Group, Waltham Centre for Pet Nutrition, Freeby Lane, Waltham-on-the-Wolds, Leicestershire, LE14 4RT, United Kingdom

Introduction

Obesity is a major risk factor for disease and rapid weight loss is often clinically essential. However, no evidence-base exists for safe weight loss management in Equidae. This study evaluated the efficacy of controlled feed restriction alone in promoting weight loss and characterised markers for tracking weight loss, health and welfare in obese horses and ponies

Methods and materials

Twelve animals in body condition score (BCS) 7-8.8/9, (1=emaciated to 9=obese) were included in a 16 week programme. Animals were of mixed age, breed (Shetland pony to Warmblood) and gender and demonstrated no other abnormalities. Two, equally-sized groups were established by random assignation and animals were individually housed on wood-shavings.

Feedstuffs were offered to 1.25% of body mass (BM) as dry matter (DM) daily. Group 1 (477±80 kg, 9±1.6 years), were offered grass hay (DE, 7.5 MJ/kgDM) to 1.15% of BM with 0.1% as a nutrient 'balancer' meal (DE, 13.8 MJ/kgDM; BUCKEYE®). Group 2 (484±73 kg, 10±1.9 years) received 0.8% of BM as a chaff-based, complete feedstuff (DE, 8.5 MJ/kgDM; SPILLERS®) and the remainder as hay. Hay was fed from doubled, small mesh nets.

Body fat contents, calculated following D_2O dilution (Dugdale *et al.*, in press), and insulin sensitivities (Eiler *et al.*, 2005) were determined at the outset and end of the study. BCS, BM and ultrasound-derived measures of subcutaneous and retroperitoneal fat depths were recorded weekly.

Results and discussion

Animals remained healthy and no signs of abnormal behaviour were expressed. Feeds were completely consumed and diets performed equally in achieving constant weekly rates of decrease from outset BM (0.49±0.06%) and BCS (0.07/9±0.02 points). These rates were loosely associated (R^2=0.48). Weight loss was unremarkable in 4/12 animals, emphasising individual responses to dietary restriction. Despite a trend towards a reduction in total body fat content and depths of accessible fat deposits, basal insulin concentrations decreased (39.6±29.1 mU/l to 13.2±1.6 mU/l). Even further dietary restriction may be warranted to achieve significant weight loss in 'resistant' animals when increased exercise is not possible.

References

Dugdale A.H.A., Curtis G.C., Cripps P.J., Harris P.A., Argo C.McG. In Press. Effect of dietary restriction on body condition, composition and welfare of overweight and obese pony mares. Equine Veterinary Journal.

Eiler, H, Frank, N, Andrew, FM, Oliver, JW and Fecteau, KA 2005. Physiological measurements of blood glucose homeostasis via combined intravenous glucose and insulin tolerance testing in Horses. Am. J. Vet. Res. 66, 1598-1604.

The effect of water temperature on loss of water-soluble carbohydrates from hay soaked in water for up to 16 hours

A.C. Longland[1], C. Barfoot[2] and P.A. Harris[3]
[1]Equine and Livestock Nutrition Services, Tregaron, Ceredigion, Wales, United Kingdom
[2]Mars Horse Care UK Ltd, Old Wolverton, Buckinghamshire, United Kingdom
[3]Equine Studies Group Waltham Centre for Pet Nutrition, Waltham-on-the-Wolds, Leicestershire, United Kingdom

Introduction

Over consumption of water-soluble carbohydrates (WSC) has been associated with the onset of laminitis. In an attempt to reduce the WSC content of hay, and thereby the intakes of WSC, grass hay is often soaked in water prior to feeding to equines prone to laminitis. Animals prone to laminitis are often fed hay throughout the year, and the ambient temperature of the water used to soak their hay may be considerably less than 10 °C in winter and more than 20 °C in summer. However, the effect of temperature of the soak water on the rate or extent of WSC loss from hays has not been reported. The aim of this study, therefore, was to determine the effect of soaking hay at different temperatures on WSC loss from hay.

Methods and materials

Six, two kg segments of three meadow hays ranging in WSC content from 154-219 g WSC/kg DM, were separately submerged in 24 l of water, for 1, 3 and 16 hrs. Duplicate segments of each hay were submerged in water at ambient temperatures of 8 °C, 16 °C or in water with an initial temperature of 49 °C. Hay samples (ca. 100 g) were dried and milled prior to WSC analysis. Results were analysed by ANOVA.

Results and discussion

Average proportional losses of total WSC from the three hays 1, 3 and 16 hours post-submergence were 0.16, 0.26 and 0.33 at 8 °C, 0.31, 0.33 and 0.43 at 16 °C and 0.36, 0.35 and 0.46 at an initial water temperature of 49 °C. Overall, soaking at 8 °C generally resulted in significantly (P<0.05) reduced losses of WSC than from the same hays soaked at 16 °C or at an initial temperature of 49 °C at 1 and 3 hours post-submergence and for one hay significantly (P<0.05) reduced WSC losses at 16 hours compared with the same hay soaked at 16 °C. Thus, soaking hays at cold temperatures may result in lower losses of WSC compared to when the same hays are soaked under warmer conditions.

Part 6. Functional nutritional ingredients

Functional nutritional ingredients: science behind the claims for health

C.E. Dunnett[1] and I. Vervuert[2]
[1]Independent Equine Nutrition, P.O. Box 319, Mildenhall, Suffolk, IP28 6SY, United Kingdom
[2]Institute of Animal Nutrition, Nutrition Diseases and Dietetics, Faculty of Veterinary Medicine, University of Leipzig, 04159 Leipzig, Germany

Abstract

Dietary feed supplements to promote health are widely used in equine nutrition. However, many of the commercially available products are promoted with little scientific basis for the assertions made on their labels or in other marketing material. The aim of this paper is to discuss some of the more commonly encountered nutraceuticals in the equine diet with reference to their potential health benefits. Evaluation of the therapeutic potential of nutraceuticals for horses is difficult due to the limited species specific scientific evidence of efficacy. However, this does not preclude the existence of beneficial properties.

Keywords: nutraceutical, supplement, functional ingredient

Introduction

The preventative and ameliorative effect of diet on disease in horses remains a continued focus of research and an area that is predicted to be significantly enhanced through the use of emerging investigative technologies such as nutrigenomics. Whilst macro nutrition is recognised to have a significant role to play in diseases such as laminitis, exertional rhabdomyoloysis and developmental orthopaedic disease, the role of nutraceutical ingredients for health is more controversial. However, many of the commercially available dietary supplements are being promoted with little scientific basis for the assertions made on their labels. Besides the lack of data on efficacy of many dietary supplements in horses, there is growing concern about doping relevant substances like capsaicin as a natural ingredient from chilli pepper. McKeever and Britton (2002) summarized some of the relevant questions which should guide a decision on the use of dietary supplements to promote health and exercise performance in horses (Table 1).

Whilst there have been many other reviews published concerning the potential benefits of functional or nutraceutical ingredients in horses, these have mainly focussed on a particular area of health such as the digestive tract (Julliand, 2006) or joint function (McIlwraith, 2008). The aim of this review is to discuss a wide range of functional ingredients that are commonly used in the manufacture of feed supplements referencing the science available to assess their potential health benefits.

The term nutraceutical was developed from the words 'nutrition' and 'pharmaceutical' in the early 90's and whilst it is often used in the marketing of equine complementary feeds or 'supplements', it has no regulatory definition. In equine nutrition, nutraceuticals are commonly described as feed ingredients that may have a beneficial effect on the health of horses.

We have exploited feed ingredients such as plants, herbs and other dietary micronutrients, for their beneficial effect on disease and other ailments in our horses for generations. The medicinal properties of plants have also in some instances led to some effective treatments in established medicine.

Omega-3 polyunsaturated fatty acids

The proposed role of omega-3 fatty acids in maintaining joint and skin health, and in supporting immune function, fertility and respiratory health makes them an attractive nutraceutical ingredient

Table 1. Summary of testing the efficacy of feed supplements according to McKeever and Britton (2002).

Topic	Question
Biochemical or physiological basis	Does the claim make physiological sense? Is there contradiction of previous knowledge? Is there scientific data available in other animal species?
Utilization of the respective substance in horses	Pharmacokinetics and pharmakodynamics of the substance in the horse? Route of administration? Bioavailability? Dose-dependent effect?
Testing procedure of the substance or product in horses	Efficacy of the product? Accuracy of experimental methods? Blinded randomized placebo or control performed studies? Number of subjects? Appropriate parameters?
Statistical methods	Correct interpretation of the data?

for horses. Linseed meal, which is a rich source of the omega-3 fatty acid alpha-linolenic acid, is used fairly extensively in proprietary horse feed and supplements. However, although alpha-linolenic acid is a precursor of the longer chain more bioactive omega- 3's, eicosapentanoic acid (EPA) and eicosahexanoic acid (DHA), the efficiency of conversion is very low, estimated to be only about 5% in humans (Gerster, 1998). Nutraceutical ingredients that provide a more concentrated source of either or both EPA and DHA such as fish oils e.g. tuna oil and salmon oil, as well as some plant sources in the form of algae (Kris-Etherton *et al.*, 2000) are now more commonly seen in equine products.

It has been established that the circulating level of DHA and EPA is affected by diet. Supplementation of mares with DHA and EPA (10-40 g/day for 28 days) increased the level in plasma in a dose dependent manner following only 7 days of supplementation (King *et al.*, 2008). However, the authors noted that the plasma concentration of DHA and EPA appeared to be affected by the availability of fresh forage and that the plasma level returned to baseline within 42 days following cessation of supplementation. Contradictory to this study, (Bergero *et al.*, 2002) found no significant effect on blood serum and skin fatty acid levels in horses using 20 g of an omega-3 and omega- 6 rich oil from Purple Viper´s Bugloss seeds.

There is not a large amount of data to support the health benefits of omega- -3 fatty acids in horses, although single studies of the effect of omega-3 fatty acid supplementation on immune response, respiratory health, seasonal pruritis, sperm quality and glycaemic and insulinaemic responses, have been previously published.

Sperm cell membranes have a high DHA content, which is associated with membrane fluidity and sperm quality. In sub-fertile men, where sperm motility and sperm count is reduced, the level of DHA and the ratio of total omega-3: omega-6 fatty acids found in sperm cells is reduced compared to normal controls (Aksoy *et al.*, 2006). The DHA content of a stallion's diet may be low unless there is significant access to fresh grazing. Traditional concentrate rations are normally rich in omega-6 fatty acids, and typically low in omega-3 fatty acids including DHA. The effect of DHA supplementation on sperm quality and motility in stallions is not unequivocal. Semen from stallions supplemented with DHA for 14 weeks showed a 3 fold increase in DHA concentration and a 50% increase in the ratio of DHA to the omega-6 fatty acid docosapentanoic acid (DPA) (Brinsko *et al.*, 2005). There was no significant effect of supplementation on motility of fresh semen, but the motion characteristics of cooled semen were suggested to improve, especially in those stallions with marginal fertility under AI conditions. In contrast, supplementation with 29 g of long chain omega-3 PUFA for 90 days failed to alter motility characteristics of either fresh or cold stored semen, despite a significant 46% increase in daily sperm output (Forsyth *et al.*, 2006).

There has also been increased interest in the omega- 3 fatty acid composition of the diet with respect to respiratory disease. In humans, a protective role for omega-3 fatty acids against human asthma has been suggested (McKeever and Britton, 2004), although clinical trials have been equivocal, see review (Raviv and Smith, 2010). Pulmonary function was not altered in horses with recurrent airway obstruction, a condition with clinical similarities to human asthma, when they were supplemented with seal blubber oil, a rich source of long chain omega-3 fatty acids (Khol-Parisini *et al.*, 2007). However, the omega-3: omega-6 ratio in plasma and leukocyte plasma membranes was increased and the leukocyte counts in epithelial lung lining fluid were also reduced in the omega-3 supplemented horses. This may suggest an effect of omega- 3 supplementation on pulmonary inflammation. This idea was supported by (Hall *et al.*, 2004) as the ratio of omega-3: omega-6 in plasma was associated with an increased production of less inflammatory eicosanoids in neutrophils obtained from healthy horses. However a major flaw of the study of (Khol-Parisini *et al.*, 2007) was the failure to express leukocyte counts as a percentage of epithelial lung lining fluid as the volume of lung lining fluid obtained varied between the horses.

The beneficial effect of dietary omega- 3 fatty acids on anti-inflammatory conditions such as arthritis in humans is supported by a meta analysis of controlled trials investigating the pain relieving effects of omega- 3 fatty acids (Goldberg and Katz, 2007). Supplementation of arthritic horses with 15 g and 19.8 g of EPA and DHA respectively for 90 days suggests some benefit as the concentration of white blood cells in synovial fluid and concentration of inflammatory marker PGE_2 in plasma was reduced (Manhart *et al.*, 2009). However, a lower intake of omega- 3 fatty acids (5.95 g for 75 days) failed to show any significant effect on lameness score, although there was a reported trend for an increase in trotting stride length (Woodward *et al.*, 2007).

In humans, the inclusion of omega- 3 fatty acids into the diet is of added interest as they may be beneficial for glucose homeostasis and insulin sensitivity (Riserus, 2008; Vessby, 2000). Although the mechanisms are still unclear, it has been suggested that the quality of dietary fat namely omega-3 fatty acids affects the fatty acid composition of cell membranes, and thus cell membrane function (Lombardo and Chicco, 2006; Riserus, 2008). The fatty acid composition of the cell membranes may affect several cellular functions, such as the translocation of glucose transporters, cell membrane fluidity and ion permeability (Taouis *et al.*, 2002). Furthermore, data suggests there are regulatory effects on gene expression and enzyme activity (Taouis *et al.*, 2002). As in humans, there is some evidence that chronic adaptation to high-glycaemic diet may lead to insulin resistance in horses (Treiber *et al.*, 2005). The addition of fish oil to a starchy meal, however, did not affect acute glucose and insulin responses (Vervuert *et al.*, 2010). To what extent long-term fish oil supplementation may improve insulin sensitivity in healthy or in insulin-resistant horses needs further clarification.

Bioavailable silicon

Research has continued into the potential benefits of one of the less recognised trace elements silicon on bone strength and integrity. Silicon is a natural constituent of plants and provides structure and rigidity to cell walls. It therefore forms a natural part of the horse's diet, however the availability in feed ingredients may be limited. Silicon plays a role in the development of new bone and is also important for the calcification process. Research in the early nineties suggested that supplementation with silicon from sodium aluminium silicate, otherwise known as sodium zeolite A (SZA) (2% of dietary intake) may reduce bone resorption (Frey *et al.*, 1992). In addition, feeding pigs a Si containing supplement reduced the overall osteochondrosis score (Frantz *et al.*, 2008) and subsequent data reported a significant decrease in injury rates in quarter horses fed a similar form and amount of silicon (Nielsen *et al.*, 1993).

Latterly it has also been established that this silicon in the form of SZA is available to foals via the milk of supplemented mares hereby providing Si to the suckling foal (Lang *et al.*, 2001). The effect of SZA on osteochondrosis (OCD) has also been investigated, however, there is no current evidence to suggest a role for silicon in the healing of OCD lesions (Turner *et al.*, 2007), although the effect of silicon supplementation on the development of OCD has not been investigated.

The form of SZA, a chalk-like powder, and the level of intake used in these earlier studies (about 200 g per day for a 500 kg horse) makes its practical use as a supplement for top dressing difficult. However, concerns regarding the impact of the relatively high aluminium content of SZA on calcium availability were not supported in balance studies as calcium apparent digestibility and retention was actually increased following supplementation with SZA (O'Connor *et al.*, 2008). An alternative source of silicon that has a low aluminium content, oligomeric orthosilicic acid (OSA) has also recently been investigated for its potential use in horses. Preliminary data showed a trend towards increased plasma silicon concentration and an improved retention of silicon with OSA supplementation (~30 ml per day) compared to a control and SZA supplemented group (200 g per day) (O'Connor *et al.*, 2008). However, up to date there is no established Si requirement for horses.

Pre and probiotics

Numerous and varied nutraceuticals are marketed that target the equine digestive system. These include probiotics (which may consist of live bacteria or live yeast), prebiotics or fructo-oligosaccharides (ScFOS) that are usually derived from either chicory or sugar beet and finally mannanoligosaccharides (MOS) that are often derived components of yeast cell walls. The premise behind the use of all these additives is to maintain, or in some instances to re-establish, the population of beneficial bacteria and other microflora within the digestive tract.

Live yeasts

The live yeast *Saccharomyces cerevisiae* is one of the more extensively studied probiotic ingredients available, with the majority of the published work in horses to date having used the strain NCYC 1026. Live yeast improves fibre fermentative capacity and other aspects of digestion (Hill *et al.*, 2001). There is also growing evidence to suggest that *Saccharomyces cerevisiae* ameliorates the potentially detrimental changes to the hindgut population and environment that are associated with feeding rations that are high in starch and low in digestible fibre (Julliand, 2006), or as the result of anthelmintic treatment (Goachet *et al.*, 2004). Any protection afforded by live yeast containing feeds or supplements is likely mediated through a beneficial effect on the microbial balance including an increase in the ratio of lactate utilizing: lactate producing bacteria, as well as an increase in the level or activity of cellulolytic bacteria. Stabilisation of the microbial population is paralleled by moderation of the characteristic fall in caecal and colonic pH following feeding (Medina *et al.*, 2002; Moore and Newman, 1994). This work would suggest that live yeast containing feeds and supplements may well be useful for horses with an increased risk of gastrointestinal disturbance that can be related to an aberration in the hindgut environment and microbial population.

The direct effect of live yeast on issues such as colic, laminitis and diarrhoea remains largely unreported in the literature, although such ingredients are commonly used prophalactically by horse owners. Another strain of yeast *Saccharomyces boulardii* has, however, been reported to significantly decrease the severity and duration of acute enterocolitis in a group of hospitalised horses (Desrochers *et al.*, 2005).

Previously, there has been some scepticism as to whether viable yeast cells reach their site of action in the GI tract. Certainly the recent work conducted by Gobert *et al.* (2006) confirms the transit of *Saccharomyces cerevisiae* (NCYC 1026) to the caecum and colon. These authors showed that this live yeast does not colonise the gut and is in fact rapidly removed within 72 hours once supplementation is ceased. In addition, the diversity of results obtained in horses to date under *in vitro* or *in vivo* conditions is equivocal (Table 2); therefore the efficacy of yeast supplementation remains somewhat speculative.

Table 2. Summary of literature about yeast supplementation in horses under in vitro *or* in vivo *conditions.*

Yeast	Outcome	Authors
In vitro experiments		
Saccharomyces cerevisiae, dosage 4.5 x 10^9 CFU/g	Increase in neutral detergent fiber digestibility	(Santos *et al.*, 2008)
Saccharomyces cerevisiae, dosage not specified	Decrease in ammonium, increase in acetate	(Lattimer *et al.*, 2007)
Saccharomyces cerevisiae, dosage 2 x 10^7 CFU/g	No difference in pH or short chain fatty acids	(Köpke *et al.*, 2007)
In vivo experiments		
Saccharomyces cerevisiae, dosage 4.5 x 10^9 CFU/g	Increase in acid detergent fiber digestibility in grain or forage based diets	(Jouany *et al.*, 2008)
Saccharomyces cerevisiae, dosage not specified	Reduction in aerobic bacteria No change in anaerobic bacteria, lactic acid bacteria or *lactobacillus*	(Santos *et al.*, 2008)
Saccharomyces cerevisiae, dosage 4.5 x 10^9 CFU/g	Increase in pH and decrease in lactate in grain based diets	(Medina *et al.*, 2002)
Saccharomyces cerevisiae, dosage not specifies	Increase in dry matter digestibility, crude protein and neutral detergent fiber digestibility	(Glade, 1991)

Prebiotics

The potential benefit of prebiotic ingredients for horses exposed to an increased risk of colic was investigated by (Volter, 1999) who reported a reduced incidence of colic in a group of 126 horses supplemented with short chain fructooligosaccharide (ScFOS). These findings were repeated in a subsequent trial where the degree of efficacy was reported to relate to the quantity of prebiotic administered (Julliand, 2006). The mechanism of the proposed effect on the incidence of colic in horses remains unresolved, although recent studies suggest that ScFOS may help to limit the detrimental effects of a sudden change to a high starch diet on the hindgut microflora by reducing the accumulation of lactate in the colon (Respondek *et al.*, 2008). Previously it was considered that the action of ScFOS was primarily in the hindgut, however, (Respondek *et al.*, 2005) suggests that they are also effective in the stomach. Whilst there was no significant effect on the microbial population reported here, the gastric pH was significantly higher with ScFOS supplementation, which could be relevant for gastric health.

MOS reputedly binds pathogenic bacteria within the digestive tract, thus reducing their ability to adhere to the gut wall and initiate disease. Whilst there are studies to support this action in other species, there is very little species specific work that has been carried out on MOS in horses. A single study, however, has reported that the incidence of diarrhoea where treatment was justified was significantly reduced in the foals of mares supplemented with MOS prior to and following foaling (Ott, 2005).

Chondroprotective ingredients

In horses, pain and lameness are associated with a diversity of lower limb conditions including joint disease. Joint disease is commonly diagnosed in equine medicine with a large number of cases being classified as osteoarthritis (OA). OA is a complex and painful disease process that culminates in the degradation of articular cartilage. Treatment strategies aim to reduce inflammation and pain and conventionally include acute pharmacological intervention with intra-articular corticosteroids and long-term use of non-steroidal anti-inflammatory drugs (NSAIDs). Chronic side effects associated with the latter may partly explain the popularity of alternative pharmacotherapies involving oral administration of nutraceuticals, as these are perceived as a benign treatment for equine OA. Purported benefits of nutraceuticals include their NSAID dose-reducing effect in OA cases when used in conjunction with these drugs, and their potential disease or structure modifying properties.

Currently available equine nutraceuticals in this area make extensive use of one or more common 'active' components, often referred to as chondroprotective agents. Gosh *et al.* (1992) summarized the main required characteristics of chondroprotective substances as follows:
- increase in collagen and proteoglycan synthesis of chondrocytes;
- increase in hyaluronic acid synthesis in synovial cells;
- inhibition of enzymes which destroy cartilage;
- inhibition of fibrin synthesis in synovial and subchondral vessels.

Chondroitin sulphate and hyaluronate are components of joint cartilage itself, whereas glucosamine, methylsulphonylmethane (MSM), glutamine and glucuronate are precursors in the formation glucosaminoglycan chains within proteogylycans. Other substances which are involved in cartilage metabolism are summarized in Table 3 (Mautone *et al.*, 2000).

Glucosamine

Glucosamine is probably the most commonly used chondroprotective ingredient in equine feeds and supplements. However, there is still controversy surrounding its efficacy. A recent *in vitro* study investigating the effect of clinically relevant concentrations of glucosamine on equine chondrocytes and synoviocytes reported that glucosamine had no impact on either proteoglycan synthesis or matrix metalloproteinase either under inflammatory or non inflammatory conditions. Although the authors suggested some local anti-inflammatory effects mediated via a decrease in IL-1 stimulated PGE2 and microsomal PGE2 synthase (Byron *et al.*, 2008). It has also recently been suggested that the concentration of glucosamine attained in synovial fluid following oral administration is significantly higher in inflamed compared to normal joints, which may mediate a therapeutic effect (Meulyzer *et al.*, 2009). Additionally despite the lower delivery of glucosamine from glucosamine sulphate, this

Table 3. Substances which are involved in cartilage metabolism.

Substance	Importance
Manganese, vitamin A	Co-factor of glucosaminoglycan synthesis
Copper, vitamin C, gelatine	Collagen synthesis
Omega- 3 fatty acids	Decrease in inflammatory response
Selenium, vitamin E	Antioxidants

source resulted in a greater and more prolonged elevation in glucosamine concentration in plasma and synovial fluid following both IV and oral administration of an equivalent dose (Meulyzer *et al.*, 2008).

The initial low bioavailability of glucosamine in horses (Du *et al.*, 2004; Laverty *et al.*, 2005) has subsequently been suggested to represent a significant underestimate, as the chemical analysis did not account for the binding of glucosamine to serum proteins (Pearson and Lindinger, 2008). Consequential to the lack of conclusive data on the effective dosage of oral glucosamine supplementation, there is a considerable range in the delivered level of glucosamine and chondroitin sulphate in commercial products. For example, Oke *et al.*, (2006) reported a range of glucosamine concentrations from 1.7-29.6 mg/50 mg product. According to information provided on the product labels, the average recommended daily dose ranged from 1800 to 12,000 mg glucosamine for an average sized mature horse (Oke *et al.*, 2006).

There have now been many published trials investigating the efficacy of glucosamine or glucosamine containing supplements on joint function, however, according to Pearson and Lindinger (2008) many of these are confounded due to experimental design, or other factors meaning that firm conclusions can still not be drawn on the efficacy of glucosamine for joint health. For example, the oral administration of glucosamine hydrochloride and chondroitin sulphate improved stride characteristics in old horses; however the treatment group included 15 horses whereas the control group was represented by only 5 horses. In addition, the allocation to the respective supplementation group, which was dependent on osteoarthritis history, was not randomised (Forsyth *et al.*, 2006). Oral treatment with a glucosamine hydrochloride and chondroitin sulphate compound improved lameness score in horses, however placebo group and management of the horses during supplementation period (e.g. rest) was not defined in that study.

In contrast, a randomized, blinded-controlled study, using a carpitis model induced by intra-articular injection of Freund´s Complete Adjuvant in horses, the oral supplementation of sulphated glycosaminoglycans did not improve lameness score, stride length or other parameters related to carpal joint injury (White *et al.*, 1994).

Herbal analgesic anti-inflammatories

Pain and inflammation associated with pathological conditions in horses are commonly treated with NSAID, such as phenylbutazone, flunixin and ketoprofen. These drugs are effective for the control of mild to moderate pain, but present potential side effects, notably gastric ulceration, when used chronically.

A number of herbal products are marketed as complementary alternatives to conventional therapeutic agents, some of which contain extracts of Willow bark or other *Salix* species. Such plants contain naturally high levels of salicylates and have been used for many years for pain, inflammation and fever. In contrast to humans however, salicylates are relatively ineffective analgesic anti-inflammatories in horses. Other herbal products available include extracts or dried forms of Devil's Claw, Feverfew, Turmeric, and *Boswellia* sp.

Devil´s Claw (*Harpagophytum procumbens*) is widely used in commercial products for the treatment of painful and osteochondrotic conditions (Brien *et al.*, 2006). Devil´s Claw is an extract obtained from the root of the *Harpagophytum procumbens*, a member of the sesame family found in the Kalahari region in South Africa. It has been shown that this remedy has anti-inflammatory and analgesic effects making it of particular use in the treatment of osteoarthritis (see review Brien *et al.*, 2006). A clear mechanism for anti-inflammatory action has yet to be established, although it is purported to inhibit the arachidonic, the cyclo-oxygenase, and the lip-oxygenase pathways (Tippler *et al.*, 1996). The potential active constituents are supposed to be iridoid glycosides and harpagoside and possibly phenol derivates (Lanhers *et al.,* 1992). From the available clinical studies in humans, it is yet not possible to conclude that Devil´s Claw is effective or not. Major flaws of the performed studies in humans include preparations of Devil´s Claw in terms of extract ratio and the concentrations of active constituents, classification of osteoarthritis and other methodological limits like blinding or parameters to assess efficacy (Brien *et al.,* 2006). Beside the high number of supplements containing Devil´s Claw in the treatment of painful and osteochondrotic conditions in horses, equine clinical studies have not been published.

Feverfew (*Tanacetum parthenium*) is a popular herbal remedy used in humans for the treatment of migraine, fever and arthritis and it is reported to have analgesic, anti-inflammatory and anti-pyretic properties. Parthenolide is believed to be the active constituent and *in vitro* experiments have shown it to be a potent inhibitor of platelet aggregation. Animal studies suggest a prophylactic effect in rheumatoid arthritis and inhibition of oedema.

Curcumin and boswellic acid have been shown to inhibit the production of inflammatory mediators. Curcumin, the major pharmacologically active component of Turmeric, is reported to block the production of prostaglandins and thromboxane by selectively inhibiting COX-2 (Maheshwari *et al.,* 2006), whereas boswellic acid reduces the production of leukotrienes through its inhibitory action on lipoxygenases (Banno *et al.,* 2006). Although herbal anti-inflammatory products such as these are marketed, no trials have been published that describe efficacy or toxicity in horses.

Capsaicin is the biochemical constituent that gives the pungent characteristics of *Capsicum* species (chilli peppers). Its principal pharmacological actions are the induction of excitation of the nociceptive phase, familiar to us as the hot/burning sensation after ingestion, and a subsequent analgesic anti-inflammatory action of longer duration, as reviewed by (Hayman and Kam, 2008). The potential therapeutic usefulness of capsaicin has recently been evaluated in horses. In studies, capsaicin was shown to relax tracheal smooth muscle *in vitro* (Zhu *et al.,* 1997), and topical perineural capsaicin application was shown to have a significant analgesic effect in an experimental reversible foot lameness model (Kathy *et al.,* 2003). Recently, the use of dietary capsaicin to benefit joint health and immune function in horses has also been investigated. Despite encouraging results from an *in vitro* study utilising macrophage cultures and an *in vivo* study in mice, dietary capsaicin at dose rates of 50 or 100 mg/day for 28 days in the form of dried jalapeno powder had no observable beneficial effects on synovial PGE2 production, nor on acute phase and humoral responses to antigenic challenge (Hardin *et al.,* 2007). Capsaicin and its botanical derivatives appear to have some potential therapeutic efficacy through topical application. It should be noted however, that in the UK capsaicin-containing products intended for topical therapeutic use offered without veterinary marketing authorisation are limited to an active ingredient level no greater than 0.012%. Furthermore it must be underlined that capsaicin is a doping relevant substance.

Immunomodulatory ingredients

The immune system in horses is complex and relies upon many functional elements to deliver a comprehensive defence against infection and disease. The impact of diet on immunity in horses is a relatively new and emerging area of research and although the number of studies undertaken in this field are quite limited, there are a few ingredients worthy of discussion.

Glutamine

Glutamine is one of the most abundant amino acids within the horse and it has a number of significant physiological roles ascribed to it. Amongst these is the provision of a fuel supply for cells of the gastro-intestinal tract and the immune system. It is largely accepted in human medicine that glutamine has a major impact on the functionality of the immune system and meta-analysis of clinical studies has concluded that glutamine supplementation has a beneficial effect on infectious complications and reduces mortality when included in parenteral nutrition (see review) (Melis *et al.*, 2004). Reduced glutamine availability and impaired immune response are associated with prolonged intensive training and competition, which may lead to viral infection (Castell and Newsholme, 2001). Similarly, in horses, even a mild viral challenge is associated with a major depletion of plasma glutamine (>30%, (Routledge *et al.*, 1999). Dietary supplementation in horses with both glutamine and a glutamine-rich water-stabilised peptide effectively increased plasma glutamine concentration (Harris *et al.*, 2006). However, the efficacy of dietary glutamine supplementation has not as yet been evaluated in horses during intense training or periods of infection.

Echinacea

This plant, which is indigenous to North America, is described in herbal texts as being beneficial during chronic viral and bacterial infections and where immunosuppression is identified. The rationale for the use of Echinacea was strengthened by the publishing of a meta-analysis of 14 randomised controlled trials showing an overall 65% reduction in the incidence of the common cold in human subjects, where Echinacea was provided as a preventative ingredient in their diet. This effect was improved when subjects were fed a combination of Echinacea and vitamin C (Shah *et al.*, 2007).

In horses, the authors of a double-blind placebo controlled trial, where a standardized powdered root extract of *Echinacea angustifolia* was fed for 42 days, reported an improvement in the infection fighting capacity of white blood cells, indicated by an increase in circulating lymphocyte count. Phagocytic ability and migration capacity of neutrophils was also suggested to increase (O'Neill *et al.*, 2002). However, the conclusions drawn by these authors have subsequently been challenged (Ralston, 2007).

Ginseng

Another plant that frequently features in equine supplements claiming immune support is Ginseng. Ginseng is a perennial plant that grows primarily in the Northern hemisphere and is characterised by the presence of ginsenosides. There are several different types of true ginseng and most are described as being 'adaptogenic', which means that they increase the body's resistance to stress, anxiety and fatigue. The use of ginseng by human athletes is relatively well documented with a few studies reporting improved resistance to infection and or improvements in performance, although overall the results seem to be quite mixed (Senchina *et al.*, 2009). In horses, there is little species specific evidence to support its use currently. The immune response to 24 hours of road transport in horses was unaffected when a supplement that contained *Eleutherococcus senticosus* (Siberian Ginseng) amongst other ingredients was fed (Stull *et al.*, 2004). However, in a more recent trial, horses fed a relatively low dose of ginseng *(Panax quinquefolium)* (35 mg/kg body weight, 1.7 mg/kg total ginsenosides)) showed an improved antibody response to vaccination for equine herpes virus (EHV-1) compared to a control group of horses.

Nucleotides

Another group of ingredients that are worthy of mention, but which have not been used extensively in horses are nucleotides. Nucleotides are found ubiquitously through the body, as they form part of the basic structure of both DNA and RNA. Cells that have a high rate of turnover, such as those of the immune system and the enterocytes in the digestive tract, have a higher requirement for nucleotides. Nucleotides are found naturally in the diet but there is potential for further supplementation. Nucleotides have been added to infant milk replacers and to diets for young animals with beneficial effects reported on response to vaccine and immune status (Maldonado *et al.*, 2001). To date there are no published trials in horses exploring the potential immunological benefits of nucleotides, although two studies have been carried out to look at the effect of ribose, which is a sugar required for nucleotide synthesis, on exercise performance, however, the results from these trials were inconclusive (Kavazis *et al.*, 2004; Kavazis *et al.*, 2002).

Conclusion

The available evidence to support the many health claims used in relation to feed supplements is commonly lacking, making evaluation of their beneficial properties difficult. This is primarily due to the low number of species specific studies carried out on the many available nutraceutical ingredients. Additionally, where data is available, the small number of animals used often restricts the statistical power of these studies making statistical significance more elusive. Furthermore, despite the increasing use of nutraceutical ingredients in human products, their use in horses is restricted by the existing veterinary marketing and feeding stuffs legislation, which precludes the presentation of 'feed ingredients' as being medicinal and also requires evidence of both efficacy and safety which may legally limit the future use of efficacious nutraceuticals in horses.

References

Aksoy, Y., H. Aksoy, K. Altinkaynak, H.R. Aydin and A. Ozkan, 2006. Sperm fatty acid composition in subfertile men. Prostaglandins Leukotrienes and Essential Fatty Acids 75, 75-79.

Banno, N., T. Akihisa, K. Yasukawa, H. Tokuda, K. Tabata, Y. Nakamura, R. Nishimura, Y. Kimura and T. Suzuki, 2006. Anti-inflammatory activities of the triterpene acids from the resin of Boswellia carteri. Journal of Ethnopharmacology 107, 249-253.

Bergero, D., N. Miraglia, M. Polidori, Ziino and D. Gagliardi, 2002. Blood serum and skin fatty acid levels in horses and the use of dietary polyunsaturated fatty acids. Animal Research 51, 157-163.

Brien, S., G.T. Lewith and G. McGregor, 2006. Devil's Claw (Harpagophytum procumbens) as a Treatment for Osteoarthritis: A Review of Efficacy and Safety. The Journal of Alternative and Complementary Medicine 12, 981-993.

Brinsko, S.P., D.D. Varner, C.C. Love, T.L. Blanchard, B.C. Day and M.E. Wilson, 2005. Effect of feeding a DHA-enriched nutriceutical on the quality of fresh, cooled and frozen stallion semen. Theriogenology 63, 1519-1527.

Byron, C.R., M.C. Stewart, A.A. Stewart and H.C. Pondenis, 2008. Effects of clinically relevant concentrations of glucosamine on equine chondrocytes and synoviocytes *in vitro*. American Journal of Veterinary Research 69, 1129-1134.

Castell, L.M. and E.A. Newsholme, 2001. The relation between glutamine and the immunodepression observed in exercise. Amino Acids 20, 49-61.

Desrochers, A.M., B. Dolente, M.F. Roy, R.C. Boston and S. Carlisle, 2005. Efficacy of saccharomyces boulardii for treatment of horses with acute enterocolitis. Journal of the American Veterinary Medical Association 227, 954-959.

Du, J., N. White and N.D. Eddington, 2004. The bioavailability and pharmacokinetics of glucosamine hydrochloride and chondroitin sulfate after oral and intravenous single dose administration in the horse. Biopharmaceutics and Drug Disposition 25, 109-116.

Forsyth, R.K., C.V. Brigden and A.J. Northrop, 2006. Double blind investigation of the effects of combined Glucosamine Hydrochloride (GHCL) and Chondroitin Sulphate (CS) on stride characteristics of veteran horses. Equine Veterinary Journal Supplement. 36, 622-625.

Frantz, N.Z., G.A. Andrews, M.D. Tokach, J.L. Nelssen, R.D. Goodband, J.M. DeRouchey and S.S. Dritz, 2008. Effect of dietary nutrients on osteochondrosis lesions and cartilage properties in pigs. American Journal of Veterinary Research 69, 617-624.

Frey, K.S., G.D. Potter, T.W. Odom, D.M. Senor, V.D. Reagan, V.H. Weir, J. Elslander, S.P. Webb, E.L. Morris, W.B. Smith and K.E. Wiegand, 1992. Plasma silicon and radiographic bone density in weanling Quarter Horses fed sodium zeolite A. Journal of Equine Veterinary Science 12, 291-295.

Gerster, H., 1998. Can adults adequately convert alpha-linolenic acid to eicosapentaenoic acid and docosahexaenoic acid. International Journal for Vitamin and Nutrition Research 68, 159-173.

Glade, M.J., 1991. Effects of dietary yeast culture supplementation of lactating mares on the digestibility and retention of the nutrients delivered to nursing foals via milk. Journal of Equine Veterinary Science 11, 323-329.

Goachet, A.G., J.M. Ricard, E. Jacotot, M. Varloud and V. Julliand, 2004. Effet de l'administration orale de trois anthelminthiques sur microfloe colique cheval. Congrès de l'Association Vétérinaire Équine Française pp.

Gobert, J., G. Bertin and V. Julliand, 2006. Digestive fate of Saccharomyces cerevisiae CBS 493 94, fed at 3 different concentrations to horses. Reproduction Nutrition Development 46 (Suppl 1).

Goldberg, R.J. and J. Katz, 2007. A meta-analysis of the analgesic effects of omega-3 polyunsaturated fatty acid supplementation for inflammatory joint pain. Pain 129, 210-223.

Gosh, P., M. Smith and C. Wells, 1992. Second-line agents in osteoarthritis. In: Dixon, J. and D. Furst (ed.) Second-line agents in the treatment of rheumatic diseases. Marcel Dekker Inc, New York. 363-427 pp.

Hall, J.A., R.J. Saun, S.J. Tornquist, J.L. Gradin, E.G. Pearson and R.C. Wander, 2004. Effect of Type of Dietary Polyunsaturated Fatty Acid Supplement (Corn Oil or Fish Oil) on Immune Responses in Healthy Horses. Journal of Veterinary Internal Medicine 18, 880-886.

Hardin, S., L. Potts, J. Waggoner, C. Löest and J. Turner, 2007. Dietary Capsaicin Does Not Alter Synovial Concentrations of Prostaglandin E2 or the Acute Phase Response in Aged Horses after Antigenic Challenge. Journal of Equine Veterinary Science 27, 494-498.

Harris, R.C., P.A. Harris, N.B. Routledge, J.R. Naylor and A.M. Wilson, 2006. Plasma glutamine concentrations in the horse following feeding and oral glutamine supplementation. Equine veterinary journal. Supplement, 637-642.

Hayman, M. and P.C.A. Kam, 2008. Capsaicin: A review of its pharmacology and clinical applications. Current Anaesthesia and Critical Care 19, 338-343.

Hill, J., S.V. Tracey, M. Willis, L. Jones and A.D. Ellis, 2001. Yeast culture in equine nutrition and physiology. In: Jacques, L.a. (ed.) Proceedings of Alltech's 17th Annual Symposium. Nottingham University Press, Kentucky, USA. 97-114 pp.

Jouany, J.P., J. Gobert, B. Medina, G. Bertin and V. Julliand, 2008. Effect of live yeast culture supplementation on apparent digestibility and rate of passage in horses fed a high-fiber or high-starch diet. Journal of Animal Science 86, 339-347.

Julliand, V., 2006. Pre and probiotics: potential for equine practice. In: Association, E.H.a.N. (ed.) 3rd European Equine Health and Nutrition Congress. Merelbeke, Belgium. 1-6 pp.

Kathy, K.S., H.F. Jonathan, A.G. Stephen, E.G. Thomas and G.J. Benson, 2003. Effects of Topical Perineural Capsaicin in a Reversible Model of Equine Foot Lameness. Journal of Veterinary Internal Medicine 17, 563-566.

Kavazis, A.N., J. Kivipelto, H.S. Choe, P.T. Colahan and E.A. Ott, 2004. Effects of ribose supplementation on selected metabolic measurements and performance in maximally exercising Thoroughbreds. Journal of Animal Science 82, 619-625.

Kavazis, A.N., J.S. Sobota, J. Kivipelto, M.B. Porter, P.T. Colahan and E.A. Ott, 2002. Ribose supplementation in maximally exercising Thoroughbreds. Equine Veterinary Journal Supplement. 34, 191-196.

Khol-Parisini, A., R. van de Hoven, S. Leinker, H.W. Hulan and J. Zentek, 2007. Effects of feeding sunflower oil or seal blubber oil to horses with recurrent airway obstruction. Canadian Journal of Veterinary Research 71, 59-65.

King, S.S., A.A. Abughazaleh, S.K. Webel and K.L. Jones, 2008. Circulating fatty acid profiles in response to three levels of dietary omega-3 fatty acid supplementation in horses. Journal of Animal Science 86, 1114-1123.

Köpke, K., B. Graveley, M. Bochnia, S. Klein, I. Vervuert, D. Cuddeford and M. Coenen, 2007. Effects of yeast on *in vitro* fermentation of horse diets. In: Coenen, M. and I. Vervuert (ed.) 11th Congress European Society of Veterinary and Comparative Nutrition. Merkur Druck- und Kopierzentrum GmbH, Leipzig, Germany. 62 pp.

Kris-Etherton, P.M., D.S. Taylor, S. Yu-Poth, P. Huth, K. Moriarty, V. Fishell, R.L. Hargrove, G. Zhao and T.D. Etherton, 2000. Polyunsaturated fatty acids in the food chain in the United States. American Jouranal of Clinical Nutrition 71, 179S-188.

Lang, K.J., B.D. Nielsen, K.L. Waite, G.M. Hill and M.W. Orth, 2001. Supplemental silicon increases plasma and milk silicon concentrations in horses. Journal of Animal Science 79, 2627-2633.

Lanhers, M.-C., J. Fleurentin, F. Mortier, A. Vinche and C. Younos, 1992. Anti-Inflammatory and Analgesic Effects of an Aqueous Extract of Harpagophytum procumbens. Planta Medica 58, 117-123.

Lattimer, J.M., S.R. Cooper, D.W. Freeman and D.L. Lalman, 2007. Effect of yeast culture on *in vitro* fermentation of a high-concentrate or high-fiber diet using equine fecal inoculum in a Daisy II incubator. Journal of Animal Science 85, 2484-2491.

Laverty, S., J.D. Sandy, C. Celeste, P. Vachon, J. Marier and A.H.K. Plaas, 2005. Synovial fluid levels and serum pharmacokinetics in a large animal model following treatment with oral glucosamine at clinically relevant doses. Arthritis and Rheumatism 52, 181-191.

Lombardo, Y.B. and A.G. Chicco, 2006. Effects of dietary polyunsaturated n-3 fatty acids on dyslipidemia and insulin resistance in rodents and humans. A review. The Journal of Nutritional Biochemistry 17, 1-13.

Maheshwari, R.K., A.K. Singh, J. Gaddipati and R.C. Srimal, 2006. Multiple biological activities of curcumin: A short review. Life Sciences 78, 2081-2087.

Maldonado, J., J. Navarro, E. Narbona and A. Gil, 2001. The influence of dietary nucleotides on humoral and cell immunity in the neonate and lactating infant. Early Human Development 65, S69-S74.

Manhart, D.R., B.D. Scott, P.G. Gibbs, J.A. Coverdale, E.M. Eller, C.M. Honnas and D.M. Hood, 2009. Markers of Inflammation in Arthritic Horses Fed Omega-3 Fatty Acids. The Professional Animal Scientist 25, 155-160.

Mautone, G., M. Donohoe, T.E. McAlindon, M.P. LaValley and D.T. Felson, 2000. Efficacy of Glucosamine and Chondroitin for Treatment of Osteoarthritis. Journal of the American Medical Association 284, 1241-1248.

McIlwraith, C.W., 2008. The use of nutraceuticals for equine joint disease. In: Robinson, N.E. and K.A. Sprayberry (ed.), Current therapy in equine medicine. Elsevier Health Sciences. 94-96 pp.

McKeever, K.H. and J. Britton, 2002. Does it work? Testing the efficacy of feed supplements. In: Advances in Equine Nutrition, Ed: Pagan, J.D., Nottingham University Press, Nottingham. pp 65-68.

McKeever, T.M. and J. Britton, 2004. Diet and Asthma. Am. J. Respir. Crit. Care Med. 170, 725-729.

Medina, B., I.D. Girard, E. Jacotot and V. Julliand, 2002. Effect of a preparation of *Saccharomyces cerevisiae* on microbial profiles and fermentation patterns in the large intestine of horses fed a high fiber or a high starch diet. Journal of Animal Science 80, 2600-2609.

Melis, G.C., N. ter Wengel, P.G. Boelens and P.A.M. van Leeuwen, 2004. Glutamine: recent developments in research on the clinical significance of glutamine. Current Opinion in Clinical Nutrition and Metabolic Care 7, 59-70.

Meulyzer, M., P. Vachon, F. Beaudry, T. Vinardell, H. Richard, G. Beauchamp and S. Laverty, 2008. Comparison of pharmacokinetics of glucosamine and synovial fluid levels following administration of glucosamine sulphate or glucosamine hydrochloride. Osteoarthritis and Cartilage 16, 973-979.

Meulyzer, M., P. Vachon, F. Beaudry, T. Vinardell, H. Richard, G. Beauchamp and S. Laverty, 2009. Joint inflammation increases glucosamine levels attained in synovial fluid following oral administration of glucosamine hydrochloride. Osteoarthritis and Cartilage 17, 228-234.

Moore, B.E. and K.E. Newman, 1994. Influence of feeding yeast culture (Yea-Sacc 1026) on cecum and colon pH of the equine. Journal of Animal Science 72 (Suppl 1), 261.

Nielsen, B.D., G.D. Potter, E.L. Morris, T.W. Odom, D.M. Senor, J.A. Reynolds, W.B. Smith, M.T. Martin and E.H. Bird, 1993. Training distance to failure in young racing quarter horses fed sodium zeolite A. Journal of Equine Veterinary Science 13, 562-567.

O'Connor, C.I., B.D. Nielsen, A.D. Woodward, H.S. Spooner, B.A. Ventura and K.K. Turner, 2008. Mineral balance in horses fed two supplemental silicon sources. Journal of Animal Physiology and Animal Nutrition 92, 173-181.

O'Neill, W., S. McKee and A.F. Clarke, 2002. Immunological and haematinic consequences of feeding a standardised Echinacea (*Achinacea angustfolia*) extract to healthy horses. Equine vet. J. 34, 222-227.

Oke, S., A., Aghazadeh-Habashi, J.S. Weese and F. Jamali, 2006. Evaluation of glucosamine levels in commercial equine oral supplements for joints. Equine Veterinary Journal 38, 93-95.

Ott, E.A., 2005. Influence of Bio-Mos, a mannaoligosaccharide supplement, on the immune system of the mare and neonatal foal. In: Jacques, L. (ed.), Nutritional biotechnology in the feed and food industry. Proceedings of Alltech's 21st annual symposium. Nottingham University Press, Lexington, Kentucky, USA, p. 447-454.

Pearson, W. and M. Lindinger, 2008. Critical review of research evaluating glucosamine-based nutraceuticals for treatment of joint pain and degenerative joint disease in horses. In: Zimmermann, N. (ed.), Proceedings of the 6th Mid-Atlantic Nutrition Conference. Maryland Feed Industry Council, Timonium, Maryland, USA, p. 184-191.

Ralston, S.L., 2007. Evidence-Based Equine Nutrition. Veterinary Clinics of North America – Equine Practice 23, 365-384.

Raviv, S. and L. Smith, 2010. Diet and asthma. Current Opinion in Pulmonary Medicine 16, 71-76.

Respondek, F., A.G. Goachet and V. Julliand, 2008. Effects of dietary short-chain fructooligosaccharides on the intestinal microflora of horses subjected to a sudden change in diet. Journal of Animal Science 86, 316-323.

Respondek, F., A.G. Goachet, F. Rudeaux and V. Julliand, 2005. Effect of short chain fructo-oligosaccharides on the microbial and biochemical profile of different segments of the gastrointestinal tract in horses. In: Lauk, H.D., B. Huskamp and E. Deegen (ed.), Pferdeheilkunde. Hippiatrika, Hanover, Germany. 69-70 pp.

Riserus, U., 2008. Fatty acids and insulin sensitivity. Current Opinion in Clinical Nutrition and Metabolic Care 11, 100-105

Routledge, N.B., R.C. Harris, P.A. Harris, J.R. Naylor and C.A. Roberts, 1999. Plasma glutamine status in the equine at rest, during exercise and following viral challenge. Equine Veterinary Journal Supplement 30, 612-616.

Santos, A.S., C.V.M. Guedes, L.M.M. Ferreira, L. Evangelista, G. Gaspar, G. Bertin and A.M. Rodrigues, 2008. The effect of *Saccharomyces cerevisiae* CBS 493.94 level on *in vitro* fermentation using faecal inoculum from horses fed diets with different fibre concentrations In: Saastamoinen, M.T. and W. Martin-Rosset (ed.), Nutrition of the exercising horse. Eaap publication no. 125. Wageningen Academic Publishers, Wageningen, the Netherlands, p. 361-363.

Senchina, D.S., N.B. Shah, D.M. Doty, C.R. Sanderson and J.E. Hallam, 2009. Herbal supplements and athlete immune function--what's proven, disproven, and unproven? Exercise Immunology Review 15, 66-106.

Shah, S.A., S. Sander, C.M. White, M. Rinaldi and C.I. Coleman, 2007. Evaluation of echinacea for the prevention and treatment of the common cold: a meta-analysis. Lancet Infectious Diseases 7, 473-480.

Stull, C.L., S.J. Spier, B.M. Aldridge, M. Blanchard and J.L. Stott, 2004. Immunological response to long-term transport stress in mature horses and effects of adaptogenic dietary supplementation as an immunomodulator. Equine Veterinary Journal 36, 583-589.

Taouis, M., C. Dagou, C. Ster, G. Durand, M. Pinault and J. Delarue, 2002. N-3 Polyunsaturated fatty acids prevent the defect of insulin receptor signaling in muscle. American Journal Physiology Endocrinology and Metababolism 282, E664-671.

Tippler, B., T. Syrovets, D. Loew and T. Simmet, 1996. Harpagophytum procumbens: Wirkung von extrakten auf die Eicosanoid biosyn-these in Ionophor A 23187- stimuliertem menschlichem Vollblut. In: D, L. and R. N (ed.) Phyto-pharmaka II. Forschung und klinische Anwendung. Steinkopff-Verlag, Darmstadt. 95-100. pp.

Treiber, K.H., R.C. Boston, D.S. Kronfeld, W.B. Staniar and P.A. Harris, 2005. Insulin resistance and compensation in Thoroughbred weanlings adapted to high-glycemic meals. Journal of Animal Science 83, 2357-2364.

Turner, K.K., B.D. Nielsen, C.I. O'Connor and D.S. Rosenstein, 2007. Silicon supplementation and osteochondrotic lesions in 2-year-old Standardbreds: a preliminary study. Equine and Comparative Exercise Physiology 4, 53-58.

Vervuert, I., S. Klein and M. Coenen, 2010. Short-term effects of a moderate fish oil or soybean oil supplementation on postprandial glucose and insulin responses in healthy horses. The Veterinary Journal 184, 162-166.

Vessby, B., 2000. Dietary fat and insulin action in humans. British Journal of Nutrition 83, S91-S96.

Volter, R., 1999. Alimentation du cheval 2nd edition, France Agricole Paris. 75 pp.

White, G.W., E.W. Jones, J. Hamm and T. Sanders, 1994. The Efficacy of Orally Administered Sulfated Glycosaminoglycan in Chemically Induced Equine Synovitis and Degenerative Joint Disease. Journal of Equine Veterinary Science 14, 350-353.

Woodward, A.D., B.D. Nielsen, C.I. O'Connor, C.D. Skelly, S.K. Webel and M.W. Orth, 2007. Supplementation of dietary long-chain polyunsaturated omega-3 fatty acids high in docosahexaenoic acid (DHA) increases plasma DHA concentration and may increase trot stride lengths in horses. Equine and Comparative Exercise Physiology 4, 71-78.

Zhu, F.X., X.Y. Zhang, M.A. Olszewski and N.E. Robinson, 1997. Mechanism of capsaicin-induced relaxation in equine tracheal smooth muscle. American Journal of Physiology. Lung Cell Molecular Physiology 273, L997-1001.

Evaluation of nutritional functional ingredients for improvement of digestive tract health and performance

A. Zeyner
Chair for Nutrition Physiology and Animal Nutrition, University of Rostock, Justus-von-Liebig-Weg 8, 18059 Rostock, Germany

Abstract

The variety of supplements on the equine market seems to be unmanageably high. In contrast to the large number, scientific evidence from studies in horses that support the claims in question is rather seldom. To make a selection, only additives are discussed here aiming to enhance digestion and absorption in the lower gut, to stabilize large bowl fermentation, and to improve athletic performance. Of these, in turn, a further selection includes those supplements that are otherwise often forgotten because their effects, despite indeed investigated, are not as spectacular. Special attention is paid exogenous digestive enzymes (α-amylase, lipase, phytase) and lecithin as possible promoters of digestion in the small intestine, probiotics which may prevent digestive disorders and enhance microbial fermentation in the large intestine, and amino acids which may balance the dietary amino acid profile and compensate for limited synthesis of muscle protein caused by a low availability of individual amino acids.

Keywords: feed supplements, digestive enzymes, lecithin, probiotics, amino acids

Introduction

Definition, approval and handling of additives for use in animal nutrition within the European Union are clearly regulated by the Council Regulation (EC) No 1831/2003. The publication of approved additives takes place in constantly updated community registers. However, in practical horse feeding substances are also in use, which do not meet these regulations, both in terms of their type and recommended dosage and the intended purpose. Thus, taking supplements here such additives are understood that are not only fed to meet a known requirement of an individual nutrient, but to follow a further target. This can be done by feeding clearly more than requirement figures, if existing, recommend, or by giving substances for which no demand is known. Nevertheless, the following purposes for administering supplements shall be treated here:

1. Improving digestion and absorption in the lower gut.
2. Stabilizing large bowel fermentation.
3. Improving athletic performance.

Selected supplements addressing the individual purpose and their expected effects are listed in Table 1. However, the number of additives commonly used in practice is even more extensive. Further requests could be that supplements increase fertility, calm down nervous horses, and so on.

Supplements to improve digestion and absorption in the lower gut (digestive enzymes, lecithin)

Auto-enzymatic digestion in the lower gut is required to function properly as a precondition for an effective use of easily available nutrients (hydrolyzable carbohydrates, protein, fat) and further for protection of the large intestine against high influxes of nutrients like starch or fat, which may interfere negatively with microbial fermentation (for review see Kienzle *et al.*, 1992ab; Julliand *et al.*, 2006; Zeyner, 2008).

Table 1. Important objectives for the administration of feed additives in horses and selected supplements used for these purposes.

Objective for administration	Supplements (selection)
1. Improving digestion and absorption in the lower gut	Digestive enzymes and lecithin
2. Stabilizing large bowel fermentation	Probiotic microorganisms (*Saccharomyces cerevisiae* CBS 493.94 and NCYC Sc 47, *Escherichia coli* strain Nissle 1917, *Lactobacillus rhamnosus* GG1), prebiotics, enzymes involved in cell wall digestion
3. Improve athletic performance by ergogenic effects (including anabolic properties)	Amino acids, carnosine (via β-alanine and L-histidine), β-hydroxy-β-methylbutyrate, ubiquinone, L-carnitine, dichloroacetate and 2-chloropropionat, creatine, ribose, γ-oryzanol, antioxidants

Modern rations often contain high amounts of starch even though this is not always justified energetically. In some cases starch is partially replaced by fat. Fats and oils are usually added to the feed to increase energy density, which offers an advantage when appetite limits the provision of adequate energy to maintain condition or when a reduced intake of hydrolyzable carbohydrates is advocated from a clinical standpoint (Dunnett, 2002). Strictly speaking starch or fat may overwhelm the capacity of digestion by body-own enzymes in the small intestine, even if this capacity is within the physiological range. This may advocate the use of supplements like digestive enzymes or emulsifying substances, with the hope to support digestion.

Of primary interest is an effective digestion of starch before the end of the small intestine. However, the ability of horses to degrade starch by endogenous enzymes is remarkably low (Kienzle, 1988; Kienzle *et al.*, 1992ab, 1994, 1997). In fact, the activity of body-own amylase in the chyme of the small intestine reaches 5-40 U/g only, but it seems to be inducible by the ingestion of hydrolyzable carbohydrates, at least within the mentioned limits (Kienzle, 1988; Radicke, 1990; Kienzle *et al.*, 1992a,b; Landes, 1992; Kleffken, 1993; Illenseer, 1994; Kienzle *et al.*, 1994; Rottmann, 1994; Heintzsch, 1995; Kienzle *et al.*, 1997; Landes and Meyer, 1998). Depending on the amount, origin and processing of cereal grains small intestinal digestibility of starch ranges between ≈20% and >90% (Kienzle *et al.*, 1992a,b; Julliand *et al.*, 2006; Zeyner, 2008) and can be critical low under certain circumstances. Apart from thermal processing applying amylase as a feed additive may enhance precaecal digestion (Meyer *et al.*, 1995). Indeed, Kleffken (1993), Kienzle *et al.* (1994) and Heintzsch *et al.* (1995) demonstrated that the addition of α-amylase to the diet considerably increases amylase activity in the jejunal chyme. Despite only 4% of the added amylase reached the end of the jejunum, feeding this enzyme to ground corn improved preileal digestibility of starch from about 50% to 60% (Kleffken, 1993; Meyer *et al.*, 1995). Similarly adding an enzyme mix with α-amylase, xylanase, β-glucanase and pectinase to a barley-based diet enhanced preileal digestibility of starch by 10% absolutely (Heintzsch *et al.*, 1995). Feeding α-amylase alone or together with amyloglucosidase to a diet based on steam-rolled triticale elevated the glycaemic response in repeated measurements (Richards *et al.*, 2004). Amyloglucosidase alone failed to have such an effect. Nevertheless, whereas the glycaemic response to the control feed increased, with time, the response to the α-amylase added diet decreased. This obvious adaptation to high-starch feeding by more effective endogenous starch degradation seems to have its counterpart on the resorptive level. That's how SGLT1 expression enhanced, with time, 2-fold in the duodenum and 3.3-fold in the ileum of horses, fed high vs. low (6.0 vs. <1.0 g/kg bwt/d) amounts of hydrolyzable carbohydrates (Dyer *et al.*, 2009). Further research is needed to decide whether horses well adapted to high-starch feeding benefit from applying α-amylase as feed additive.

Fat is the energy-concentrate *per se* that can partly be used instead of starch. The digestibility of fats in the horse is generally considered to be high (for review see Kronfeld *et al.*, 2004; Zeyner, 2008). Nonetheless, any practically relevant amounts of dietary fats or oils remarkably increase faecal fat excretion indicating that considerable amounts of dietary fat may reach the hind gut (Zeyner, 2002, Zeyner *et al.*, 2003). This is undesirable, first because precaecally undigested fat does not provide energy to the host and second because it may hinder microbial fibre fermentation (Jansen *et al.*, 2000; Zeyner *et al.*, 2000; Zeyner, 2002, Zeyner *et al.*, 2002). It is conceivable that feeding exogenous lipase enhances small intestinal digestibility of fat. Related studies are scarce. Verthein (1981) did not find any effect of an enzyme mix with cellulase, protease, amylase, amyloglucosidase and lipase on total tract digestibility of fat. However, that could not be expected first because the experimental diet did not contain supplemental fat and second because the measurements did not include partial digestibility. Interestingly, Heintzsch *et al.* (1995) found an increase of preileal digestibility of fat by 10% absolutely in horses fed a mixed feed with 10% crude fat in the dry matter following addition of an enzyme mix without lipase, α-amylase, xylanase, β-glucanase, and pectinase.

However, since horses do not have a gall bladder adding an emulsifying substance like lecithin may also improve precaecal fat digestion. Several studies revealed that added lecithins of different origin (corn and soy bean) are well accepted by horses (Holland *et al.*, 1993, 1995; Zeyner and Lengwenat, 1997; Zeyner *et al.*, 1997; Holland *et al.*, 1998; Zeyner *et al.*, 1999). Further, horses consuming 1.33 g of soy bean oil per kg of body weight (bwt) and day could be protected from fat-induced depression of fibre digestion when purified soy lecithin was added in an amount of 10%, but not 5%, of the dietary soy bean oil (Zeyner *et al.*, 2002). However, preventing depression of fibre digestion by limiting the amount of dietary fat seems to be more effective.

Another interesting point is the breakdown of phytate-complexes because, due to its molecular structure, phytate can fix not only phosphorus but other elements (e.g. calcium, magnesium, zinc), too. Moreover, dietary protein and amino acids may also form complexes with phytate. Rumen microbes are able to produce phytase which makes nutrients captured in the phytate structure available for

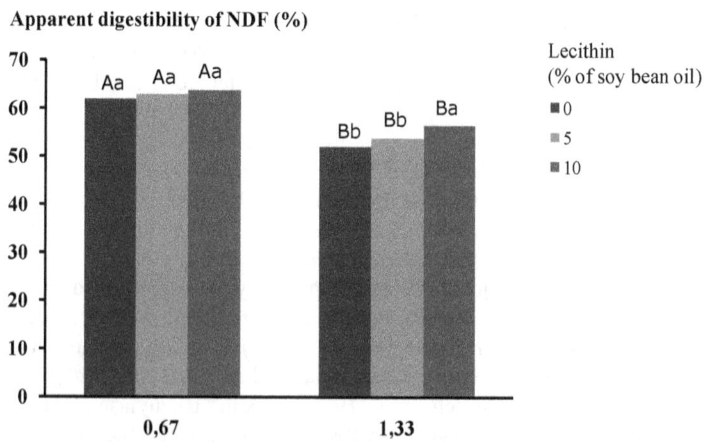

Figure 1. Impact of various concentrations of soy lecithin[1] in diets providing different amounts of soybean oil on apparent digestibility of neutral detergent fibre (Zeyner et al., 2002).
[1]Purified soy lecithin with 97% phospholipids (from that 23% phosphatidylcholine, 20% phosphatidyl-etholamine, and 14% phosphatidylinositol; ± pooled s.d. = 2.69); [AB] different superscripts indicate with P<0.05 significantly different means between the respective columns of both levels of soy bean oil; [ab] different superscripts indicate with P<0.05 significantly different means within a given level of soy bean oil)

absorption in the small intestine. Hind gut microbes are said to have the same potential. This may be of value for phosphorus, because this mineral can be in net absorbed in the small intestine and in the hind gut as well (Meyer, 1992). In contrast, calcium and most trace elements are predominantly absorbed in the small intestine. Phytate-phosphorus seems to be partly available to horses whereby sodium phytate is obviously better digestible (Schryver *et al.*, 1971; Matsui *et al.*, 1999) than phosphorus from natural feedstuffs (Hintz *et al.*, 1973; Van Doorn *et al.*, 2004). However, there is yet no clear evidence that added phytase indeed enhances phosphorus availability in horses. Interestingly, calcium digestibility seems to benefit more from supplementation of phytase to diets with different amounts of phosphorus and various phosphorus sources (monocalcium phosphate *vs.* wheat and rice bran) than phosphorus digestibility (Van Doorn *et al.*, 2004).

Supplements to stabilize large bowel fermentation (with emphasis on probiotics)

Stabilization of fermentation in the large intestine is mainly required to prevent hind gut acidosis caused by high-starch feeding and to prevent or treat diarrhoea. Of minor importance is the attempt to improve fibre fermentation by oral administration of digestive enzymes (e.g. cellulase, hemicellulase, xylanase) or to support enterocytes energetically by butyrate from the fermentation of supplied prebiotics.

Probiotics as feed additives for horses are highly popular. They can be defined as viable microbes which, if delivered in sufficient amounts, act as active germs in the gut and, thus, yield positive effects on the host`s health (see Julliand and Zeyner, 2009). Different modes of action have been reported for probiotic strains in the digestive tract; for instance competition with pathogens for nutrients and for places to adhere to the intestinal wall. However, the underlying mechanisms are not fully understood. More sophisticated mechanisms are currently being discussed, like direct antagonistic effects against specific pathogenic microbes, influences on the microbial formation of short chain fatty acids and lactic acid *via* synergism with other microbes, stimulating the rate of glucose transport through brush border vesicles, reducing paracellular permeability, and immunestimulation (for review see Breves, 2004; Julliand and Zeyner, 2009). It can be speculated that some probiotics may act in the foregut and the terminal tract as well, and that individual microbes may have complementary effects.

However, in several farm animals it has been demonstrated that probiotics can be useful to prevent diarrhoea or to improve performance. Benefit and risk of probiotics for horses have been summarized by Julliand and Zeyner (2009). Precondition for the development of an active effect in the terminal intestine is the ability to survive the passage through the gut. This ability has only been investigated for *Lactobacillus rhamnosus* strain GG1, *Escherichia coli* strain Nissle 1917 and *Saccharomyces cerevisiae* CBS 493 94. Interestingly, results for these highly different bacteria were quite similar in adult horses, indicating the ability to survive the gastrointestinal passage but not for permanent implantation or multiplication in this environment (Newman and Spring, 1993; Moore *et al.*, 1994; Weese *et al.*, 2003; Zeyner *et al.*, 2003; Julliand, 2005; Gobert *et al.*, 2006; Jouany *et al.*, 2009). In foals, due to the provisionally unstable microbial community within the gut, a transient colonisation is principally conceivable when the probiotic microorganism is started to be given early after birth. Nonetheless, the current recommendation to supply probiotics every day should be maintained for the time being.

In horses, there are only few documented effects of probiotics. Thus, *Escherichia coli* strain Nissle 1917 has been shown to have the potential to lower frequency and severity of first foal's heat, when started to be administered immediately after birth (Zeyner and Neuhaus, cited from Julliand and Zeyner, 2009). Effects of probiotic yeasts (*Saccharomyces cerevisiae* CBS 493 94 and NCYC Sc 47) in horses are much better investigated than those of bacteria. Probiotic yeasts seem to act as gut flora stabilizers and digestibility enhancers. It has been demonstrated that adding *Saccharomyces cerevisiae* can alter the count of total anaerobic bacteria and fermentation conditions in the terminal tract and may diminish starch-induced pH-decrease there (Spring *et al.*, 1995; Medina *et al.*, 2002; Julliand, 2005). Further, the activity of bacterial enzymes involved in plant cell wall digestion may be elevated (Jouany *et al.*, 2009). Horses receiving viable yeast show indeed a better fibre digestibility and an overall improved digestibility of energy, nitrogen and phosphorus (Glade *et al.*, 1991a; Jouany *et al.*, 2008). These favourable effects may explain, at least in part, a better milk quality and increased nutrient retention in sucking foal's when mares received *Saccharomyces cerevisiae* (Glade *et al.*, 1991b). However, there are a number of studies which show no effect on physiological parameters when adding yeast to horses' diet (for detailed discussion see Dunnett and Vervuert, 2010, this publication).

Despite such promising results, any microorganism that is expected to have probiotic properties should be carefully checked for safety *in vivo*. This has been impressively highlighted in the case of *Lactobacillus pentosus* WE7 (Weese and Rousseau, 2005). The strain isolated from equine faeces showed inhibitory capacities against pathogens (*Salmonella spp.*, *E. coli*, *S. zooepidemicus*, *C. difficile*, *C. perfringens*). *In vivo* effects have been studied in 153 foals. Despite the favourable preconditions *in vitro*, foals of the administered group developed significantly more depression, anorexia, and colic bouts than foals of the control group. A further point that needs to be investigated is whether probiotic microbes which are beneficial for other species, like lactic acid producing bacteria, are likewise favourable for horses.

The horse's digestive tract does normally not need the help of added probiotic microorganisms, but it may benefit from probiotics under certain critical circumstances. Then supplementation should only be done with a clearly defined aim and over a limited duration.

Supplements to increase athletic performance (with emphasis on amino acids)

There are a variety of substances on the market which promise to improve athletic performance in horses. Whether the expectations on the substance in question is justifiable or not, often more pharmacologically than nutritionally relevant effects are expected. Maybe the performance-enhancing potential of improved feeding and training strategies are sometimes overlooked, consciously or unconsciously ignored or at least underestimated. Therefore standard practice is to use single feedstuffs reclassified as supplements (e.g. glucose for 'glycogen-loading'; sodium bicarbonate for intracellular acid-base regulation under anaerobic stress). Supplements available on the equine market are often used without secure knowledge of their impact, and/or without sufficient understanding of side-effects and safety.

In a narrower sense additives to improve athletic performance are hoped to fulfil at least one of the following goals:
- elevating the muscle glycogen pool and improving glycogen availability;
- intensifying oxidative metabolism;
- supplying precursors for ATP metabolism;
- improving the intracellular acid-base regulation; and
- support the development of skeletal muscles tissue.

In a wider sense supplements are used in the hope to influence the immune response, to support cartilage maturation and to improve hoof horn quality, lung function, antioxidant metabolism and animal behaviour, just to name a few.

Supplements claimed to be advantageous for exercising horses have been examined and discussed at full length by Harris (2008). The most common ergogenic additives, partly with supposed anabolic properties, were mentioned in Table 1 (amino acids, carnosine, β-hydroxy-β-methylbutyrate, ubiquinone, L-carnitine, dichloroacetate and 2-chloropropionat, creatine, ribose, γ-oryzanol). Although many supplements should be beneficial from a theoretical point of view, clear scientific evidence based on studies in the equine is predominantly missing. Under the above quoted additives, L-carnitine seems to have a particularly interesting potential. However, advantages like protecting and maybe improving muscle function through buffering the mitochondrial concentration of acetylCoA (Foster and Harris, 1987; Harris *et al.*, 1987), mitigation of damage caused by hypoxia (Spiering *et al*., 2007), and metabolic adaptation of muscle fibres to the type of exercise in question (Rivero *et al.*, 2002) may be more relevant than an expected increase in fat oxidation in the mitochondria (for review see Zeyner and Harmeyer, 1999; Harris, 2008).

Recently amino acids are of increased interest. Although various amino acid preparations with performance enhancing claims are on the equine marked, evidence for their efficacy is scarce (O'Connor *et al.*, 2002). Expected effects of added amino acids in this regard are the expression of an age appropriate growth and muscle anabolism (Ott *et al.*, 1979; Graham *et al.*, 1994, Staniar *et al.*, 2001; Koslowski and Liebert, 2009; Koslowski *et al.*, 2009), improving performance and compensation for exercise-induced muscle catabolism, respectively (Graham-Thiers *et al.*, 2001, 2003; Graham-Thiers and Kronfeld, 2005; Hackl *et al.*, 2009; Van den Hoven *et al.*, 2009), influencing behaviour (O'Reilly, 2006), delaying central fatigue during endurance exercise through branched-chain amino acids induced reduction of undesired tryptophan effects (Bigard *et al.*, 1996; Blomstrand *et al.*, 1996; Casini *et al.*, 2000; Stefanon *et al.*, 2000; Trottier *et al.*, 2002; Grimmett and Sillence, 2005; Nery *et al.*, 2005), and providing precursors for the synthesis of so called ergogenic substances; for example carnosin (Powell *et al.*, 1991; Harris *et al.*, 2006; Kendrick *et al.*, 2008) and L-carnitine (for review see Bremer, 1983; Zeyner and Harmeyer, 1999).

The supply of amino acids that is strictly required for an optimal development of skeletal muscles is of particular interest. Because amino acids seem to be absorbed from the small intestine only (Wysocki and Baker, 1975; McMeniman *et al.*, 1987; Slade *et al.*, 1971; Schmitz *et al.*, 1990; Schubert, 1992) essential amino acids must be ingested from the feed, but they may also be extracted from microbes present in the lower gut. However, the latter has not yet been quantified meaning that the current assumption that horses benefit exclusively from orally ingested essential amino acids should be maintained for the time being. There is a lack of studies addressing amino acid requirements in horses. Based on balance studies and broken-line analysis, NRC (2007) gives a lysine requirement for maintenance of 0.036 g/kg body weight and day as minimum and 0.054 g/kg body weight and day as optimum. It is interesting to note that, despite the relatively higher need for high-quality crude protein in growing horses, the relationship between crude protein and lysine is assumed to result in lysine being 4.3% of the crude protein required for both, maintenance and growth (NRC, 2007). No responsible information exists on the horse's need for other essential amino acids. Amino acid ratios in the ideal protein or, if not determined as in horses, the amino acid profile of the product in question (e.g. muscle or milk) could be helpful to derive recommendations. The situation is complicated in fact because only one study exists dealing with the precaecal digestibility of several amino acids in horses (De Almeida *et al.*, 1998).

Thoroughbred foals and yearlings offered a pasture supplement with synthetic lysine and threonine grew at the same or greater rates than animals on the control supplement (Staniar *et al.*, 2001), supporting the idea that protein quality for fast-growing horses will be improved by supplementation with amino acids that are probably limiting to protein synthesis. This is quite plausible for horses with a remarkably high weight gain. Nevertheless, almost grown 2.5 years old Warmblood-type stallions in light training developed higher increase in muscle circumference (*M. extensor digitorum communis*) when synthetic lysine was added (Koslowski and Liebert, 2009; Koslowski *et al.*, 2009; Table 2). A similar effect has always been demonstrated in young but already grown horses (<10 years) and aged horses (>20 years) as well, all receiving regular light exercise (Graham-Thiers and

Table 2. Differences in withers height, muscle development and blood urea concentration between 2.5 years old Warmblood-type stallions fed either a diet supplemented with 17 g/d L-lysine (Lysine+)[1] or a control diet (Lysine-)[2] during 70 days of training[3] (Koslowski and Liebert, 2009; Koslowski et al., *2009).*

Parameters		Lysine+ (n=6)	Lysine- (n=6)
Δ Withers height	(cm)	3.2^a	2.3^b
Δ Muscle circumference[4]	(mm)	19.5^a	7.0^b
Δ Blood urea	(mg/dl)	-2.3^b	-0.59^a

Δ = d70 minus d0 of youth training; [1] totally 60.7 g/d lysine; [2] totally 43.7 g/d lysine; [3] 25 min/d trot; [4] circumference of the *M. extensor digitorum communis;* [a,b] different superscripts within a raw indicate significantly different means ($P<0.05$).

Kronfeld, 2005). The horses of both groups maintained muscle mass better when synthetic lysine and threonine were supplemented. Interestingly, lysine provided by the basal diets in both studies already fulfilled NRC recommendations (NRC, 2007). The point was that the amino acid profile of the individual supplemented diets more closely resembled the muscle amino acid ratio of horses reported by Bryden (1991). More research is needed to be able to exploit the potential for developing and maintaining muscle mass and probably for muscle regeneration in athletic horses of different ages.

Assessing a possible extra need for amino acids caused by exercise is highly complicated because there are no amino acids retained in any product of athletic work that could be taken into account for quantitative considerations. Indeed, whether supplementation with amino acid and protein mixtures in exercised horses can be helpful is not well researched. Casini *et al.* (2000) and Stefanon *et al.* (2000) did not achieve any clear effect on performance following exercise in trotters receiving branched chain amino acids. The same was observed in tryptophan supplemented horses undergoing repeated bouts of treadmill tests (Vervuert *et al.*, 2005). However, it is justified to assume that exercise of certain intensity induces muscle catabolism which in turn would be accompanied by shifts in amino acid metabolism. The degree and characteristic of such metabolic shifts are probably specific for individual amino acids and may further be influenced by factors like intensity and duration of the exercise performed, athletic fitness of the horse, metabolic availability of fuels for energy, and postprandial status. Corresponding alterations in plasma amino acids following exercise or caused by regular training have repeatedly been measured (Miller and Lawrence, 1988, King and Suleiman, 1998; Pösö *et al.*, 2001; Essen-Gustavsson and Jensen-Waern, 2002; Hackl *et al.*, 2009; Van den Hoven *et al.*, 2009). However, plasma amino acids reflect the balance between appearance in the blood as the sum of intake and tissue release and disappearance by tissue uptake and excretion (Cynober, 2002). This opens a wide field for interpretation. Coenen *et al.* (2006) demonstrated that equine plasma amino acid levels can be elevated by feeding a gelatin supplement. In humans, anabolic effects of amino acid supplementation as well as a faster recovery after exercise have been reported by Carli *et al.* (1992) and Chandler *et al.* (1994). To reduce fatigue, improve stamina and achieve quicker recovery following heavy exercise, individual plasma amino acid patterns are used in human beings as a basis for personalized amino acid supplements (Spona, 1998).

It would be more helpful to identify the point when a net efflux of amino acids from the muscle indicates the onset of muscle catabolism. Investigations on amino acid dynamics in the muscle of exercised horses are rare (Essen-Gustavsson and Jensen-Waern, 2002; Matsui *et al.*, 2006; Van den Hoven *et al.*, 2009). Matsui *et al.* (2006) applied radio-labelled phenylalanine to investigate amino acid metabolism in horse muscle. They found a mixture of 10 amino acids given intravenously after strenuous exercise being able to reduce protein breakdown and increase protein synthesis in the hind limb. This result is confirmed by Van den Hoven *et al.* (2009) who found that a mixture of 18 amino acids administered orally shortly after exercise increased the intramuscular pool of amino acids. Although excessive nitrogen supply is undesired in athletic horses administering certain amino acids immediately after exercise may have the potency to compensate for depressed regeneration of muscle protein caused by critically low availability of individual amino acids.

In this area further investigation is worthwhile. Particularly the amount and composition of amino acid mixtures depending on various types and intensities of exercise should be addressed.

Conclusion

Currently there are many, frequently not sufficiently investigated, supplements on the market, claiming to improve athletic performance or health. Supplements for the regulation of digestion are of less interest. Among these for example α-amylase and lecithin, respectively, may improve the precaecal digestion of starch and fat. However, similar effects would possibly be achieved simply by feeding rations which are more strictly restricted in starch and fat. Focusing on the hindgut, probiotic yeasts seem to be able to protect the hindgut pH against the acidifying effect of rations high in starch and to elevate fiber fermentation. Bacteria which may have probiotic properties in horses have yet to be studied more thoroughly. Among the supplements which are said to promote athletic performance, L-carnitine and amino acids show some potential which warrants further investigation.

References

Bigard, A.X., P. Lavier, L. Ullmann, H. Legrand, P. Douce and C.Y. Guezennec, 1996. Branched-chain amino acid supplementation during repeated prolonged skiing exercise at altitude. Int. J. Sports. Nutr. 6, 295-306.

Bryden, W.L., 1991. Amino acid requirements of horses estimated from tissue composition. Proc. Nutr. Soc. Aust., p. 53.

Blomstrand, E., S. Ek and E.A. Newsholme, 1996. Influence of ingesting a solution of branched-chain amino acids on plasma and muscle concentrations of amino acids during prolonged submaximal exercise. Nutrition 12, 485-490.

Bremer, J., 1983. Carnitine – metabolism and functions. Physiol. Rev. 63, 1420-1480.

Breves, G.., 2004. Probiotische Mikoorganismen und ihre zellulären Wirkungsmechanismen. In: Mikrobiologie und Tierernährung. 20. Hülsenberger Gespräche. Lübeck, June 9-11, 2004: 9-18.

Carli, G., M. Bonifazi, L. Lodi, C. Lupo, G. Martenelli and A. Viti, 1992. Changes in exercised-induced hormone response to branched chain amino acids administration. Europ. J. Appl. Physiol. and Occupational Physiol. 64, 272-277.

Casini, L., L. Gatta, L. Magni and B. Colombani, 2000. Effect of prolonged branched chain amino acid supplementation on metabolic responses to anaerobic exercise in standardbreds. J. Equine Vet. Sci. 20, 120-123.

Chandler, R.M., H.K. Byrne, J.G. Patterson and J.L. Ivy, 1994. Dietary supplements affect the anabolic hormones after weight-training exercise. J. Appl. Physiol. 76, 839-845.

Coenen, M., K. Appelt, A. Niemeyer and I. Vervuert, 2006. Study of gelatin supplemented diet on amino acid homeostasis in the horse. Equine Vet. J. (Suppl.) 36, 606-610.

Cynober, L.A., 2002. Plasma amino acid levels with a note on membrane transport: characteristics, regulation, and metabolic significance. Nutrition 18, 761-766.

De Almeida, F.C., S.C. Valadares Filho, A.C. Queiroz, M.I. Leão, J.F.C. Silva and P.R. Cecon, 1998. Digestibilidade aparente e verdadeira pré-cecal e total da proteína em dietas com diferentes níveis protéicos em eqüinos. R. Bras. Zootec. 27(3), 521-529.

Dunnett C.E., 2002. Dietary lipid form and function. Proc. Xth Equine Nutr. Conf., Kentucky, USA.

Dunnett C.E. and I. Vervuert, 2010. Functional nutritional ingredients: Science behind the claims for health. In: Ellis, A.D., A.C. Longland, M. Coenen and N. Miragli (eds.), The impact of nutrition on the health and welfare of horses. EAAP publication no. 128, Wageningen Academic Publishers, Wageningen, the Netherlands, p. 241-253.

Dyer, J., M. Al-Rammahi, L. Waterfall, K.S.H. Salmon, R.J. Geor, L. Boure, G.B. Edwards, C.J. Proudman and S.P. Shirazi-Beechey, 2009. Adaptive response of equine intestinal Na^+/glucose cotransporter (SGLT1) to an increase in dietary soluble carbohydrate. Pflügers Archiv – European J. Physiol. 458, 419-430.

Essen-Gustavsson, B. and M. Jensen-Waern, 2002. Effect of an endurance race on muscle amino acids, pro- and macroglycogen and triglycerides. Equine Vet. J. (Suppl.) 34, 209-213.

Foster, C.V. and R.C. Harris, 1987. Changes in the free and bound carnitine in muscle with maximal sprint exercise in the Thoroughbred horse. In: Gillespie, J.R. and N.E. Robinson (editors). Equine Exercise Physiology 2. ICEEP Publications, Davis (California): 332-340.

Glade, M.J., 1991a. Dietary yeast culture supplementation of mares during late gestation and early lactation: effects on dietary nutrient digestibilities and fecal nitrogen partitioning. J. Equine Vet. Sci. 11, 10-16.

Glade, M.J., 1991b. Effects of dietary yeast culture supplementation of lactating mares on the digestibility and retention of the nutrient delivered to nursing foals via milk. J. Equine Vet. Sci. 11, 323-329.

Gobert, J., G. Bertin and V. Julliand, 2006. Digestive fate of *Saccharomyces cerevisiae* CBS 493.94, fed at three different concentrations to horses. Reprod. Nutr. Develop. 46 (Suppl. 1), S95.

Graham, P.M., E.A. Ott, J.H. Brendemuhl and S.H. TenBroeck, 1994. The effect of supplemental lysine and threonine on growth and development of yearling horses. J. Anim. Sci. 72, 380-386.

Graham-Thiers, P.M. and D.S. Kronfeld, 2005. Amino acid supplementation improves muscle mass in aged and young horses. J. Anim. Sci. 83, 2783-2788.

Graham-Thiers, P.M., D.S. Kronfeld, C. Hatsell, K. Stevens and K. McCraigt, 2003. Amino acid supplementation improves muscle mass in aged horses. Proc. 18th Equine Nutr. Physiol. Soc. Symp., East Lansing (MI), June 4-7, 2003: 134.

Graham-Thiers, P.M., D.S. Kronfeld, K.A. Kline and D.J. Sklan, 2001. Dietary protein restriction and fat supplementation diminish the acidogenic effect of exercise during repeated sprints in horses. J. Nutr. 131, 1959-1964.

Grimmett, A. and M.N. Sillence, 2005. Calmatives for the excitable horse: a review of L-tryptophan. Vet. J. 170, 24-32.

Hackl, S., R. Van den Hoven, M. Zickl, J. Spona and J. Zentek, 2009. The effects of short intense exercise on plasma free amino acids in standardbred trotters. J. Anim. Physiol. Anim. Nutr. 93, 165-173.

Harris, P.A., 2008. Ergogenic aids in the performance horse. Ergogenic aids in the performance horse. In: M.T. Saastamoinen and W. Martin-Rosset (eds.), Nutrition of the exercising horse. EAAP publication no. 125, Wageningen Academic Publishers, Wageningen, the Netherlands, p. 373-398.

Harris, P., M. Coenen, D.L. Frape, L.B. Jeffcott and H. Meyer, 2006. Equine nutrition and metabolic disease. In: Higgins, A. and J. Snyder (eds.), Equine Manual. Elsevier Saunders, London: 151-222.

Harris, R.C., C.V.L. Foster and E. Hultman, 1987. Acetylcarnitine formation during intense muscular contraction in humans. J. Appl. Physiol. 63, 440-442.

Heintzsch A., 1995. Effekte einer Enzymmischung (α-Amylase, Xylanase, β-Glucanase, Pectinase) als Futteradditiv auf die praeileale Verdaulichkeit stärkereicher Rationen beim Pferd. Tierärztl. Hochschule, Hannover, Doctoral thesis.

Hintz, H.F., A.J. Williams, J. Rogoff and H.F. Schryver, 1973. Availability of phosphorus in wheat bran when fed to ponies. J. Anim. Sci. 36, 522-525.

Holland, J.L., D.S. Kronfeld, G.A. Rich, K.A. Kline, J.P. Fontenot, T.N. Meacham and P.A. Harris, 1998. Acceptance of fat and lecithin containing diets by horses. Appl. Anim. Behaviour Sci. 56, 91-96.

Holland, J.L., T.N. Meacham and D.S. Kronfeld, 1995. Digestibility of lecithin containing diets by horses. Proc. 14th Equine Nutr. Physiol. Symp., Ontario (California), January 19-21, 1995: 80-81.

Holland, J.L., T.N. Meacham, D.S. Kronfeld and W. Tiegs, 1993. Acceptance of lecithin containing diets by horses. Proc. 13th Equine Nutr., Physiol. Symp., Gainsville (Florida), January 21-23, 1993: 64-65.

Illenseer, M., 1994. Praeileale Verdaulichkeit von Hafer-, Kartoffel- und Maniokrationen beim Pferd. Tierärztl. Hochschule, Hannover, Doctoral thesis.

Jansen W.L., J. Van der Kuilen, S.N.J. Geelen and A.C. Beynen, 2000. The effect of replacing nonstructural carbohydrates with soybean oil on the digestibility of fibre in trotting horses. Equine Vet. J. 32, 27-30.

Jouany, J.-P., J. Gobert, B. Medina, G. Bertin and V. Julliand, 2008. Effect of live yeast culture supplementation on apparent digestibility and rate of passage in horses fed a high-fiber or high-starch diet. J. Anim. Sci. 86, 339-347.

Jouany, J.-P., B. Medina, G. Bertin and V. Julliand, 2009. Effect of live yeast culture supplementation on hindgut microbial communities and their polysaccharidase and glycoside hydrolase activities in horses fed a high-fiber or high-starch diet. J. Anim. Sci. 87, 2844-2852.

Julliand, V., 2005. Impact of nutrition on the microflora of the gastro-intestinal tract in horses. In: A. Lindner (editor), Applied equine nutrition: 85-103.

Julliand, V., A. De Frombelle and M. Varloud, 2006. Starch digestion in horses: The impact of feed processing. Livest. Sci. 100, 44-52.

Julliand, V. and A. Zeyner, 2009. The pros and cons of probiotics. In: Robinson N.E. and K.A. Sprayberry (eds.), Current Therapie in Equine Medicine. 6th revised edition. St. Louis (Missouri): Saunders Elsevier: 83-86.

Kienzle, E., 1988. Eine in-vitro Methode zum Vergleich der Abbaubarkeit verschiedener Stärkevarianten durch Pankreasamylase. Z. Tierphysiol. Tierernährg. Futtermittelkd. 60, 12-13.

Kienzle, E., J. Pohlenz and S. Radicke, 1997. Morphology of starch digestion in the horse. J. Vet. Med. A 44, 207-221.

Kienzle, E., S. Radicke, E. Landes, D. Kleffken, M. Illenseer and H. Meyer, 1994. Activity of amylase in the gastrointestinal tract of the horse. J. Anim. Physiol. Anim. Nutr. 72, 234-241.

Kienzle, E., S. Radicke, S. Wilke, E. Landes and H. Meyer, 1992a. Praeileale Stärkeverdauung in Abhängigkeit von Stärkeart und –zubereitung. Pferdeheilkunde (Sonderausgabe), 103-106.

Kienzle, E., S. Radicke, S. Wilke, E. Landes and H. Meyer, 1992b. Untersuchungen zur praeilealen Verdaulichkeit von Mais- und Haferstärke beim Pferd in Abhängigkeit von der Zerkleinerung des Getreides. 46. Tagung der Gesellschaft für Ernährungsphysiologie (Kurzfassung der Vorträge): 89-90.

King, N. and M.S. Suleiman, 1998. Effect of regular training on the myocardial and plasma concentrations of taurine and amino acids in thoroughbred horses. Amino Acids 15, 241-251.

Kleffken, D., 1993. Praeileale Verdauung von Getreidestärke (Gerste / Mais) in Abhängigkeit von Zubereitung, Rauhfutterangebot und Amylasezusatz beim Pferd. Tierärztl. Hochschule, Hannover, Germany, Doctoral thesis.

Koslowski, D. and F. Liebert, 2009. Investigations on amino acid supply in diets for young stallions and effects of lysine supplementation on growth and selected blood parameters. Proc. Soc. Nutr. Physiol. 18, 32.

Koslowski, D., F. Liebert and J. Thielebein, 2009. Investigations on muscle development (*M. extensor digitorum communis*) depending on lysine supply in diets for young stallions. Proc. Soc. Nutr. Physiol. 18, 35.

Kronfeld, D.S., J.L. Holland, G.A. Rich, T.N. Meacham, J.P. Fontenot, D.J. Sklan and P. A. Harris, 2004. Fat digestibility in *Equus caballus* follows increasing first-order kinetics. J. Anim. Sci. 82, 1773-1780.

Landes, E., 1992. Amylaseaktivität sowie Konzentration organischer Säuren im jejunal- und Caecumchymus des Pferdes nach Hafer- und Maisfütterung. Hannover: Tierärztl. Hochschule, Doctoral thesis.

Landes E. and H. Meyer, 1998. Einfluß von Fetten auf die Futteraufnahme sowie mikrobielle Umsetzungen im Magen und Dünndarm des Pferdes. Pferdeheikunde 14 (1), 51-58.

Matsui, T., H. Murakami, H. Yano, H. Fujikawa, T. Osawa and Y. Asai, 1999. Phytate and phosphorus movements in the digestive tract of horses. Equine Vet. J. 26 (Suppl.), 46-50.

Matsui, A., H. Omura, Y. Asai, T. Takahashi, A. Hiraga, K. Okamura, H. Tokimura, T. Sogino, T. Obitsu and K. Taniguchi, 2006. Effects of amino acids and glucose administration following exercise on the turnover of muscle protein in the hind limb femoral region of thoroughbreds. Equine Vet. J. (Suppl.) 36, 611-616.

McMeniman, N.P., R. Elliot, S. Groenendyk and K.F. Dowsett, 1987. Synthesis and absorption of cystein from the hindgut of the horse. Equine Vet. J. 19, 192-194.

Medina, B., I.D. Girard, E. Jacotot and V. Julliand, 2002. Effect of preparation of *Saccharomyces cerevisiae* on microbial profiles and fermentation patterns in the large intestine of horses fed a high fiber or a high starch diet. J. Anim. Sci. 80, 2600-2609.

Meyer, H., 1992. Intestinaler Wasser- und Elektrolytstoffwechsel des Pferdes. Übers. Tierernährg. 20, 135-166.

Meyer, H., S. Radicke, E. Kienzle, S. Wilke, D. Kleffken and M. Illenseer, 1995. Investigations on preileal digestion of starch from grain, potato and manioc in horses. J. Vet. Med. Series A 42, 371-381.

Miller, P.A. and L.M. Lawrence, 1988. The effect of dietary protein level on exercising horses. J. Anim. Sci. 66, 2185-2192.

Moore, B.E., K.E. Newman, P. Spring and V.E. Chandler, 1994. Effect of yeast culture (Yea-Sacc 1026) on microbial population and digestion in the cecum and colon of the equine. J. Anim. Sci. 72 (Suppl. 1), 252.

Nery, J., A. Assenza, W. Pinna, A. Sfuncia and D. Bergero, 2005. A study of horse's blood constituents and weight variations during endurance competitions. Pferdeheilkunde 21, 103-104.

Newman, K.E. and P. Spring, 1993. A comparison of two different strains of yeast in their ability to alter ruminal and cecal fermentation. J. Anim. Sci. 71 (Suppl. 1), 289.

NRC, 2007. Nutrient Requirements of Horses. 6th rev. ed. (edited by National Research Council of the National Academies). Washington, D.C., The National Academies Press.

O'Connor, C.I., B.D. Nielsen, H.C. Schott and H.M. Clayton, 2002. Effects of weight carrying, exercise and a myo-anabolic supplement on growth and muscle. Equine Vet. J. (Suppl.) 34, 178-181.

O'Reilley, B., 2006. Reducing cribbing frequency in horses through dietary supplementation of tryptophan and calcium carbonate. Auburn University, Alabama, Master thesis.

Ott, E.A., R.L. Asquith, J.P. Feaster and F.G. Martin, 1979. Influence of protein level and quality on growth and development of yearling foals. J. Anim. Sci. 49, 620-626.

Pösö, A.R., B. Essen-Gustavsson and S.G. Persson, 1993. Metabolic response to standardised exercise test in standardbred trotters with red cell hypervolaemia. Equine Vet. J. 25, 527-531.

Powell, D.M., L.M. Lawrence, J. Novakofski, L.R. Moser and M.J. Biel, 1991. Effect of dietary L-histidine supplementation on muscle carnosine and buffering capacity in horses. Proc. 12th Equine Nutr. Physiol. Symp., Calgary, June 06-08, 1991: 115-119.

Radicke S., 1990. Untersuchungen zur Verdauung von Mais- und Haferstärke beim Pferd. Tierärztl. Hochschule, Hannover, Doctoral thesis.

Richards, N., M. Choct, G.N. Hinch and J.B. Rowe, 2004. Examination of the use of exogenous alpha-amylase and amyloglucosidase to enhance starch digestion in the small intestine of the horse. Anim. Feed Sci. Technol. 114: 295-305.

Rivero, J., H.P. Sporleder, E. Quiroz-Rothe, I. Vervuert, M. Coenen and J. Harmeyer, 2002. Oral L-carnitine combined with training promote changes in skeletal muscle. Equine Vet. J. (Suppl.) 34, 269-274.

Rottmann, J., 1994. Untersuchungen zur Verdaulichkeit (insgesamt und praeileal) von Maissilage und Maiskolbensilage beim Pferd. Tierärztl. Hochschule, Hannover, Doctoral thesis.

Schmitz, M., F. Ahrens and H. Hagemeister, 1990. Beitrag der Absorption von Aminosäuren im Dickdarm zur Proteinversorgung bei Pferd, Rind und Schwein. J. Anim. Physiol. Anim. Nutr. 64, 12-13.

Schryver, H.F., H.F. Hintz and P.H. Craig, 1971. Phosphorus metabolism in ponies fed varying levels of phosphorus. J. Nutr. 101, 1257-1263.

Schubert R., 1992. Utilization of ^{15}N-urea for intestinal synthesis of bacterial protein and for milk protein in pony mares. Pferdeheilkunde (Sonderausgabe), 137-139.

Slade, L.M., R. Bishop, J.G. Morris and D.M. Robinson, 1971. Digestion and absorption of 15N-labelled microbial protein in the large intestine of the horse. British Vet. J. 127, XI-XIII.

Spiering, B.A., W.J. Kraemer, J.L. Vingren, D.L. Hatfield, M.S. Fragala, J.-Y. Ho, C.M. Maresh, J.M. Anderson and J.S. Volek, 2007. Responses of criterion variabe to different supplemental doses of L-Carnitine L-tartrate. J. Strength Conditioning Res. 21, 259-264.

Spona, J., 1998. Substituieren – aber individuell. Internet J. Geriatrics Gerontology 4, 18-19.

Spring, P., K.E. Newman and C. Wenk, 1995. Einfluss einer Hefekultur auf die mikrobielle Aktivität im Dick- und Blinddarm des Pferdes *in vitro* und *in vivo*. 5. Symposium Vitamine und Zusatzstoffe in der Ernährung von Mensch und Tier (Tagungsband). Jena, 28-29 September, 1995: 500-505.

Staniar, W.B., D.S. Kronfeld, J.A. Wilson, L.A. Lawrence, W.L. Cooper and P.A. Harris, 2001. Growth of Thoroughbreds fed a low-protein supplement fortified with lysine and threonine. J. Anim. Sci. 79, 2143-2151.

Stefanon, B., C. Bettini and P. Guggia, 2000. Administration of branched chain amino acids to standardbred horses in training. J. Equine Vet. Sci. 20, 115-119.

Trottier, N.L., B.D. Nielsen, K.J. Lang, P.K. Ku and H.C. Schott, 2002. Equine endurance exercise alters serum branched chain amino acid and alanine concentrations. Equine Vet. J. 34, 168-172.

Van den Hoven, R., A. Bauer, S. Hackl, M. Zickl, J. Spona and J. Zentek, 2009. Changes in intramuscular amino acid levels in submaximally exercised horses – a pilot study. J. Anim. Physiol. Anim. Nutr. 94, 455-464.

Van Doorn, D.A., H. Everts, H. Wouterse and A.C. Beynen, 2004. The apparent digestibility of phytate phosphorus and the influence of supplemental phytase in horses. J. Anim. Sci. 82, 1756-1763.

Verthein B., 1981: Auswirkungen einer Enzymgabe auf die Futterverdaulichkeit beim Pferd. Tierärztliche Hochschule, Hannover, Doctoral thesis.

Vervuert, I. and M. Coenen, 2005. Metabolic responses to oral tryptophan supplementation before exercise in horses. J. Anim. Physiol. Anim. Nutr. 89, 140-145.

Weese, J.S., M.E.C. Anderson, A. Lowe and G.J. Montheith, 2003. Preliminary investigation on the probiotic potential of *Lactobacillus rhamnosus* strain GG in horses: fecal recovery following oral administration and safety. Can. Vet. J. 44; 299-302.

Weese, J.S. and J. Rousseau, 2005. Evaluation of *Lactobacillus pentosus* WE7 for prevention of diarrhea in neonatal foals. J. Am. Vet. Med. Assoc., 226, 2031-2034.

Wysocki, A.A. and J.P. Baker, 1975. Utilization of bacterial protein from the lower gut oft he equine. Proc. 4th Equine Nutr. Physiol. Symp., Pomona, January 16-18, 1975: 21-43.

Zeyner A., 2002. Ernährungsphysiologische Wirkungen eines Austausches von stärkereichen Komponenten durch Sojaöl in der Reitpferdeernährung. Georg-August-University, Göttingen, Habilitation thesis.

Zeyner, A., 2008. Energy providing nutrient sources. In: Saastamoinen, M.T. and W. Martin-Rosset (eds.), Nutrition of the exercising horse. EAAP publication no. 125. Wageningen Academic Publishers, Wageningen, the Netherlands, p. 277-294.

Zeyner, A., N. Albers, W. Schrödl, G. Vallentin, H. Fuhrmann and M. Krüger, 2003. Probiotic Escherichia coli strain Nissle 1917 in adult horses: Tolerance and presence of viable bacteria in the faeces: In: Schubert, R., G. Flachowsky, G. Jahreis and R. Bitsch, 9. Symposium Vitamine und Zusatzstoffe in der Ernährung von Mensch und Tier. Jena, September 24-25, 2003: 420-423.

Zeyner, A., C. Behnisch and J. Gropp, 1997. Untersuchungen zur fettreichen Fütterung (Sojaöl, Sischöl, Lecithin) von Pferden. VDLUFA-Schriftenreihe 46, 323-326.

Zeyner, A., C. Behnisch and J. Gropp, 2003. Investigations on the palatability and digestibility of fish oil containing diets in adult riding horses. Proc. 7th Congr. Europ. Soc. Vet. Comp. Nutr. (ESVCN). Hannover, October 03-04, 2003: 75.

Zeyner, A., I. Buchholz and F.J. Schweigert, 1999. Untersuchungen zum Einfluß von Fettgaben auf die Konzentration an β-Carotin, Vitamin A und Vitamin E im Blutserum von Pferden. 2. Mitteilung: Sojaöl und Sojalecithin. In: Schubert, R., G. Flachowsky, R. Bitsch and G. Jahreis (eds), Vitamine und Zusatzstoffe in der Ernährung von Mensch und Tier. Jena: 358-361.

Zeyner, A. and J. Harmeyer, 1999. Metabolic functions of L-carnitine and its effects as feed additive in horses. Arch. Anim. Nutr. 52, 115-138.

Zeyner, A. and O. Lengwenat, 1997. Untersuchungen zur oralen Gabe von mittelkettigen Triglyceriden und Sojalecithin sowie von L-Carnitin an Galopprennpferde. In: Schubert, R., G. Flachowsky, R. Bitsch and G. Jahreis (edS.), Vitamine und Zusatzstoffe in der Ernährung von Mensch und Tier. Jena: 304-309.

Zeyner, A., A. Schindler, A. Dittrich and J. Gropp, 2000. Untersuchungen zur Akzeptanz und Verdaulichkeit von fettreichem Futter beim Pferd. Proc. Soc. Nutr. Physiol. 9, 50.

Zeyner, A., A. Schindler, S. Grass and A. Dittrich, 2002. Effect of lecithin on the digestibility of energy and nutrients in rations with different amounts of added dietary fat in horses. Proc. Soc. Nutr. Physiol. 11, 38.

Lack of effect of linseed-based feed additive on gastric ulcers in horses

S. Hyyppä[1], S. Särkijärvi[1], M.T. Saastamoinen[1] and T. Eskonen[2]
[1]MTT Agrifood Research Finland, Equines, Opistontie 10 A 1, 32100 Ypäjä, Finland
[2]Department of Clinical Veterinary Sciences, Faculty of Veterinary Medicine, P.O. Box 57, 00014 University of Helsinki, Finland

Abstract

Gastric ulceration is a common problem in performance horses. The 16 experimental animals were in training and had gastric ulcers. The basal feed consisted of haylage/hay and concentrate, the forage-to-concentrate ratio being on average 75:25 on a dry matter basis. Linseed-based (65%) feed (1 g/kg live weight per day) was given to the treatment group. Evaluation of the treatment effect was based on gastroscopy performed at the beginning of the trial and then twice at an interval of 6 weeks. The linseed-based feed did not decrease the number or severity of gastric lesions compared to the control group.

Keywords: gastric ulcer, linseed

Introduction

Ulceration of the non-glandular mucosa of the equine stomach is recognized as a common problem in performance horses. The main cause of ulcers is considered to be prolonged exposure to gastric acids in the absence of feed and saliva to neutralize the acids. The clinical signs of ulcers are not always apparent or they may be fairly non-specific such as dull hair coat, behavioral changes, poor athletic performance, poor appetite, weight loss, excessive recumbency and frequent colic. A definitive diagnosis can be made with gastroscopy. (Bell *et al.*, 2007).

Currently omeprazole, an acid pump inhibitor, is regarded as the drug of choice in the treatment of gastric ulcers. Oral administration of omeprazole has been shown to be very effective in both treatment and prevention of gastric ulceration, provided that daily dosing is maintained (Andrews *et al.*, 1999; Merritt *et al.*, 2003; Lester *et al.*, 2005). However, the recommended prolonged use of this medication makes the treatment fairly expensive.

Attempts have been made to find an alternative, less expensive product than omeprazole that offers acceptable treatment and prevention of gastric ulcers. A variety of products, including mucilaginous compounds, pectins, magnesium oxides, other acid buffers, and unsaturated fatty acids, have been proposed to have mucosal protective factors. So far the efficacy of pectin-lecithin complexes has varied (Venner *et al.*, 1999; Murray and Grady, 2002). Antacids are generally considered to be effective for 1-2 hours to neutralize existing acid. However, the need for large and frequent doses makes their use relatively impractical in the horse.

When whole linseeds are cooked, or crushed linseeds are soaked in warm water, they exude large quantities of a mucilaginous material. It is possible that the mucus could form a protective layer on the non-glandular mucosa of the equine stomach. Linseed has also a high concentration of unsaturated fatty acids, including alpha-linolenic acid and linoleic acid. These may have the ability to suppress gastric acid production (Das, 1998). This study tested the effect of a linseed-based feed on gastric ulceration in horses.

Material and methods

The experimental animals were 11 Finnhorses and 5 Standardbred horses (weight 400-580 kg) in active physical training and in apparently good health. All horses had ulcers in the non-glandular region of the stomach in the vicinity of the *margo plicatus*, as determined by gastroscopy at the beginning of the trial. Lesion severity was scored on a scale of 0-5. Animals were divided into a control group and a treatment group based on the similarity of and the severity of ulceration of horses. The basal feed of the horses consisted of haylage/hay and concentrate (oats and pelleted commercial concentrate), the forage-to-concentrate ratio being on average 75:25 on a dry matter basis. In total horses ate 5.1-8.6 kg dm/day roughage, which was given in equal amounts three times a day 30 min before the concentrates. Linseed-based (65%) feed was added to the diet of treatment horses at 1 g/kg live weight per day. Ingredients on a fresh weight basis in the pelleted experimental feed were: linseed (650 g/kg), beet pulp (150 g/kg), dry garlic (100 g/kg), dry carrot (50 g/kg) and molasses (50 g/kg). The feed was given in three equal portions with the concentrates at 06:30, 12:30 and 17:30. The portion of the linseed based feed for the subsequent feed was prepared by soaking feed in warm (45 °C) water (1.5 litres per 500 g feed) at the time of the previous feeding. Mucoid material was evident in the soaked feed prior to feeding.

During the 12 weeks study period gastroscopy was performed twice at week 6 and 12. The evaluation of the treatment effect was based on these findings.

Results and discussion

During the study the general health of all horses was apparently good and they remained in training throughout the trial. Despite their apparent good health, most horses exhibited grade 1-3 ulcers at every examination point (Table 1). The linseed-based feed did not decrease the number or severity of gastric lesions compared to the control group. At the end of the trial three horses had lower ulcer scores than at the first gastroscopy. This indicated a fairly low rate of spontaneous healing of ulcers

Table 1. Gastric ulceration of the treatment and the control groups based on gastroscopy performed at the beginning, at 6 weeks and at 12 weeks (Scores 0-5; 0 = healthy, 5 = very severe).

Treatment group	Beginning of the trial	6 weeks	12 weeks
Pair 1	1	1	1
Pair 2	1	1	1
Pair 3	1	0	0
Pair 4	2	2	2
Pair 5	2	1	2
Pair 6	1	1	1
Pair 7	1	3	3
Pair 8	1	1	1
Control group			
Pair 1	1	1	1
Pair 2	1	1	1
Pair 3	1	1	1
Pair 4	2	1	0
Pair 5	2	0	3
Pair 6	1	2	1
Pair 7	1	1	1
Pair 8	1	0	0

in actively trained horses. The severity of the ulcers seemed to reflect the training intensity, as the most severe ulcerations of these horses were found mainly in those horses that competed in races.

Although poor appetite is one of the symtoms of gastric ulcers, the palatability of the linseed product was quite good. Only two of the treatment horses (Pairs 4 and 8) occasionally left a portion of their concentrates uneaten. No side effects from the product were recorded. However, the linseed-based feed did not decrease the number or severity of gastric lesions compared to the control group.

In conclusion the linseed-based feed failed to restore gastric health. This suggests that, if medication costs are to be avoided, better feeding management should be employed.

References

Andrews, F.M., R.L. Sifferman, W. Bernard, F.E. Hughes, J.E. Holste, C.P. Daurio, R. Alva and J.L. Cox, 1999. Efficacy of omeprazole paste in the treatment and prevention of gastric ulcers in horses. Equine Vet. J. Suppl. 29, 81-86.

Bell, R.J., T.D. Mogg and J.K. Kingston, 2007. Equine gastric ulcer syndrome in adult horses: a review. N. Z. Vet. J. 55, 1-12.

Das, U.N., 1998. Hypothesis: cis-unsaturated fattyacids as potential anti-peptic ulcer drugs. Prostaglandins, leukotrienes and essential fatty acids. 58, 377-380.

Lester, G.D., R.L. Smith and I.D. Robertson, 2005. Effects of treatment with omeprazole or ranitidine on gastric squamous ulceration in racing Thoroughbreds. J. Am. Vet. Med. Assoc. 227, 1636-1639.

Merritt, A.M., L.C. Sanchez, J.A. Burrow, M. Church and S. Ludzia, 2003. Effect of GastroGard and three compounded oral omeprazole preparations on 24 h intragastric pH in gastrically cannulated mature horses. Equine Vet. J. 35, 691-695.

Murray, M.J. and T.C. Grady, 2002. The effect of a pectin-lecithin complex on prevention of gastric mucosal lesions induced by feed deprivation in ponies. Equine Vet. J. 34, 195-198.

Venner, M., S. Lauffs and E. Deegen, 1999. Treatment of gastric lesions in horses with pectin-lecithin complex. Equine Vet. J. Suppl. 29, 91-96.

Effect of linseed based feed supplementation on sand excretion in horses

S. Särkijärvi, S. Hyyppä and S. Saastamoinen
MTT Agrifood Research Finland, Animal Production Research/Equines, Opistontie 10 A 1, 32100 Ypäjä, Finland

Abstract

The effect of linseed based feed on the sand removal was monitored radiographically in 16 Finnhorse mares. Linseed based (65%) feed was added to the diet of treatment horses at the level of 1 g/kg live weight and sand removal was investigated for 11 weeks. The excretion of sand from the gastrointestinal (GI) tract was very slow. In the final evaluation, the measured area of sand accumulations was diminished in average from 145 to 41 cm^2 in the treatment group and in the control group from 158 to 95 cm^2. The relative clearances were in average 0.77 in the treatment group and 0.56 in the control group. The result indicated clear tendency (P=0.095) in the benefit of the treatment group. Linseed based feed seemed to enhance the sand removal from the digestive tract of a horse.

Keywords: faecal sand, linseed, radiography, sand excretion

Introduction

Sand ingestion can result from grazing on scarce pasture on a sandy soil or eating feeds from the ground (Husted *et al.*, 2005). However, some horses do eat sand deliberately. Ingested sand typically accumulates in the ventral colon and the right dorsal colon, because in the colon the larger diameter and reduced flow allows sand to settle. Intestinal sand can cause bowel irritation leading to a variety of symptoms like weight loss, poor performance and intermittent diarrhea. In severe cases sand can cause impaction colic, when surgical intervention is sometimes necessary to save the horse (Bertone *et al.*, 1988; Specht and Colahan, 1988).

Only a tentative diagnosis can be made from a history in an environment with sandy soil. Neither testing for faecal sand gives a definitive diagnosis. Intestinal sand accumulation may be detected with ultrasound, but radiography is the only method that gives a fair estimate of the volume of sand in the intestine (Keppie *et al.*, 2008; Ruohoniemi *et al.*, 2001).

A variety of products, like psyllium, mineral oil and magnesium sulphate, have been used to prevent sand accumulation or to remove it from the gut. So far, they all have a questionable effect on sand removal (Ruohoniemi *et al.*, 2001). When linseeds (whole or crushed) are soaked in water, mucus similar to psyllium mucilloid are formed. The linseed mucus, fibre fractions and possible prebiotic effect, could be helpful in removing the intestinal sand and restoring the gut motility. In this study, the effect of linseed based feed on the sand removal was monitored.

Material and methods

Sixteen Finnhorse mares with radiographically evident sand accumulation(s) in the cranioventral abdomen were included in the study. The horses were randomly selected from a healthy population that is known to exhibit the problematic behaviour of sand eating. The horses were 4 to 21 years old and had mean live weight (LW) 614 kg (s.d. 42 kg). The presence of sand in faeces was examined by mixing faecal material with water (Colahan, 1987). The cranioventral abdomen was radiographed from the right side of the horse as described in (Korolainen and Ruohoniemi, 2002). The amount of sand and the shape and location of the sedimentation were recorded. The size of sand accumulation was evaluated visually and scored from 0-4 (Korolainen and Ruohoniemi, 2002). In addition, the two-dimensional area (cm^2) of the accumulation was measured from the radiographs. On the grounds of the similarity (size and shape) of sand accumulations, the horses formed pairs after first examination. The horses of each pair were then randomly allotted to control group and treatment group.

The basal feeding of both groups consisted of haylage/hay and oats in a ratio of 80:20 on a dry matter basis. Linseed based feed was added to the diet of treatment group at the level of 1 g/kg live weight and served daily in three equal meals. Each portion was soaked in hot water minimum of two hours before feeding. Ingredients in the pelleted experimental feed were: linseed (650 g/kg), beet pulp (150 g/kg), dried garlic (100 g/kg), dried carrot (50 g/kg) and molasses (50 g/kg). All the horses were exposed to a light daily exercise in paddock or by riding. Further sand ingestion from the paddocks was prohibited by winter conditions. Feed allowances were calculated to meet the Finnish nutrient requirements for light work (MTT, 2006).

The resolution of sand was investigated radiographically in 2-3 week intervals in total of eleven weeks. The evaluation of the treatment effect was based on the change in the measured area (cm^2) in relation to the original size of the accumulation. Statistical analysis was carried out using the MIXED procedure of the SAS system (SAS 9.2).

Results and discussion

In general, the health of the horses during the experiment was good. The LW of the horses remained constant and didn't differ between groups. There were no evident symptoms to implicate sand accumulations in either of the groups, except for some horses having mild short-term diarrhea. However, there was no immediate need for treatment. Although the largest sand accumulations were around 400 cm^2 e.g. 14x28 cm, no serious symptoms (impaction colic) were experienced.

Sand was present in all of the samples examined in faecal sand test during the whole experiment irrespectively of the group. The amount of sand detected at the sand test did not seem to correlate with the size of sand accumulation visible in the radiographs or the clearance of sand from the digestive tract. This finding is in agreement with the results noted before (Ruohoniemi *et al.*, 2001, Edens and Cargile, 1997).

In the beginning of the experiment, there was no difference between the groups in the size of sand accumulations (P=0.59). In the final evaluation, the measured area of sand accumulations was diminished in average from 145 to 41 cm^2 in the treatment group and in the control group from 158 to 95 cm^2. The relative clearances were in average 0.77 in the treatment group and 0.56 in the control group. The result indicated clear tendency (P=0.095) in the benefit of the treatment group. In overall, the excretion of sand from the GI tract was very slow and not predictable. One horse in the control group failed to remove any sand from the digestive tract in the 11 weeks study.

Conclusions

Sand accumulations were easily detectable in radiographic evaluation. Sand test was not a reliable tool in predicting sand clearance. Although some horses had radiographically evident large sand accumulations, they were asymptomatic, which shows a poor correlation between radiological and clinical findings. Linseed based feed seemed to enhance the sand removal from the GI tract of a horse in a long term use. However, since the excretion of sand was very slow, efforts should be made to prevent sand eating with better management.

References

Bertone, J.J., J.L. Traub-Dargatz, R.W. Wrigley, D.G. Bennett and R.J. Williams, 1988. Diarrhea associated with sand in the gastrointestinal tract of horses. J. Am. Vet. Med. Assoc. 193, 1409-1412.

Colahan, P., 1987. Sand colic. In: Robinson, N.E. and W.B. Saunders Co. (eds.), Current Therapy in Equine Medicine II. Philadelphia. pp. 55-58.

Edens, L.M. and J.L. Cargile, 1997. Medical management of colic. In: Robinson, N.E. and W.B. Saunders Co. (eds.), Current Therapy in Equine Medicine IV. Philadelphia. pp. 182-191.

Husted, L., M.S. Andersen, O.K. Borggaard, H. Houe and S.N. Olsen, 2005. Risk factors for faecal sand excretion in Icelandic horses. EquineVet. J. 37, 351-355.

Keppie, N.J., D.S. Rosenstein, S.J. Holcombe and H.C. Schott II, 2008. Objective radiographic assessment of abdominal sand accumulation in horses. Vet. Radiol. Ultrasound 49, 122-128.

Korolainen, R. and M. Ruohoniemi, 2002. Reliability of ultrasonography compared to radiography in revealing intestinal sand accumulations in horses. Equine. Vet. J. 34, 499-504.

MTT, 2006. Rehutaulukot ja ruokintasuositukset (Feed tables and feeding recommendations). [online]. Jokioinen: MTT Agrifood Research Finland. Published 14.2.2006, [cited 4.3.2010]. Available at: http://www.mtt.fi/rehutaulukot/. URN:NBN:fi-fe20041449.

Ruohoniemi, M., R. Kaikkonen, M. Raekallio and L. Luukkanen, 2001. Abdominal radiography in monitoring the resolution of sand accumulations from the large colon of horses treated medically. Equine Vet. J. 33, 59-64.

Specht, T.E. and P.T. Colahan, 1988. Surgical treatment of sand colic in equids: 48 cases (1978-1985). J. Am. Vet. Med. Assoc. 193, 1560-1564.

Organic versus inorganic zinc supplementation: effect on markers of zinc status

C.E. Dunnett and M. Dunnett
Independent Equine Nutrition, P.O. Box 319, Mildenhall, Suffolk, IP28 6SY, United Kingdom

Abstract

In a crossover design, eight ponies were supplemented with two forms of zinc. The aim of this study was to investigate the effect of zinc supplementation and source on indicators of zinc status in grass fed ponies. Plasma and whole blood Zn concentrations were not good indicators of dietary Zn intake. Zinc supplementation had a significant effect on hair growth rate, which may limit its use to evaluate zinc status.

Keywords: inorganic, organic zinc supplementation

Introduction

Zinc (Zn) is an important dietary trace mineral, as it is required for the function of over 100 metalloenzymes. Traditionally, inorganic sources of supplementary Zn such as sulphate have predominated. Whilst there is evidence for improved absorption of organic Zn in other species, the results to date in horses have been contradictory. The aim of this study was to investigate the effect of zinc supplementation and source on indicators of zinc status in grass fed ponies.

Material and methods

In a crossover design, eight ponies with a mean age of 4 (±0.8) years and bodyweight 240 (±51) kg were supplemented for 2 months with one of two forms of zinc supplementation (periods 2 and 4), with a 2 month washout period between (period 3) and at the end (period 5) of the periods of supplementation. Prior to the study the ponies with accustomed to a grass based diet for a period of 2 months (period 1). The ponies were randomly divided into two groups (group O/IN or group IN/O) and assigned to either of two forms of Zn supplementation. During each washout period the ponies returned to their grass based diet.

Group O/IN were supplemented with an organic form of Zn (Bioplex®, Alltech Inc) during period 2 and an inorganic form of Zn (zinc sulphate) during period 4. Group IN/O was supplemented with an inorganic form of Zn (zinc sulphate) during period 2 and an organic form of Zn (Bioplex®, Alltech Inc) during period 4.

The ponies were fed a grass based diet with an analysed Zn content of 27 mg/kg dry matter, which is below the current NRC requirements (NRC, 2007). Based on an estimated voluntary dry matter (dm) intake of 2.5% of bodyweight, the basal diet provided approximately 162±35 mg of Zn per head per day or 0.68 mg of Zn per kg bodyweight per day. Zn supplementation of both groups provided an additional 375 mg Zn/head/day above that of the basal diet. In addition, the Zn supplements were fed together with approximately 200 g of grass pellets (Zn, 40 mg/kg dm) and 100 g of soaked sugar beet (Zn, 28 mg/kg dm). The total intake of Zn during the two periods of supplementation was therefore estimated to be 93±12 mg Zn per Kg DM or 2.3±0.3 mg Zn per kg bodyweight.

Health was monitored throughout this feeding study and samples of plasma, whole blood and hair were taken at the end of each feeding period for assessment of Zn status. Plasma and whole blood samples were analysed for Zn concentration by inductively coupled plasma optical emission spectroscopy (ICP-OES). Hair samples were analysed for Zn concentration by inductively coupled plasma mass spectrometry (ICP-MS). Mane hair growth rate was also measured.

Once the normality of the data was established an analysis of variance (ANOVA) for repeated measures was carried out and this was followed when significance of (P<0.05) was reported by paired T tests to establish significant differences between groups. Significant results were reported at P<0.05.

Results

Mean plasma and whole blood Zn concentrations prior to supplementation (end of period 1) were 8.3±1.9 and 36.7±9.2 μmol/l, respectively. Plasma and whole blood Zn concentration varied throughout the feeding study, but there was no clear effect of Zn supplementation (Table 1). There was also no significant effect of Zn source on plasma or whole blood Zn concentration.

Mean growth of mane was 27.24±3.11, 42.94±3.89, 32.40±10.07, 38.61±8.12 and 33.57±4.78 mm for periods 1-5, respectively. Growth of mane hair was significantly increased following zinc supplementation during both periods 2 and 4 (P<0.01, P<0.05) when compared to the baseline level observed following the period of acclimatisation (period 1). Hair growth did not return to the baseline level during either of the two washout periods (3 and 5). Although hair growth was numerically lower during periods 3 and 5 compared to the previous period of supplementation, this was not significantly different. However, hair growth during the two washout periods (3 and 5) was also not significantly different from that the baseline level (period 1). There was also no apparent effect of the source of supplementary Zn (Figure 1) on hair growth.

Mean concentration of Zn in mane hair prior to supplementation (end of period 1) was 138±10.8 μg/g. Hair Zn concentration fluctuated during the study with no clear relationship with Zn supplementation or any significant increase in response to Zn supplementation irrespective of the source of Zn (Table 1).

Discussion

Plasma and whole blood Zn concentrations were similar to those reported previously (Schryver *et al.*, 1980; Wichert *et al.*, 2002a) and were found within the normal range, albeit at the lower end (Lewis, 1995) despite the likely marginal zinc intake whilst on the basal diet. Whilst there was some variation in Zn concentration in both plasma and whole blood there was no clear relationship to Zn supplementation, irrespective of Zn source. This is in agreement with earlier research in ponies that reported no significant increase in serum Zn concentration following supplementation with Zn at 250

Table 1. Concentration of zinc (Zn) in plasma, whole blood Zn μmol/l (mean ± SD) NS and hair μg/g (mean ± SD) NS in response to supplementary zinc in organic or inorganic form.

		Period 1 No Supplement	Period 2 Supplement	Period 3 No Supplement	Period 4 Supplement	Period 5 No Supplement
Plasma Zn	O/IN[1]	8.15±2.26	8.35±2.49	8.10±0.96	7.70±0.73	8.60±1.23
	IN/O[2]	8.40±1.66	8.43±0.67	8.80±0.26	7.53±1.01	8.90±2.04
Whole Blood Zn	O/IN	34.37±10.80	40.80±5.09	44.70±4.16	45.03±4.00	42.60±10.92
	IN/O	40.30±8.06	47.90±9.19	49.17±1.55	42.00±2.88	47.80±5.68
Hair Zn	O/IN	138.47±6.30	193.93±83.99	147.09±15.27	125.61±49.11	136.79±13.97
	IN/O	136.76±16.97	143.71±32.05	128.34±28.71	132.44±24.97	135.73±55.02

[1] Group O/IN received organic Zn during period 2 and inorganic Zn during period 4; [2] group IN/O received inorganic Zn during period 2 and organic Zn during period 4.

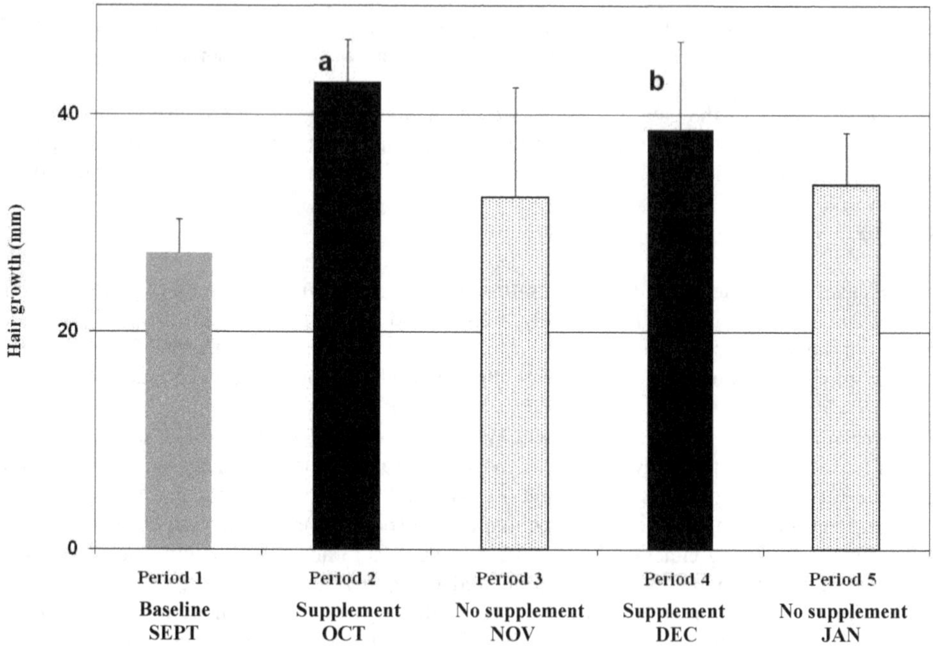

Figure 1. Mean (± SD) mane hair growth prior to, during and after zinc (Zn) supplementation (375 mg/head/day).
[a] Significant difference from period 1 (P<0.01); [b] Significant difference from period 1 (P<0.05).

mg/head per day, although an increase was apparent with a much higher intake of 520 mg/head/day (Schryver *et al.*, 1980). However, this may simply reflect the timing of blood samples in relation to supplementation, as previously serum response to a discrete administration of Zn was dose dependent when peak serum concentration and area under the curve was examined (Wichert *et al.*, 2002b).

In contrast, Zn supplementation had a highly significantly effect on hair growth, which increased during both periods of supplementation in this study. Seasonal effects were discounted, as monthly hair growth in horses of various breeds has previously been shown to be linear over a calendar year with no observable seasonal effects (Dunnett, 2005). Furthermore, the rate of growth has also been shown to be lowest near to the withers, which was the site of sampling in this study (Dunnett, 2005). This is perhaps not surprising as Zn is a key mineral involved in the process of keratinisation and plays a catalytic, structural and regulatory role mediated via a large number of Zn dependent enzymes (Tomlinson *et al.*, 2004). However, despite the known association between Zn supplementation and keratin synthesis in horses, an increase in the rate of hair growth in response to Zn supplementation has not been previously reported, although it has been described in dogs (Lowe *et al.*, 1994). The significant increase in hair growth during the periods of Zn supplementation is likely to have been influenced by the marginal zinc intake in these ponies outside of the periods of supplementation. Prolonged Zn deficiency is associated with a reduction in hair growth in other species (Neathery *et al.*, 1973). In addition, changes to the mechanical properties of horse hair in terms of diameter, elasticity and strength has previously been reported in response to a combination of zinc and copper supplementation (Kania *et al.*, 2009).

Zn concentration in mane hair was in agreement with other authors e.g. Asano *et al.* (2005). Whilst there was some variation in the Zn concentration in mane hair during this study, there was no clear relationship with either Zn supplementation or Zn source, which is in contrast to previous work in dogs that reported an increase in hair Zn concentration with Zn supplementation, which was greatest with an organic versus inorganic source of Zn (Lowe *et al.,* 1994). However, the significant increase in hair growth in these ponies in response to Zn supplementation may have masked an increased Zn uptake into hair, as the zinc is incorporated into an increased volume of hair. This relationship to hair growth rate is apparent in humans, where the zinc concentration in hair is paradoxically increased when hair growth rate is suppressed, e.g. with severe malnutrition (Erten *et al.,* 1978). The effect of hair growth rate on the deposition of zinc in dogs is likely to be less relevant due to an overall much lower hair growth rate. The effect of hair growth rate on hair zinc concentration could be corrected for by using a static component of hair such as melanin, which is unaffected by hair growth rate.

Conclusion

Plasma and whole blood Zn concentrations were not good indicators of dietary Zn intake in grass fed ponies. The usefulness of the Zn concentration in equine hair requires further evaluation as an indicator of Zn status, due to a highly significant effect of Zn supplementation on hair growth rate where historical zinc intake may have been marginal.

Acknowledgements

The authors would like to acknowledge Zoe Stevenson of Alltech Inc for her technical input and also to Alltech Inc for their funding of this study.

References

Asano, K., K. Suzuki, M. Chiba, K. Sera, R. Asano and T. Sakai, 2005. Twenty-eight element concentrations in mane hair samples of adult riding horses determined by particle-induced X-ray emission. Biological Trace Element Research 107, 135-140.

Dunnett, M., 2005. The diagnostic potential of equine hair: A comparative review of hair analysis for assessing nutritional status, environmental poisoning and drug use and abuse. In: Pagan, J.D. (ed.), Advances in Equine Nutrition III. Nottingham University Press, Nottingham. 85-106 pp.

Erten, J., A. Arcasoy, A.O. Cavdar and S. Cin, 1978. Hair zinc levels in healthy and malnourished children. American Journal of Clinical Nutrition 31, 1172-1174.

Kania, M., D. Mikolajewska, K. Marycz and M. Kobielarz, 2009. Effect of diet on mechanical properties of horse hair. Acta of Bioengineering and Biomechanics 11, 55-57.

Lewis, L.D., 1995. Equine Clinical Nutrition: Feeding and Care, Williams and Wilkins, Baltimore. 25-60 pp.

Lowe, J.A., J. Wiseman and D.J.A. Cole, 1994. Zinc source influences zinc retention in hair and hair growth in the dog. Journal of Nutrition 124, 2575S-2576S.

Neathery, M.W., W.J. Miller, D.M. Blackmon, F.M. Pate and R.P. Gentry, 1973. Effects of Long Term Zinc Deficiency on Feed Utilization, Reproductive Characteristics, and Hair Growth in the Sexually Mature Male Goat. Journal of Dairy Science 56, 98-105.

NRC, 2007. Nutrient Requirements of Horses 6th Edition, The National Academies Press, Washington DC. 54-68 pp.

Schryver, H.F., H.F. Hintz and J.E. Lowe, 1980. Absorption, excretion and tissue distribution of stable zinc and 65 zinc in ponies. J. Anim Sci. 51, 896-902.

Tomlinson, D.J., C.H. Mulling and T.M. Fakler, 2004. Formation of keratins in the bovine claw: roles of hormones, minerals, and vitamins in functional claw integrity. J. Dairy Sci. 87, 797-809.

Wichert, B., T. Frank and E. Kienzle, 2002[a]. Zinc, copper and selenium intake and status of horses in Bavaria. Journal of Nutrition 132.

Wichert, B., K. Kreyenberg and E. Kienzle, 2002b. Serum response after oral supplementation of different zinc compounds in horses. J. Nutr. 132, 1769S-1770.

Comparative effect of organic and inorganic selenium supplementation on markers of selenium status

M. Dunnett and C.E. Dunnett
Independent Equine Nutrition, P.O. Box 319, Mildenhall, Suffolk, IP28 6SY, United Kingdom

Abstract

In a crossover design, eight ponies were supplemented with two forms of selenium (Se). The aim of this study was to investigate the effect of Se supplementation and source on Se status in grass fed animals, where basal Se intake may be marginal. Plasma and whole blood Se were shown to be sensitive indicators of acute changes in dietary Se intake. Increased incorporation of Se into mane hair following 2 months of supplementation with an organic (Sel-Plex®) compared to inorganic (selenite) selenium source supports the previously suggested assertion of increased availability of organic selenium.

Keywords: organic, inorganic selenium supplementation

Introduction

Adequate dietary selenium (Se) is essential to support optimum functioning of the antioxidant, immune and reproductive systems in horses. The low natural abundance of Se in most pastures, grains and forages makes dietary supplementation essential (Surai, 2006). The narrow margin of safety for sodium selenite makes organic Se an attractive alternative for supplementation in horses. The aim of this study was to investigate the effect of Se supplementation and source on Se status in grass fed animals using measured parameters in both blood and hair.

Materials and methods

In a crossover design, eight ponies with a mean age of 5 (±1.7) years and bodyweight 270 (±57) kg were supplemented for 2 months with one of two forms of Se (periods 2 and 4), with a 2 month washout period between (period 3) and at the end (period 5) of the periods of supplementation. Prior to the study the ponies with accustomed to a grass based diet for a period of 2 months (period 1). The ponies were randomly divided into two groups (group O/IN or group IN/O) and assigned to either of two forms of Se supplementation. During each washout period the ponies returned to their grass based diet.

Group O/IN was supplemented with an organic form of Se (Sel-Plex®, Alltech Inc) during period 2 and an inorganic form of Se (sodium selenite) during period 4.

Group IN/O was supplemented with an inorganic form of Se (sodium selenite) during period 2 and an organic form of Zn (Sel-Plex®, Alltech Inc) during period 4.

The ponies were fed a grass based diet and the Se supplementation of both groups provided an additional 1.5 mg Se/head/day above that of the basal diet. In addition, the Se supplements were fed together with approximately 200 g of grass pellets and 100 g of soaked sugar beet.

Samples of plasma, whole blood and hair were taken at the end of each dietary period (1-5) and ponies were assessed for general health by a veterinarian throughout the study. Plasma and whole blood Se concentration was analysed by inductively coupled plasma optical emission spectroscopy (ICP-OES), plasma selenomethionine (SeMet) concentration was analysed by liquid chromatography-mass spectrometry (LC-MS). A small section of mane hair, close to the withers, was shaved at the start of the study and was then clipped at the end of each study period to ensure hair samples did not break and were representative of the period of interest. Hair samples were washed and extracted according to established methodology to remove surface contamination (Chyla and Zyrnicki, 2000) and hair Se concentration was analysed by inductively coupled plasma mass spectrometry (ICP-MS).

Normality of the data was established and an analysis of variance (ANOVA) for repeated measures was carried out. When significance of ($P<0.05$) was reported with ANOVA, a paired T Test was used to establish significant differences between groups. Significant results were reported at $P<0.05$.

Results

There was a significant effect of Se supplementation and period on plasma and whole blood Se concentration ($P<0.001$), but no statistically significant effect of Se source (Table 1). Mean plasma and whole blood Se concentration prior to supplementation (end of period 1) was 1.8±0.38 and 2.3±0.23 μmol/l, respectively. Mean plasma Se was significantly higher following Se supplementation during both supplementation period 2 (2.4±0.28 μmol/l, $P<0.001$) and period 4 (2.21±0.29 μmol/l, $P<0.01$), when compared to the previous baseline feeding period.

Whole blood Se was also significantly higher following Se supplementation during supplementation period 2 (3.10±0.19 μmol/l, $P<0.001$) and period 4 (3.38±0.31 μmol/l, $P<0.001$), when compared to the previous baseline feeding period. Plasma and whole blood Se concentration declined rapidly on cessation of supplementation.

There was no statistically significant effect of selenium source on either plasma or whole blood Se concentration following supplementation. However, there was a trend towards higher plasma Se concentration following supplementation with the organic compared to the inorganic Se source (7 out of 8 ponies), but this failed to reach statistical significance ($P=0.06$) (Table 1).

There was no clear change in plasma total SeMet in response to Se supplementation irrespective of Se source (Table 1). Whilst there were small fluctuations in plasma total SeMet concentration throughout the study, ANOVA failed to reveal any significant effects or interactions with respect to this parameter. There was no correlation between SeMet and total Se in plasma ($r^2=0.1$).

Table 1 Concentration of plasma and whole blood selenium (Se) μmol/l and plasma selenomethionine (SeMet) μmol/l (mean±SD) in response to supplementary selenium (1.5 mg/head/day) with organic (Sel-plex) or inorganic forms (sodium selenite).

		Period 1 No supplement	Period 2 Supplement	Period 3 No supplement	Period 4 Supplement	Period 5 No supplement
Plasme Se	O/IN	1.77±0.28	2.45±0.33[b]	1.31±0.35[a]	2.06±0.28[a]	1.82±0.24
	IN/O	1.76±0.51	2.34±0.27[b]	1.50±0.55[a]	2.35±0.23[a]	2.11±0.31
Whole Blood Se	O/IN	2.29±0.13	3.14±0.24[b]	2.41±0.29[a]	3.26±0.34	2.75±0.53[a]
	IN/O	2.35±0.33	3.07±0.14[b]	2.62±0.41[a]	3.51±0.26	3.04±0.66[a]
Plasma Se Met	O/IN	0.54±0.08	0.57±0.16	0.55±0.13	0.52±0.04	0.53±0.05
	IN/O	0.41±0.10	0.50±0.08	0.52±0.08	0.52±0.08	0.48±0.02

[1] Group O/IN eceived organic Se during period 2 and inorganic Se during period 4; [2] group IN/O received inorganic Se during period 2 and organic Se during period 4; [a] represents a significant difference from period 1 (both groups combined) $P<0.05$; [b] represents a significant difference from period 1 (both groups combined) $P<0.001$.

Hair Se concentration was affected by dietary Se intake and statistical analysis revealed a significant effect of both treatment (P<0.01) and period (P<0.001). Hair Se concentration increased in response to Se supplementation irrespective of Se source. Mean Se concentration was significantly higher at the end of the first period of supplementation (period 2,580±95 vs. 344±104 ng/g) (P<0.001). Additionally, Se concentration did not return to the baseline level following either non-supplementation periods (period 3 and 5) and remained significantly higher than period 1, respectively (469±138, P<0.05; 456±98 ng/g, P<0.05).

Hair Se concentration was also influenced by Se source and post supplementation hair Se concentration was greater when all of the ponies were fed organic Se compared to inorganic Se, respectively (mean 626±114 vs. 497±101 ng/g, P<0.01) see Figure 1. There was, however, no effect of Se source on the decline in Se concentration in hair 2 months following cessation of Se supplementation.

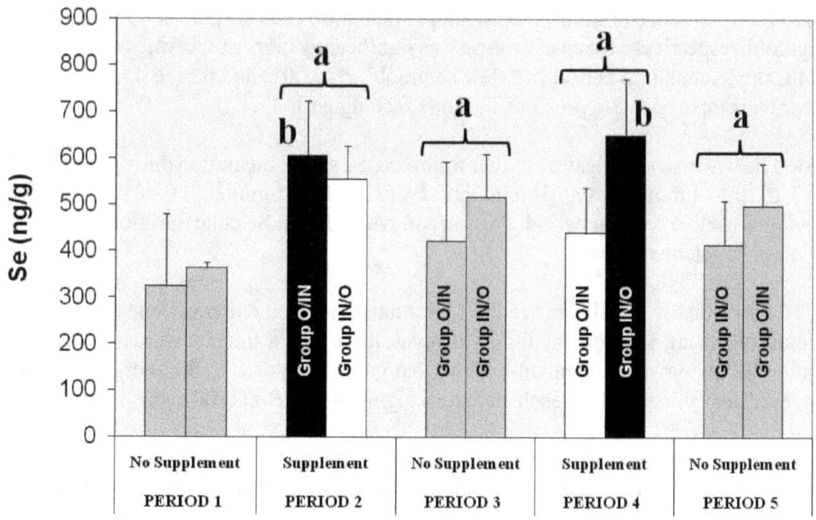

Figure 1. Mean concentration of selenium (Se) in equine hair (ng/g ± S.D) prior to and after Se supplementation (Sel-plex or sodium selenite) for two months (1.5 mg/head/day).
O/IN ponies fed organic Se (Sel-Plex®) during period 2 and inorganic Se (sodium selenite) during period 4; IN/O ponies fed inorganic Se (sodium selenite) during period 2 and organic Se (Sel-Plex®) during period 4; [a] Significantly different from period 1 (both (P<0.001) (pony groups combined); [b]Se-plex group was significantly higher than sodium selenite Se (P<0.01).

Discussion

Prior to Se supplementation the plasma and whole blood Se concentration in this group of grass fed ponies was at the lower end of the previously described normal range (Stowe, 1998; Stowe and Herdt, 1992). However, the level of Se in the basal diet was probably low resulting in a marginal dietary Se intake. Plasma and whole blood Se was sensitive to the frequent changes in Se supplementation and provided a good indicator of dietary Se intake, although in contrast to (Calamari *et al.*, 2009) there was no significant effect of Se source, which may reflect the shorter period of supplementation as well as differences in basal Se intake. SeMet represented about 28% of the total plasma Se concentration, but the correlation between the two was weak. Despite SeMet being the major form of Se in Se enriched yeast, the plasma concentration of SeMet was unchanged by organic Se supplementation. However, this may be explained by the association of Se primarily as selenocysteine within selenoprotein P, a major circulatory transporter of Se (Hoffmann *et al.*, 2007).

Improved absorption of organic Se during this study in these grass fed ponies is credible, as the incorporation of Se into mane hair was increased significantly in response to organic compared to inorganic Se supplementation and in addition there was also a trend towards increased plasma Se concentration with the former. Increased incorporation of Se into equine hair with organic Se supplementation has been previously reported (Calamari *et al.,* 2008), but the period of supplementation in this study was shorter (2 months) and was carried out in grass fed ponies with a crossover rather than block design.

The incorporation of Se into hair occurs through substitution of Se for sulphur within the disulphide bridges found in the structure of keratin. The Se content of hair remained elevated above its pre-supplementation level 2 months after cessation of Se supplementation, which suggests that the elevated uptake of Se into new hair was sustained despite cessation of supplementation. The plasma and whole blood level of Se declined rapidly once supplementation ceased suggesting that there may be some other mechanism present to sustain Se uptake into new hair. Selenoprotein P is expressed in skin and hair follicles, which may represent a less labile sources of Se (Lee *et al.,* 2008).

Conclusion

In grass fed ponies, where basal Se intake may be marginal, plasma and whole blood Se were shown to be sensitive indicators of acute changes in dietary Se intake. Increased incorporation of Se into mane hair following 2 months of supplementation with an organic (Sel-Plex®) compared to inorganic (selenite) selenium source supports the previously suggested assertion of increased availability of organic selenium. Hair analysis can also be used as a good retrospective indicator of chronic changes in dietary Se intake

Acknowledgements

The authors would like to acknowledge Zoe Stevenson of Alltech inc. for her support and also to Alltech Inc for their funding of this study.

References

Calamari, L., G. Brigati and G. Bertin, 2008. Assessment of selenium status in horses fed diet supplemented with sodium selenite or Se-yeast(ed.) Proceedings of the XIV SIVE/FEEVA Congress. Societa Italiana Veterinara Per Equini, Lido di Venezia, Italy. 351-352 pp.

Calamari, L., A. Ferrari and G. Bertin, 2009. Effect of selenium source and dose on selenium status of mature horses. Journal of Animal Science 87, 167-178.

Chyla, M.A. and W. Zyrnicki, 2000. Determination of metal concentrations in animal hair by the ICP method: Comparison of various washing procedures. Biological Trace Element Research 75, 187-194.

Hoffmann, P.R., S.C. Hoge, A.L. Ping, F.W. Hoffmann, A.C. Hashimoto and M.J. Berry, 2007. The selenoproteome exhibits widely varying, tissue-specific dependence on selenoprotein P for selenium supply. 3963-3973 pp.

Lee, S.R., J.M. Yon, I.J. Baek, M.R, C.G. Park, B.J. Lee, Y.W. Yun and S.Y. Nam, 2008. Spatiotemporal expression of the selenoprotein P genein postimplantational mouse embryos. The International Journal of Developmental Biology 52, 1005-1011.

Stowe, H.D., 1998. Selenium supplementation for horse feed. In: Pagan, J.D. (ed.), Advances in Equine Nutrition. Nottingham University Press, Nottingham, UK. pp 97-103.

Stowe, H.D. and T.H. Herdt, 1992. Clinical assessment of selenium status of livestock. Journal of Animal Science 70, 3928-3933.

Surai, P.F., 2006. Selenium in nutrition and health, Nottingham University Press, Nottingham, UK. 974 pp.

Influence of garlic supplementation on respiratory health and incidence of anaemia in horses

M.T. Saastamoinen, S. Hyyppä and S. Särkijärvi
MTT Agrifood Research Finland, Animal Production Research, Equines, Opistontie 10 A1, 32100 Ypäjä, Finland

Abstract

Garlic (*Allium sativum*) is claimed to have many beneficial properties to the health of humans and animals. It is commonly used in horses e.g. to treat respiratory diseases and infections in the horse's lungs. However, in addition to possible positive influences it may have adverse health effects. This study shows that long-term supplementation of dried garlic at a level of 32 mg/kg BW seemed to reduce the amount of tracheal symptoms and accumulation of tracheal exudates. Also the amount of neutrophil cells in the tracheal mucus was smaller in the garlic supplemented horses. Concerning the adverse effects, Hb, HcT and RBC values of the supplemented horses showed a tendency to decrease during the 83-days study period indicating slight anaemia in those individuals. Consequently, it is possible that even low garlic supplementation levels can be harmful when the supplementation period is long.

Keywords: nutrition, respiratory health, dietary supplementation

Introduction

Garlic (*Allium sativum*) has been used in the diet of humans for centuries because of its believed positive health effects. For example, it may help to clear mucus in the airways. In addition, it has many anti-microbial and anti-parasitic properties. The active components in garlic include organosulfur compounds.

Garlic is also commonly used in horses to treat respiratory diseases and infections in the horse's lungs, to give relief from the symptoms of coughs. Pearson (2003) reported a significant decrease in respiratory rate in horses supplemented with garlic. Williams and Lamprecht (2008) have reviewed studies on health influences of garlic in human and some animal species.

Although garlic has been used as a supplement in diets for horses, the dosage of garlic required for beneficial effects in horses is not known. Furthermore, there is little information on possible adverse health effects of garlic in horses. There may be a risk of anaemia when large doses of garlic are fed to horses. Pearson *et al.* (2005) reported that a daily dose of dried garlic over 200 mg/kg BW developed indications of Heinz body anaemia. The report of The National Research Council (2008) gives presumed and historical safe intakes of 90 and 15 milligrams per kilogram of body weight, and concluded that the threshold level above which the risk of an adverse event will increase significantly is likely to be between 15 and 200 mg/kg BW of dried garlic. The aim of this study was to evaluate the possible positive influence on airway health as well as possible adverse health effects of garlic supplementation in horses.

Material and methods

The effect of garlic supplementation was tested using twelve Finnhorse mares (aged between 5 to 17 years, average BW 628 kg). The horses were paired according to matched health status, management and upper respiratory tract characteristics as determined by endoscopy. One of each matched pair was the experimental animal receiving garlic, the other member of the pair was the control, receiving no garlic. The horses had access to a paddock for four hours daily and, in addition they were exercised for one hour by riding or driving, corresponding to light work.

The horses were individually fed with a hay and oat diet and supplemented with a linseed based feed at maintenance energy level, the forage-to-concentrate ratio being 80:20. In addition, the experimental group was supplemented with 20 grams of dried garlic flakes, corresponding 0.002% of the DM intake and 32 mg/kg BW, which is between limits given by The National Research Council (2008). The supplementation continued for a total of 83 days.

Upper respiratory tract (ethmoidal region, pharyngeal openings of guttural pouches, soft palate, larynx and trachea) examination by endoscopy was carried out on days 41 and 83 of the study. The findings were registered. Tracheobronchial aspirates were drawn at the time of the endoscopy and cytologic and bacteriological (neutrophil cells) evaluation was made from the tracheal mucus. Blood samples were collected on the corresponding days. The blood analysis consisted of white blood cells (WBC), red blood cells (RBC), mean cell corpuscular volume (MCV), haematocrit (HcT) and haemoglobin (Hb) contents. Also the white blood cells were differentiated.

The experimental design was a randomized block design. The data were analysed with a linear model. The differences between the treatments were tested with T-test ($P<0.05$). Categorical variables (neutrophil cells in tracheal mucus) and 0/1-variables (symptoms = 1 or no symptoms = 0) were not tested statistically, but were presented descriptively, because of the small number of observations and their subjective scoring. The classification concerning the neutrophil cells in bronchoalveolar smear samples was as follows: non or some single cells(-); single cells and a few small pool of cells (+); several large pools of cells (++); abundant pools of cells (+++); and extreme abundance of cells (++++).

Results and discussion

The garlic supplementation seemed to reduce the amount of tracheal symptoms and accumulation of tracheal exudates based on the endoscopy examination. The symptoms disappeared in three of the six horses supplemented with garlic, one horse remaining without any symptoms during the study period. The symptoms remained in two horses. Concerning the control horses, the symptoms remained in three horses, fluctuated in two horses and disappeared in one horse.

The amount of neutrophils in the tracheal mucus was smaller in the garlic supplemented horses. A large amount of neutrophil cells (+++) was found only in one sample in the supplemented group, but in four of the six horses samples in the control group. The neutrophils in the tracheal mucus of the control horses remained high or increased in three individuals during the study, but decreased in two of the supplemented horses (from +++ to +), and in one of them the amount increased (from ++ to +++).

The supplemented horses showed slightly declining Hb, HcT and RBC values during the study (Table 1). The mean (of days 41 and 83) as well as the final values of Hb, HcT and RBC in the supplemented horses were numerically (statistically non-significantly) smaller compared to the control horses. However, the final Hb value of the garlic supplemented horses was below the normal ranges for Finnhorses (MTT, 2009), indicating a slight anaemia. The other parameters were within the ranges reported for healthy Finnhorses (e.g. Pösö *et al.*, 1983; Saastamoinen, 1994). There were no differences in the differentiated white blood cells (WBC).

The supplemented amount of dried garlic in the present study (32 mg/kg BW) is within those limits (15 and 200 mg/kg BW of dried garlic) given by The National Research Council (2008). Consequently, it is possible that low supplementation levels may be harmful when the supplementation period is

Table 1. Haematology of the horses during the study period.

Parameter	Garlic supplementation			Control		
	Initial	Day 41	Day 83	Initial	Day 41	Day 83
WBC, x 10^9 /l	8.15	7.52	6.97	7.76	6.95	7.52
RBC, x 10^{12} /l	7.17	7.15	6.57	7.29	7.46	7.04
MCV, fl	50.0	49.5	49.2	50.1	50.0	49.7
HcT, %	36.9	35.4	32.3	36.4	37.2	34.9
Hb, g/l	130.5	128.3	118.0	130.7	135.5	127.3

WBC = white blood cells; RBC = red blood cells; MCV = mean cell corpuscular volume; HcT = haematocrit; Hb = haemoglobin; the differences between the supplemented and control groups were statistically non-significant.

long. The decrease in haematology values is more critical to oxidatively stressed hard-working horses than horses in light work. However, Pearson *et al.* (2005) found that the recovery from anaemia was largely complete five weeks after termination of garlic supplementation. The supplementation period in their study was 71 days, but they used only two horses.

Conclusions

The used level (32 mg/kg BW) of dried garlic fed to horses seemed to reduce the amount of tracheal symptoms and accumulation of tracheal exudates, but caused also a slight anaemia, when fed continuously for ca. three months. Thus, these findings pointed out that there may be a risk of adverse health effects even if quite low doses are fed for long periods. Further research is needed to find safe garlic doses and supplementation periods for horses. It is also important to obtain further information of the possible positive health effects of garlic.

References

MTT, 2009. Laboratory values of horses. Laboratory of MTT Agrifood Research Finland.

Pearson, W., 2003. Ethnoveterinary medicine: The science of botanicals in equine health and disease. In: Proceedings of Second annual European Equine Health and Nutrition Congress, The Netherlands, 31-40.

Pearson, W., H.J. Boermans, W.J. Bettler, B.W. McBride and M.I. Lindinger, 2005. Association of maximum voluntary dietary intake of freeze-dried garlic with Heinz body anemia in horses. Am. J. Vet. Res. 66, 457-465.

Pösö, A.R., T. Soveri and H.E. Oksanen, 1983. The effect of exercise on blood parameters in Standardbred and Finnish-bred horses. Acta Vet. Scand 24, 170-184.

Saastamoinen, M.T., 1994. Propionic acid treated grain (oats) in the diet of horses. Agric. Sci. Finl. 3, 161-168.

The National Research Council, 2008. Safety of dietary supplements for horses, dogs and cats. The National Academies Press, Washington D.C. 260 pp.

Williams, C.A. and E.D. Lamprecht, 2008. Some commonly fed herbs and other fuctional foods in equine nutrition: a review. Vet. J. 178, 21-31.

A randomised placebo controlled double blind study on the effect of subspecies of rose hip (*Rosa canina*) on the immune system, working capacity and behaviour of horses

K. Winther[1], A. Kharazmi[2], A.S.V. Hansen[3] and J. Falk-Rønne[4]
[1]Department of Clinical Biochemistry, Frederiksberg Hospital, Nordre Fasanvej 57, 2000 Frederiksberg, University of Copenhagen, Denmark
[2]Department of Microbiology, Rigshospitalet, University of Copenhagen, Blegdamsvej 9, 2100 Copenhagen Ø, Denmark
[3]Institute for Plant biology and Biotechnology, University of Copenhagen, Faculty of Life Sciences, Bülowsvej 17, 1870 Frederiksberg, Copenhagen, Denmark
[4]The Horseclinic at Lunden, Ordrup Jagtvej 201, 2920 Charlottenlund, Copenhagen, Denmark

Abstract

The aim of this study was to investigate whether rose-hip had any effects on the immune system, working capacity or behaviour of horses in exercise training. Forty-four horses, all trotters, were included in a double-blind, placebo controlled trial and were supplemented with either rose-hip (*Rosa canina*) LitoVET, or a placebo for 3 months. The results from this study suggest that rose-hip LitoVET is effective as an anti-inflammatory and anti-oxidative agent in horses and improves vitamin C status.

Keywords: rosehip, anti-oxidant, anti-inflammatory, performance

Introduction

A standardized powder produced from the seeds and husk of certain subspecies of rose-hip by HybenVital, Langeland, Denmark, has shown anti-inflammatory properties and improves the flexibility of joints in humans (Winther *et al.*, 2005). The powder also reduces pain and the consumption of pain killers such as paracetamol and NSAID in humans suffering from muscoloskeletal diseases such as osteoarthritis and rheumatoid arthritis (Warholm *et al.*, 2003; Willich *et al.*, 2010; Winther *et al.*, 2005). It has also been documented that this standardised powder, as well as a certain galactolipid named GOPO isolated from the same product, inhibits the break-down of cartilage and up-regulates the genes responsible for the generation of collagen and aggrecane, two main ingredients of cartilage (Schwager *et al.*, 2008). The present herbal remedy has therefore been suggested as a possible disease-modifying agent in joint disease.

The present study aimed to investigate whether LitoVet, a HybenVital rose hip powder, especially developed for horses, would affect the immune system, working capacity and behaviour of horses in exercise training. As rose-hip is also noted for its high content of vitamin C, the ability of a daily supplementation with rose hip to affect the plasma concentration of vitamin C in horses was also investigated. This aspect of the research was of particular interest, as it has been shown that horses can develop vitamin C depletion, as a result of strenuous exercise and that the uptake of synthetic vitamin C, added orally to their daily diet is very poor (Snow *et al.*, 1987).

Materials and methods

Seventy-four horses, that were in active trotting training were recruited to a double-blind, placebo controlled trial. All of the horses were located at the same training yard and were subject to a common training program. Trotters were used in this study as they often develop osteoarthritis early in life, as a result of their intensive training program and performance on an oval track. The horses were allocated to either of two groups supplemented with LitoVET or placebo. The allocation of horses to these groups was achieved using block randomisation. Within each block of 3 horses, two were allocated to the Litovet group and one to the placebo group. The horses were fed LitoVET (210 g daily as a dry powder added to their feed) and the placebo group were fed the same amount of a placebo powder with a similar taste, odour and colour. The mean age of the horses was 7±2.4 and 6.8±2.1 years for the LitoVET and placebo group, respectively. Mean bodyweight was 432.8±16.9 and 431.7±17.9 kg for the LitoVET and placebo group, respectively. The horses were supplemented with either LitoVET or placebo for a period of three months.

During the period of supplementation some horses were lost from both the LitoVet and placebo groups due to unrelated circumstances resulting in a smaller group of horses 44 that completed the trial, 29 horses that were fed the LitoVET and 15 horses that were fed the placebo for the period of supplementation. The horses underwent a standardised training regime during the trial that consisted of about 45 minutes of exercise 4 times per week including low intensity warm up and cool down exercise interspaced with 6-8 interval period of high intensity but sub-maximal exercise.

Venous blood samples from the jugular vein were drawn for the various analyses between 7-8 am at the same time relative to feeding and training just before the start of the trial (pre) and after 3 months of supplementation (post) with either LitoVET or placebo. The anti-inflammatory capacity was estimated using the chemotaxis of neutrophil leucocytes using a Boyden chamber (Kharazmi and Winther, 1999). Anti-oxidant capacity was estimated, by the release of antioxidative anions from neutrophils using chemiluminescense (Kharazmi and Winther, 1999). Vitamin C in serum was estimated using established photometric methodology (Hausman Lench and Lewis, 1961).

Working capacity was estimated as the time to run 1000 meters (in seconds) during actual races all conducted on the same track. Horses were placed in these races on a random basis and were competing against other horses that were not taking part in this study. Behaviour was evaluated using a questionnaire, which was completed by the staff responsible for the daily care of the horses. As the study was double-blinded, neither the staff with daily care of the horses, the veterinarians, or the trainers responsible for the exercise regime knew which horses were supplemented with the LitoVet or placebo.

Within group (LitoVET or placebo) statistical analysis of neutrophil chemotaxis and serum vitamin C was carried out using the nonparametric Wilcoxon test. This non parametric statistical test was also used to analyse the time to run 1000 metres prior to and following supplementation within the LitoVET or placebo groups. Statistical comparisons between the placebo and LitoVET supplemented groups was carried out using a non parametric Man Whitney test. The positive or negative responses to the trainer survey was analysed statistically using the non parametric sign test. All data are presented as mean ± SD.

Results

During supplementation with LitoVET, neutrophil chemotaxis declined from 30.4±14.0 to 9.0±13.5 (number of cells migrating per unit time) (P<0.004) indicating enhanced anti-inflammatory activity (Figure 1). There was no significant difference observed in neutrophil chemotaxis in the placebo group 31.5±12.1 to 33.7±12.9 (number of cells migrating per unit time). There was no placebo data for the washout period as the horses were released from the study.

LitoVET also significantly improved the anti-oxidative capacity, when estimated as chemiluminescence (P<0.05). Prior to the period of supplementation there was no significant difference in the 1000 meter run time between the placebo and LitoVET groups (P=0.263). Horses supplemented with LitoVET shortened their time to run 1000 meters significantly (78.3±2.6 to 77.2±2.4 seconds), which was equivalent to a mean decline of 1.1±1.5 seconds (P=0.02).

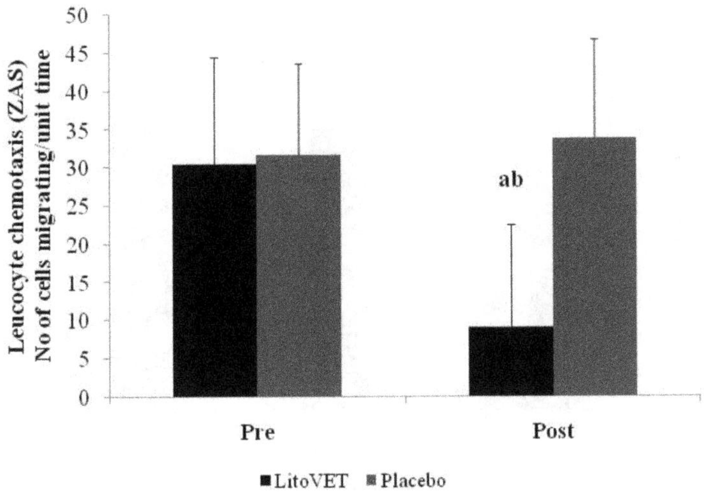

*Figure 1. Effect of supplementation with rose hip (*Rosa canina*) LitoVet (210 g daily) on leucocyte chemotaxis (ZAS) (mean ± SD) before (pre), after 3 months supplementation (post).*
[a] Significantly different from pre within group (P<0.01); [b] significantly different from corresponding group for a given time (P<0.05).

In contrast, the mean time to run 1000 meters in the placebo supplemented horses increased from 77.0±2.1 to 77.3±2.3 seconds, however this was not a statistically significant change. Whilst there was a numerical difference in time to run 1000 meters between the LitoVET and placebo group following supplementation, this did not attain statistical significance (P<0.075).

Trainers were also asked as part of a questionnaire whether the horses were more lithe or supple and easier to work the day after strenuous exercise and were asked to respond with either yes, possibly or absolutely not. In the supplemented group 19 out of 27 horses (70%) indicated that their horses were, or possibly were more supple, whereas this figure was significantly lower in the control group (8 out of 15 or 53%, P<0.05). In addition, supplementation of the diet with 210 g of LitoVet also resulted in a mean increase in the serum vitamin C level of 17% two hours following feeding, Figure 2 (30.9±4.5 vs. 36.2±5.2 μmol/l, P<0.05). There was no significant change in the serum vitamin C concentration in response to normal feeding in the placebo group of horses (29.9±3.7 vs. 30.7±3.0 vs. 29.9±3.4 μmol/l).

Discussion

The present data suggests that LitoVET works as an anti-inflammatory and anti-oxidative agent in horses and is effective at increasing the vitamin C level in serum within 2 hours of supplementation. In addition, horses supplemented with LitoVET were also reported, by their staff and trainers, to be more free moving or supple the day after strenuous exercise.

Prior to supplementation there was no significant difference in the time taken to run 1000 meters between the LitoVET and the placebo group. In contrast to the placebo group, there was a decrease in the time taken to run 1000 meters in the LitoVET group following 3 months supplementation. The anti-inflammatory and antioxidative action of LitoVET may have had a beneficial effect on tissue microtrauma or inflammation which can be associated with the exercise training (Evans, 2000), thus improving the time to complete 1000 meters through maintenance of free movement. However, whilst this data supports the possibility of a restoration of exercise performance in the LitoVET

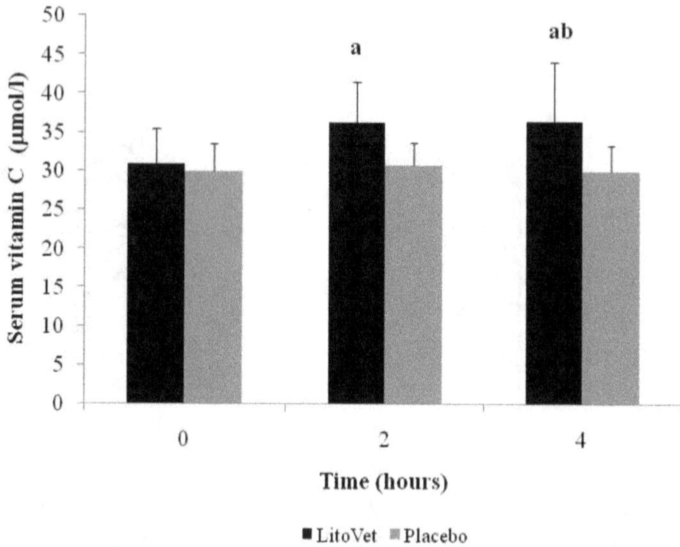

*Figure 2. Effect of supplementation with rose hip (*Rosa canina*) LitoVet (210 g single dose) on serum vitamin C (mean ± SD, µmol/l) concentration 2 and 4 hours following feeding on a single day.* [a] *Significantly different from time 0 within group (*$P<0.05$*);* [b] *significantly different from corresponding group for a given time (*$P<0.05$*).*

group, the magnitude of change in run time was small and confounding factors training effect, or differences in the competitive nature of the timed races cannot be discounted as contributory factors.

Similar effects have previously been shown in humans suffering from different types of joint diseases (Warholm *et al.*, 2003; Willich *et al.*, 2010; Winther *et al.*, 2005). It has also been reported, that rose-hip is by far the strongest antioxidant in comparison to the numerous amounts of different plants, berries and fruits that have been analysed (Halvorsen *et al.*, 2002)(Halvorsen *et al.*, 2002). The present data also indicate that rose-hip probably due largely to its vitamin C content functions as an anti-oxidant in horses.

It has been reported that sportsmen, as well as trotters are sometimes given NSAID`s to alleviate pain and stiffness the day after strenuous exercise. This medication is associated with undesirable side effects such as gastric ulceration and bleeding and its use is also precluded during racing under the prohibited substance rules of most racing jurisdiction. It is therefore of interest that in the present study this subspecies of rose-hip powder, which has been shown to have a strong anti-inflammatory capacity in humans, seems to exert a similar effect in horses but without any observable effect on gastrointestinal function, although any effects of long term administration require investigation. Equally the results of this trial need to be interpreted with care as the effects observed are likely to be dependent on the rose hip sub-species due to a varying level of the galactolipid GOPO.

The improved serum vitamin C level observed after 2 hours following supplementation is potentially beneficial, especially given the poor availability of conventionally used forms of vitamin C such as ascorbic acid in horses (Deaton *et al.*, 2003; Snow and Frigg, 1990). Whilst the present data shows potential, a larger double-blind study in horses is warranted in order to consolidate the effects of LitoVet rosehip powder.

References

Deaton, C.M., D.J. Marlin, N.C. Smith, C.A. Roberts, P.A. Harris, F.J. Kelly and R.C. Schroter, 2003. Pulmonary bioavailability of ascorbic acid in an ascorbate-synthesising species, the horse. Free Radical Research 37, 461-461.

Evans, D.L., 2000. Training and fitness in athletic horses, Rural Industries Research & Development Corporation, Kingston, Australia. pp 1-70.

Halvorsen, B.L., K. Holte, M.C.W. Myhrstad, I. Barikmo, E. Hvattum, S.F. Remberg, A.-B. Wold, K. Haffner, H. Baugerod, L.F. Andersen, O. Moskaug, D.R. Jacobs, Jr. and R. Blomhoff, 2002. A systematic screening of total antioxidants in dietary plants. Journal of Nutrition 132, 461-471.

Hausman Lench, E.R. and G.T. Lewis, 1961. Determination of ascorbic acid in blood. Analyst 50, 18-19.

Kharazmi, A. and K. Winther, 1999. Rose hip inhibits chemotaxis and chemiluminescence of human peripheral blood neutrophils *in vitro* and reduces certain inflammatory parameters *in vivo*. Inflammopharmacology 7, 377-386.

Schwager, J., N. Richard and S. Wolfram, 2008. The galactolipid GOPO mediates beneficial effects of rosehip on arthritis by reducing chemokine and interleukine production in macrophages and chondrocytes. International Journal of Rheumatic Disease 11, A96.

Snow, D.H. and M. Frigg, 1990. Bioavailability of ascorbic acid in horses. Journal of Veterinary Pharmacology and Therapeutics 13, 393-403.

Snow, D.H., S.P. Gash and J. Cornelius, 1987. Oral administration of ascorbic acid to horses. Equine Veterinary Journal 19, 520-523.

Warholm, O., S. Skaar, E. Hedman, H.M. Molmen and L. Eik, 2003. The Effects of a Standardized Herbal Remedy Made from a Subtype of *Rosa canina* in Patients with Osteoarthritis: A Double-Blind, Randomized, Placebo-Controlled Clinical Trial. Current Therapeutic Research 64, 21-31.

Willich, S.N., K. Rossnagel, S. Roll, A. Wagner, O. Mune, J. Erlendson, A. Kharazmi, H. Sörensen and K. Winther, 2010. Rose hip herbal remedy in patients with rheumatoid arthritis – a randomised controlled trial. Phytomedicine 17, 87-93.

Winther, K., K. Apel and G. Thamsborg, 2005. A powder made from seeds and shells of a rose hip subspecies (*Rosa canina*) reduces symptoms of knee and hip osteoarthritis: a randomized, double blind, placebo controlled clinical trial. Scandinavian Journal of Rheumatology 34, 302-308.

Effects of a live or heat-treated lactic-acid bacteria versus a placebo on faecal microbial communities and activities in horses

C. Faubladier[1]*, E. Jacotot*[1]*, C. Berger*[2] *and V. Julliand*[1]
[1]*ENESAD, 26 bd Dr Petitjean, 21000 Dijon, France*
[2]*Danisco, 20 rue de Brunel, 75000 Paris, France*

Abstract

Zootechnical additives could be used in order to prevent the adverse impact of starch-rich diets in the hindgut of horses, by enhancing fiber digestibility or by stabilising the gut microflora. The aim of this study was to evaluate the effect of a live or heat treated lactic-acid bacteria vs. a placebo on microbial communities and activities in the faeces of horses. Three pairs of mature French trotters (459±34 kg) were randomly assigned in a 3x3 Latin square design and received successively the three dietary treatments over three experimental periods: placebo, treatment 1 (5.10^9 CFU/day of each bacterial strain : *Propionibacteria* and *Lactobacillus plantarum*) and treatment 2 (50 g/day of cereal blend with equivalent 5.10^9 heat treated cells of *Lactobacillus rhamnosus* and *Lactobacillus farciminis*). After fourteen days of adaptation to the dietary treatment, faecal samples were collected in order to count anaerobic flora (total, lacticolytic, amylolytic, cellulolytic). Lactic acid, volatile fatty acids and pH were measured on faecal fluid samples collected at the same time. When the live lactic-acid bacteria (treatment 1) was given, the concentration of the faecal fibrolytic microflora was numerically higher for 4 out of 6 supplemented horses and its activity for the first period was significantly higher ($P\leq0.05$). In contrast, when the heat treated lactic bacteria (treatment 2) were fed, the amylolytic activity of the microflora was significantly higher in the faeces of supplemented horses ($P\leq0.05$) without disturbing the fibrolytic activity of the ecosystem. These results suggest that these two new bacterial products had positive effects on the faecal microbial communities and activities in horses.

Keywords: Propionibacteria, Lactobacillus plantarum, *probiotic*

Introduction

In horses, forage is utilized by rapid microbial degradation occurring in the hindgut and the main end-products, the volatile fatty acids (VFA), are absorbed through the intestinal mucosa and represent a major energy source.

The efficiency of the microbial degradation of forage depends greatly on feeding practice. In horses undertaking intense physical activity, diets often contain high levels of concentrate feeds rich in starch in order to satisfy their high energetic requirements. A starch intake of greater than 0.2% BW increases the quantity of starch reaching the hindgut, which can affect the microbial activity resulting in a change in the biochemical composition of the intestinal environment. This can have a negative impact on the microbial fibrolytic activity and affect fiber degradation within the hindgut (Julliand *et al.*, 2006; Zeyner, 2008).

Zootechnical additives may help to prevent the adverse impact of this type of diet, by enhancing fiber digestibility or by stabilising the gut flora. The probiotic product tested in our study is not currently registered as a zootechnical additive in the EU but is listed as a direct-fed microbial product in the US. The aim of this first study was to evaluate the effect of a live or heat treated lactic bacteria vs. a placebo on microbial communities and activities in the faeces of horses.

Material and methods

Six mature French Trotters (geldings, mean age 5±2 years; mean initial bodyweight 458.8±34.2 kg) were kept in indoor individual free-stalls (13.3 m^2) and were given access to a sandy paddock for one hour per day except on the training days. During each experimental period, horses undertook light exercise 3 days per week for 30 minutes consisting of lunging at walk (5 min) and at trot (7 min) alternatively.

The basal diet and dietary schedule corresponded to feeding practices used widely in French horse riding schools. The daily feed ration consisted of a commercial pelleted concentrate and meadow hay (Table 1). The hay represented 60% of the total daily dry matter intake (1.3 kg DM/100 kg BW) and was fed in two equal meals at 10 and 16 h. The pelleted feed (890 g DM/100 kg BW) was fed in two equal meals: 8 h and 17.30 h. The daily feed rations were calculated to cover 120% of energy requirements for maintenance and light training (Martin-Rosset and Vermorel, 2002).

Horses were randomly assigned to 3 dietary treatments within a 3x3 Latin square design and received over three experimental periods the three dietary treatments: placebo, treatment 1 (5.10^9 CFU/day of each bacterial strain: *Propionibacteria* and *Lactobacillus plantarum*) and treatment 2 (50 g/day of cereal blend with equivalent 5.10^9 heat treated cells of *Lactobacillus rhamnosus* and *L. farciminis*). Each treatment was incorporated daily into the morning meal of concentrate for 14 days before sampling. Between the two experimental periods, horses were fed the basal diet without treatment for a 23-d period. This allowed a complete wash out of the dietary treatments from the intestinal tract of the horses (Gobert *et al.*, 2006).

Faecal samples were collected from the rectum three hours after the morning meal. The first sub-sample of faeces was immediately transported to the laboratory in a CO_2-saturated flask at 38 °C and diluted under O_2-free CO_2 in an anaerobic mineral solution (Bryant and Burkey, 1953) for inoculation onto specific enumeration media.

Total viable anaerobic bacteria were determined in roll tubes under O_2-free CO_2 in a non-selective medium (Leedle and Hespell, 1980). Bacterial numbers were determined after 48 h of incubation at 38 °C from four replicate roll tubes prepared per dilutions.

Table 1. Composition and nutrient values of the hay and the pelleted feed given to horses during the experiment.

Analysed composition, DM basis, %	Hay	Pelleted feed
Dry matter (DM)	88.9	89.1
Organic matter (OM)	91.9	91.7
Crude protein (CP)	ND	12.5
Neutral detergent fiber (NDF)	64.6	42.0
Acid detergent fiber (ADF)	35.9	20.7
Acid detergent lignin (ADL)	4.6	6.0
Energy content (UFC[1]/kg DM)	0.43	0.90
Protein content (MADC[2]/kg DM)	33.7	140.3

[1] French unit of energy content (Unité Fourragère Cheval, Horse feed unit); [2] French unit of protein content (Matière Azotée Digestible Cheval, Horse Digestible Crude Protein).

Amylolytic bacteria were deeply inoculated on an adapted medium (Varloud, 2006). After a 48 h-incubation at 38 °C, bacterial colonies counts were performed on three Petri plates per dilution stained with iodine.

Lacticolytic bacteria were enumerated in roll tubes under O_2-free CO_2 in a selective medium (Mackie et Heath, 1979). Bacterial numbers were determined after 48 h of incubation at 38 °C from four replicate tubes per dilution.

Cellulolytic bacteria were enumerated in a modified broth medium (Halliwell and Bryant, 1963; Julliand *et al.*, 1999) containing filter paper cellulose strips as energy source. After two weeks of incubation at 38 °C, the most probable number (MPN) was determined from four replicate tubes per dilution (Clarke and Owens, 1983).

A second subsample was filtered (100 µm) and the pH of the filtrate was immediately measured with an electronic pH meter. The filtered content was divided into two aliquots and immediately frozen for determination of total VFA, acetate (C2), propionate (C3), butyrate (C4) and valerate (C5) concentrations by gas-liquid chromatography and determination of D- and L-lactate concentrations using an enzymatic reaction procedure quantified spectrophotometrically at 540 nm. Acetate plus butyrate to propionate ([C2 + C4]/C3) ratio was calculated according to Sauvant *et al.* (1994).

Logarithmic transformations were performed on microbial counts before statistical analysis. Data were processed by analysis of variance (ANOVA) using the GLM procedure in the SAS v8.2 software package (SAS Institute. Inc., Cary, NC). The model included the effects of the treatment, the animal, the period and the interaction between the treatment and the period

Results and discussion

Faecal samples are often used for assessing the conditions of the gut in both human and animals. Although colonic and faecal ecosystems differ both quantitatively and by their activity in horses (Da Veiga *et al.*, 2005, Faubladier *et al.*, 2006, Müller *et al.*, 2008), changes in faeces appeared to be appropriate markers for changes in the colon of horses (Julliand and Goachet, 2005).

We observed an interaction treatment x period for three parameters: the molar percentage of C2, the molar percentage of C3 and the (C2+C4)/C3 ratio. Therefore the effect of the treatment was analyzed period by period only for these three parameters.

When treatment 1 was added, the faecal microbial communities and activities were not significantly altered. During the period 1, the molar percentage of C2 and the (C2+C4)/C3 ratio increased (P=0.015 and P=0.002 respectively) whereas the molar percentage of C3 decreased (P=0.010) in faeces of horses receiving the treatment 1 compared to placebo (Δ%C2 = +17.6; Δ%C3= -12,8 and Δ(C2+C4)/C3= +1.2). This could result from a higher activity of the cellulolytic microflora concomitant with intermediate total VFA concentration. Moreover, when treatment 1 was supplemented, the faecal microbial count of cellulolytic bacteria was numerically higher for 4 horses compared to the placebo and treatment 2 possibly suggesting an improvement of fiber degradation.

When treatment 2 was supplemented, there was no change except for the faecal concentrations of total VFA (Δ= +28,3 mmol/l), total lactate (Δ= +3,6 mmol/l) and L-lactate (Δ= +2,6 mmol/l) that increased significantly compared to the placebo (P=0.023, P=0.008 and P=0.072 respectively). This could be due to higher concentration of lactate producing bacteria. Nevertheless, the balance of the faecal ecosystem was not modified. This could also be explained by a higher amylolytic microbial activity. However, the molar percentage of C2, the molar percentage of C3 and the (C2+C4)/C3 ratio were

Table 2. Effect of the treatment 1, treatment 2 and placebo on the microbial communities and activities in faeces of horses (n=6).

Item	Treatment			SEM [1]	*P*-value	
	1	Placebo	2		Treatment	Treatment x period
Total anaerobic bacteria (log10 cfu/ml)	7.3	7.1	7.4	0.3	0.119	0.334
Lacticolytic bacteria (log10 cfu/ml)	6.6	6.7	7.0	0.6	0.516	0.689
Amylolytic bacteria (log10 cfu/ml)	5.0	5.3	5.5	0.7	0.533	0.718
Cellulolytic bacteria (log10 MPN/mL)	5.6	4.7	4.8	0.9	0.236	0.325
pH	6.7	6.7	6.8	0.2	0.934	0.151
Total VFA (mmol/l)	69.7^{ab}	50.3^{a}	78.6^{b}	16.6	0.063	0.140
%C2	63.2	58.8	59.4	2.0	0.020	0.020
%C3	24.6	26.6	24.0	2.0	0.155	0.009
%C4	8.1	9.0	9.0	0.7	0.104	0.394
%C5	0.8	1.2	2.1	1.0	0.162	0.837
(C2+C4)/C3	2.9	2.9	3.1	0.2	0.264	0.008
Total lactate (mmol/l)	11.3^{ab}	9.6^{a}	13.2^{b}	1.6	0.022	0.416
D-Lactate (mmol/l)	4.8	4.3	5.1	0.6	0.168	0.325
L-Lactate (mmol/l)	6.5^{ab}	5.4^{a}	8.0^{b}	1.6	0.064	0.458
L/D	1.6	1.3	1.8	0.7	0.398	0.837

[a,b] Means that do not have a common superscript differ $P<0.05$; [1] standard error of the mean.

not modified when treatment 2 was fed whatever the period. Neither faecal pH nor the concentration of cellulolytic bacteria decreased suggesting that the fiber degradation was not disturbed.

Conclusion

We measured changes occurring in faeces of lactic acid bacteria supplemented horses. When treatment 2 was fed, the amylolytic activity of the microflora was significantly higher in the faeces of supplemented horses ($P\leq0.05$) without disturbing the fibrolytic activity of the ecosystem. In contrast, when the treatment 1 was supplemented, the concentration of the faecal fibrolytic microflora was numerically higher for 4 out of 6 supplemented horses and its activity for the first period was significantly higher ($P\leq0.05$).

These data suggested that these two new bacterial products had positive effects on the faecal microbial communities and activities in horses. They could be fed to horses as zootechnical additives in particular when rations are supplemented with starch. But treatment 1 needs first to be accepted by the EU commission.

References

Bryant M.P. and L.A. Burkey, 1953. Cultural methods and some characteristics of more numerous groups of bacteria in the bovine rumen. J. Dairy Sci. 36: 205-217.

Clarke K.R. and N.J.P. Owens, 1983. A simple and versatile micro-computer program for the determination of 'Most Probable Number'. J. Microbiol. Methods 1:133-137.

Da Veiga, L., F. Chaucheyras-Durand and V. Julliand, 2005. Comparative study of colon and faeces microbial communities and activities in horses fed a high starch diet. Pferdeheilkunde, 21, 45-46.

Faubladier C., V. Julliand, L. Da Veiga and F. Chaucheyras-Durand, 2006. A comparison of colon and faeces microbial diversity in horses using biomolecular techniques. Reproduction Nutrition Development 46 (Suppl. 1), S14.

Gobert J., G. Bertin and V. Julliand, 2006. Digestive fate of Saccharomyces cerevisiae CBS 493 94. fed at 3 different concentrations to horses. Reprod. Nutr. Dev. 46 (Suppl.1) S95-S96.

Halliwell G. and M. Bryant, 1963. The cellulolytic activity of pure culture strains of bacteria from the rumen of cattle. J. Gen. Micobiol. 32: 441-448.

Julliand, V., A. de Vaux., L. Millet and G. Fonty, 1999. Identification of Ruminococcus flavefaciens as the predominant cellulolytic bacterial species of the equine cecum. Appl. Env. Microbio. 65: 3738-3741.

Julliand, V and A.G. Goachet, 2005. Comparison of microbial ecosystems in cecal, colonic and faecal content of horses. 19th ENPS Symposium.

Julliand, V., A. De Fombelle and M. Varloud, 2006, Starch digestion in horses: the impact of feed processing. Livestock Science, 100, 44-52.

Leedle, J. and R. Hespell, 1980. Differential carbohydrate media and anaerobic replica plating techniques in delineating carbohydrate-utilizing subgroups in rumen bacteria populations. Appl. Env. Microbiol 34: 709-719.

Mackie, R.I. and A.B. Heath, 1979. Enumeration and isolation of lactate utilizing bacteria from the rumen of sheep. Appl. Environ. Microbio.38:416-421.

Martin-Rosset, W. and M. Vermorel, 2002. Evaluation et expression de la valeur énergétique des aliments et des apports énergétiques recommandés pour le cheval athlète dans le système UFC. In : Proc. 1ères journées d'étude européennes en nutrition équine "Quel système choisir en Europe pour l'évaluation des besoins nutritionnels?", Dijon, France, p. 27-68.

Müller, C.E., D. Von Rosen and P. Uden, 2008. Effect of forage conservation method on microbial flora and fermentation pattern in forage and in equine colon and faeces. Livestock Science, 119, 116-128.

Sauvant, D., P. Chapoutot and H. Archimède, 1994. La digestion des amidons par les ruminants et ses conséquences. INRA Prod. Anim. 7: 115-124.

Varloud, M., 2006. Implication des micro-organismes de l'estomac dans la digestion de l'amidon par le cheval. Doctoral thesis, INRA Paris-Grignon, France. 202 p.

Zeyner, A., 2008. Energy providing nutrient sources. In: Saastamoinen, M. and W. Martin-Rosset (eds.), Nutrition of the exercising horse. EAAP publication 125. Wageningen Academic Publishers, Wageningen, the Netherlands, p. 277-294.

Can oral intake of gamma-oryzanol (experimentally given orally as pure substance) result in doping relevant testosterone levels in the urine of mares and geldings?

A. Mößeler[1], S. Licht[1], L. Wilhelm[2] and J. Kamphues[1]
[1]Institute of Animal Nutrition, University of Veterinary Medicine Hannover Foundation, Bünteweg 2, 230559 Hannover, Germany
[2]LADR GmbH, medical supply centre, Geesthacht, Germany

Abstract

Gamma-oryzanol is a natural substance occuring in the lipid fraction of rice (*Oryza sativa*). It is a mixture of different lipohylic substances (that contains two major active molecules, phytosterols and esters of ferulic acid). Although gamma-oryzanol (GO) was listed as a prohibited substance (medication class B) by the FEI, standardised studies on the effects of this substance in horses are not available. The aim of the current study was to obtain information about the effect of gamma-oryzanol (given orally as a pure substance) on serum testosterone levels in healthy endurance horses to prove the hypothesis that gamma-oryzanol can cause elevated testosterone levels. Special interest was given to the question if gamma-oryzanol can cause testosterone levels above the critical value (20 ng testosterone/ml urine in geldings and 55 ng/ml in mares) to check if feeding rice bran oil (containing gamma-oryzanol) to horses may be doping relevant. Six healthy endurance horses (3 geldings, 3 non pregnant mares) were fed gamma-oryzanol (pure substance) over 31 days at a dosage of 2 g/500 kg bodyweight. The horses were trained daily (at a speed of 200 m/min for 1 hour) and urine samples were taken at day 0, 14 and 28 of the trial. In all samples testosterone levels were analysed by HPLC. At no time did testosterone levels in either urine or serum rise above the critical (doping threshold as determined by the FEI) in any horse. There was no effect of gender nor the duration of GO admimistration or of application as well as the stress situation on testosterone levels in serum. Therefore, oral intake of GO seemed to have no effect on serum testosterone levels in mares and geldings and the hypothesis of elevating testosterone levels in the blood could not be confirmed.

Keywords: gamma-oryzanol, doping, testosterone, rice products

Introduction

Gamma-oryzanol is a natural substance occuring in the lipid fraction of rice (*Oryza sativa*). It is a mixture of different lipohylic substances (that contains two major active molecules, phytosterols and esters of ferulic acid). Although gamma-oryzanol (GO) was listed as a prohibited substance (medication class B) by the FEI, standardised studies on the effects of this substance in horses are not available. Furthermore,are mixed feeds on the market, that contain rice bran or rice oil; both raw materials are known to have a varying content of GO, which is supposed to have anabolic effects. The anabolic effect is thought to be a result of both antioxidant and steroid-like properties. One hypothesis about the mechanism of the proposed anabolic effect is the reduced clearance of endogenous testosterone, although even this effect could not be confirmed in studies in men (1). Nonetheless there are many of companies that promote feed supplements for horses that contain gamma-oryzanol as a naturally anabolic substance. Therefore, it is used for young, growing sales horses as well as for racehorses. Because there is a lack of data there are uncertainness amongst horse riders whether they can use rice-containing products for their sports horses. There has been a case of doping in which a gelding was tested positive regarding high testosterone levels; and a relationship between the high intake of rice products (rice bran and rice oil) and these elevated levels was discussed. The aim of the current study was to obtain information about the effect of gamma-oryzanol (given orally as a pure substance) on serum testosterone levels in healthy endurance horses to prove the hypothesis that gamma-oryzanol can cause elevated testosterone levels. Special interest was given to the question if gamma-oryzanol can cause testosterone levels above the critical value (20 ng testosterone / ml urine in geldings and 55 ng / ml in mares) to check if feeding rice bran oil (containing gamma-oryzanol) to horses may be doping relevant.

Material and methods

Six healthy endurance horses (3 geldings, 3 non pregnant mares) were fed gamma-oryzanol (pure substance) over 31 days at a dosage of 2 g/500 kg bodyweight. The dosage was chosen to imitate the maximum level that could be expected if the diet is rich in rice bran containing mixed feed and rice bran oil. The horses had free access to hay and received oats and a mineral mix in order to fulfil their nutritional requirements. The horses were trained daily (at a speed of 200 m/min for 1 hour) and urine samples were taken at day 0, 14 and 28 of the trial. Additionally the horses were exposed to a competition like stress situation on day 29-31; urine samples were taken daily on that days. In all samples testosterone levels were analysed by HPLC.

Results

Testosterone levels in horses were always lower than 1.7 ng/ml (with most values <1.00 ng/ml). Urine testosterone levels were not affected by the intake of gamma-oryzanol at a dosage of 2 g/500 kg bodyweight (see Table 1). Additionally, there was no effect of gender or time on testosterone levels in blood. One mare (horse 2) had higher testosterone levels than all the other horses, but the kinetics of the levels indicates that this finding was not caused by of the ingestion of gamma-oryzanol (see Table 2).

Table 1. Urine testosterone levels (ng/ml) before and after administraion of 2 g of gamma-oryzanol/500 kg BW.

Horse	Gender	Day 0	Day 14	Day 28	Day 29	Day 30	Day 31
1	Mare	<0.02	<0.02	0.66	<0.02	<0.02	<0.02
2		<0.02	<0.02	0.65	<0.02	<0.02	<0.02
3		0.79	<0.02	1.02	<0.02	<0.02	<0.02
4	Gelding	<0.02	<0.02	<0.02	<0.02	<0.02	<0.02
5		0.31	<0.02	1.67	<0.02	<0.02	<0.02
6		0.45	<0.02	1.40	<0.02	<0.02	<0.02

Table 2. Serum testosterone levels (ng/ml) before and after administration of 2 g of gamma-oryzanol/500 kg BW.

Horse	Gender	Day 0	Day 14	Day 28	Day 29	Day 30	Day 31
1	Mare	<0.02	0.03	<0.02	<0.02	<0.02	<0.02
2		0.06	0.06	0.09	0.06	0.07	0.04
3		0.02	<0.02	<0.02	<0.02	<0.02	0.03
4	Gelding	0.03	<0.02	<0.02	<0.02	<0.02	<0.02
5		0.03	<0.02	<0.02	0.02	<0.02	0.03
6		0.03	0.03	<0.02	<0.02	<0.02	<0.02

Discussion

In the present study there was no effect of gender or the duration of GO admimistration or of pplication as well as the stress situation on testosterone levels in serum. Therefore, oral intake of GO seemed to have no effect on serum testosterone levels in mares and geldings and the hypothesis of elevating testosterone levels in the blood could not be confirmed. Special interest was given to the testosterone levels in urine – because this is the first matrix used for doping analysis. At no time did testosterone levels in either urine or serum rise above the critical (doping threshold as determined by the FEI) in any horse.

References

Fry, A.C., E. Bonner, D.L. Lewis, R.L. Johnson, M.H. Stone and W.J. Kraemer, 1997. The effects of gamma-oryzanol supplementation during resistance exercise training. Int. J. Sport Nutr. 7: 318-329.

Part 7. Nutrition, health and performance

Preventing problems whilst maximizing performance

J.E. Lindberg and A. Jansson
Swedish University of Agricultural Sciences, Department of Animal Nutrition and Management, P.O. Box 7024, 750 07 Uppsala, Sweden

Abstract

Energy-rich and palatable roughage should always be the foundation of diets for horses. The feeding strategy of the high performance horse has to be long-term, aiming at providing the energy necessary for building up adequate stores and allowing maximal performance. Moreover, the feeding strategy should be designed to support the stability of the gut microbiota. In addition, the diet should support rapid recovery after exercise and thus allow the animal to perform regularly at maximal capacity, without interfering with health. The following criteria are suggested as guidelines when composing safe and 'successful' diets to exercising horses with a high energy demand; free access to forage with a digestible energy content of 12.5 MJ or higher, the concentrate allowance should be less than 4.5 kg per day and starch intake should not exceed 1.1 g per kg body weight per meal. Supplementing the diet with fat appears to be a safe and efficient way to improve energy density and support performance.

Keywords: forage, starch, fat, cereals, soluble fiber, glycogen

Introduction

In the equine athlete, the overall performance capacity is determined by genetic potential, conditioning and nutrition. Conditioning requires a long-term strategy and besides changes in muscle metabolic properties, it includes many aspects of adaptation that are specific to the expected performance (circulation, skeleton, connective tissue, biomechanics and behaviour). The importance of nutrient availability, especially fluid and glucose, for maximal performance has also been demonstrated (Geor and McCutcheon, 1998; Geor, 2006).

The high performance horse has an energy requirement that is twice the maintenance energy requirement (NRC, 2007). Therefore, the feeding of the equine athlete should aim at providing the energy and nutrients necessary for building up adequate stores and to allow maximal performance. In addition, the diet should support rapid recovery after exercise and thus allow the animal to perform regularly at maximal capacity, without interfering with its health.

In the present paper, we will focus on feeding strategies used to provide the athletic horse with energy. In addition, we are suggesting criteria that could be used as guidelines when composing diets for athletic horses.

Digestion in horses, feeding practices and health

The gastrointestinal (GI) tract of the horse is adapted to continuous grazing of a fibre-rich diet and the horse has developed a symbiosis with the hindgut microbiota, which produces energy substrates in the form of short-chain fatty acids (SCFA). Therefore, roughage should always be the foundation of the horse diet, and grains, concentrates or other supplements should be used only to increase the energy density and to supply essential nutrients not contained in the roughage. However, a common feeding practise for performance horses has been to include large amounts of starch-rich concentrates (Mullen *et al.*, 1979; Glade, 1983; Southwood *et al.*, 1993b; Burk and Williams, 2008).

Concentrate-rich diets that contain substantial amounts of starch have been associated with a number of gastrointestinal disorders that involve disturbances of the intestinal microbiota. This is largely due to a limited capacity to digest starch in the small intestine (Kienzle, 1994), which will result in undigested starch flowing into the hindgut. This could result in colic due to excessive hindgut fermentation (Potter *et al.*, 1992a; Kienzle *et al.*, 1994), unrestrained gas production and a reduction in hindgut pH (Beyer, 1998). There are indications in the literature that the small intestinal starch digestibility in horses decreases with increasing starch intakes (Potter *et al.*, 1992a; Kienzle, 1994). Although the risk of exceeding the small intestinal starch digesting capacity will depend on the cereal starch source, earlier data suggests that there is a critical point at 2 g starch/kg body weight (BW) per meal. However, a recent study on Standardbred horses in training fed a conventional forage:concentrate diet, with a starch intake of <1 g/kg BW per day, showed a reduced microbial stability and an increase in the faecal flora of lactic acid bacteria and members of the *Streptococcus bovis/equis* complex when compared to feeding a forage-only diet (Willing *et al.*, 2009). This indicates that the previously suggested limit for starch intake (2 g starch/kg BW; Kienzle, 1994; Potter *et al.*, 1992a) to avoid digestive problems in horses and support equine health and welfare, may have to be changed.

During the years it has also been reported that some athletic horses suffer from health and behavioural disturbances that might be related to feeding strategies. In horses offered diets with a low inclusion of roughage and a high inclusion of concentrates, stereotypic behaviours have been reported (Kusunose, 1992; Gillham *et al.*, 1994; Redbo *et al.*, 1998). Moreover, gastric ulcers (Beyer, 1998) and sporadic and recurrent exertional rhabdomyolysis (MacLeay *et al.*, 1999) can be related to the use of grains and sweet-feed in equine rations. In racing Thoroughbred horses fed >4.5 kg of concentrates per day, an increased risk for rhabdomyolysis was reported (MacLeay *et al.*, 1999).

Substrate utilization during exercise

The skeletal muscle uses substrates of varying chemical composition and origin. Creatine phosphate, stored in skeletal muscle, can provide energy during the first seconds of intensive exercise and the response to dietary supplementation seems to be small or non-existent in horses (Schuback *et al.*, 2000). A continuous muscle glucose supply, via direct up-take from blood or from muscle glycogen stores, is crucial for all forms of exercise performance. Reduced pre-exercise muscle glycogen content has been shown to reduce the performance and anaerobic capacity during high-intensity work (Lacombe *et al.*, 2001). Glucose generated from muscle glycogen stores is utilized during most types of exercise and the utilization increases with increasing work intensity, being most pronounced during the first work bouts during interval work (Voiton *et al.*, 2007). Fatty acids derived from adipose tissue, plasma VLDL and intramuscular triacylglycerol are important fuels during low intensity exercise. It has been suggested that a crossover from lipid to carbohydrate dependence occurs in mammals at exercise intensities approaching 50% of VO_{2max} (Figure 1; Brooks, 1998).

Glucose supply, performance and diet

As mentioned earlier, a continuous muscle glucose supply is crucial for exercise performance and it has been shown that intravenous administration of glucose to horses running on a treadmill at 50-60% VO_{2max} increases time to fatigue (Geor, 2006). It is therefore of interest to manipulate diet carbohydrate properties and/or composition in order to optimize glucose availability during exercise. An important and relevant issue in this context is whether this should be achieved through major changes in diet ingredient composition or if it should (and can) be achieved through dietary supplementation.

In a study by Jansson *et al.* (2002) on horses exercising at 50-60% of VO_{2max} the muscle glycogen utilisation was lowered on diets with high sugar (barley syrup) content compared to diets with oats (Figure 2). This indicates that the source and availability of sugars in the diet affect glycogen utilization and could affect performance.

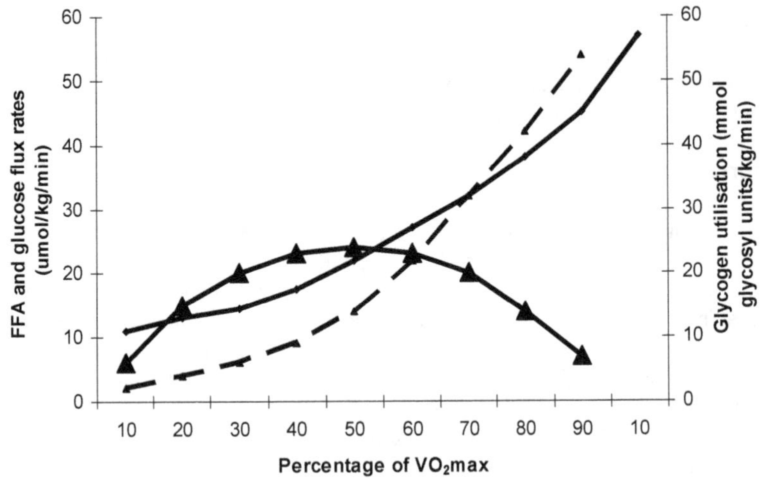

Figure 1. A schematic figure of the suggested crossover from lipid to carbohydrate dependence in skeletal muscle during exercise in mammals. The line with triangles illustrates free fatty acid flux rates, the solid line glucose flux rates and the dotted line glycogen utilization (After Brooks, 1989).

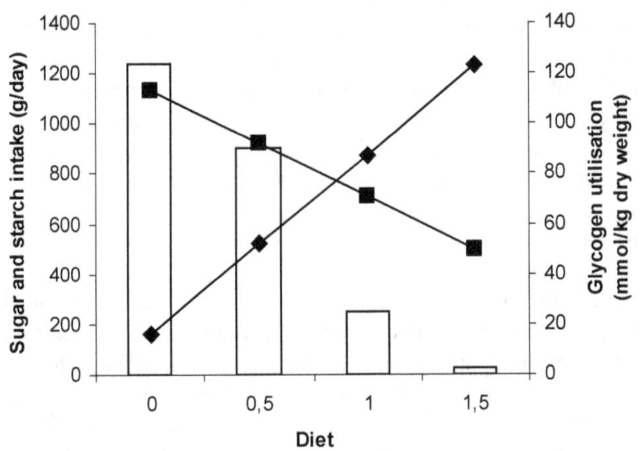

Figure 2. Sugar (line with rhombus) and starch (line with squares) intake and glycogen utilisation (bars) in four horses during exercise at 50-60% of VO_2max on fours diets including 0, 0.5, 1.0 and 1.5 kg barley syrup (after Jansson et al.*, 2002).*

Feeding horses with cereal starch, results in significant increases in plasma glucose and insulin concentrations but the responses were not clearly altered by different types of processing of either oats (Vervuert *et al.*, 2003), barley (Vervuert *et al.*, 2007) or maize (Vervuert *et al.*, 2004a). Interestingly, there appears to be only a moderate glycaemic and insulinaemic response to starch intakes <1.1 g/kg BW in horses (Vervuert *et al.*, 2009) even though highly processed cereals are used.

Supplementation of a grass meal pellet diet (0.6 kg) with glucose or fructose (0.7 g per kg BW) to trained Standardbred horses resulted in higher plasma glucose concentrations during exercise than in the basal diet, with no counter-regulation by insulin (Vervuert *et al.*, 2004b). The glycaemic response during exercise was lower with fructose than with glucose supplementation, while there were no effects on the response in blood lactate and plasma free FA (Vervuert *et al.*, 2004b). It was concluded that the use of fructose instead of glucose as a supplement of easyily available carbohydrates had no clear advantage in exercising horses. Supplementation with ribose has also been shown not to affect the exercise response. Thoroughbred geldings fed a basal grass hay and concentrate (43: 57 on as-fed basis) diet top-dressed with ribose had blood ammonia-N, plasma glucose, plasma lactic acid, VO_2, heart rate and performance (treadmill standardized exercise test) similar to those of geldings supplemented with glucose (Kavazis *et al.*, 2004).

Glycogen stores

In equines, a complete re-synthesis of muscle glycogen stores following exercise requires 48-72 h (Geor, 2006). Diet might be a means of altering the rate of glycogen synthesis but high soluble carbohydrate intakes have resulted in only a modest acceleration in muscle glycogen replenishment and a modest increase in muscle glycogen content compared to low intakes of soluble carbohydrates (Geor, 2006). The mechanisms underlying the slow glycogen replenishment after exercise in equines, as compared to humans, is still not well known (Voiton *et al.*, 2007). Probably, in horses subjected to intensive training and/or competition several days per week, it is of importance for the long-term performance capacity to have a feeding regime that support a restoration of the glycogen stores as rapidly as possible.

Reduced pre-exercise muscle glycogen content has been shown to reduce the performance and anaerobic capacity during high-intensity work (Lacombe *et al.*, 2001). Recently, oral acetate supplementation of a typical hay: grain diet was shown to enhance the rate of glycogen re-synthesis during the 4 h initial recovery period after muscle glycogen depletion (Waller *et al.*, 2008). These data suggests that acetate, which is the major SCFA produced during hindgut fermentation of dietary fiber, may be important as a substrate to support glycogen synthesis. Interestingly, Standardbred horses in training, which were fed a forage-only diet or a typical forage: concentrate diet, had elevated plasma acetate concentrations post-exercise on the forage-only diet. However, the muscle glycogen content was lower in horses on the forage-only diet (Jansson and Lindberg, unpublished).

Supplemental fat

By feeding supplemental fat it is possible to increase the energy density of the diet and to reduce the amount of starch-rich concentrate needed to meet the energy needs. In addition, it has been shown that fat inclusion alters substrate utilization and might be beneficial for performance. The fat content in pasture and in a diet based on roughage-only or roughage plus concentrate (based on ordinary feedstuffs) is low (3-4% crude fat in the dry matter) and the total amount of digestible energy (DE) coming from fat is less than 10%. Pagan *et al.* (2002), showed that consumption of a fat supplemented diet (29% of DE from fat) by mature Arabian geldings for five to ten weeks was associated with an altered metabolic response to low-intensity exercise compared with feeding the control diet (7% of DE from fat). Horses on the fat supplemented diet showed a 30% reduction in the production and utilization of glucose, a decrease in the respiratory exchange rate, a decrease in the estimated whole-body carbohydrate utilization and an increase in the whole-body rate of lipid oxidation during exercise. This indicates beneficial effects of long-term fat supplementation on substrate utilization in horses performing low-intensity exercise, which may improve exercise performance capacity.

It has been suggested that long-term feeding of supplemental fat to exercising horses will increase mobilization and speed of mobilization of free fatty acids (FA), increase speed of uptake into muscle of free FA, lower lactic acid production, have a glycogen-sparing effect and increase pre-exercise muscle glycogen levels (Potter *et al.*, 1992b; Harris and Harris, 2005). However, there is a large variation in the effects reported in the literature from different studies and only a few show direct benefits on performance traits (Geor, 2006). It is likely that much of the variations in reported effects of fat supplementation can be explained by the diversity in the study protocols and the horses used. Another aspect of feeding fat that is not very well investigated is the importance for heat production and heat balance during exercise. Theoretically, it is possible to decrease the heat load coming from the diet when more of the DE is coming from fat.

Fat sources used for supplementing the diet have varied between studies, but have mainly been derived from different plant sources (mostly corn oil and soybean oil). However, supplementation has been focused on the amount of fat added to the diet (on either a weight basis or on an energy basis) without considering fat quality in terms of the FA composition of the fat source used. It was recently shown by O'Connor *et al.* (2004), that a basal timothy-hay: concentrate diet (74: 26 on as-fed basis) supplemented with fish oil, which was fed to conditioned mature geldings for 63 days, altered exercise metabolism. The basal diet was either top-dressed with corn oil or menhaden oil at a rate of 324 mg oil per kg body weight (BW). The fish oil contained 10.6% eicosapentaenoic acid (EPA) and 8.0% docosahexaenoic acid (DHA), and was the only significant sources of EPA and DHA in the diet. During exercise (standard exercise test on a treadmill) horses receiving fish oil had a lower heart rate and tended to have lower packed cell volume than horses receiving corn oil. Serum insulin tended to be lower and glucose: insulin ratios tended to be higher during exercise in horses receiving fish oil, and plasma free FA was lower during the initial stages of the exercise test for horses receiving fish oil. This indicates that the FA composition of supplemental fat sources may be one additional factor to consider with respect to potential performance capacity in exercising horses.

Soluble fibre

Due to the fermentative capacity of the equine GI tract, dietary inclusion of soluble fibre feeds offers an attractive alternative as energy sources to starch-rich cereals. However, only a few studies have been performed on the effects of replacing starch-rich concentrates (oats mainly) with soluble fibre (beet pulp) in the diet of exercising horses (Crandell *et al.*, 1999; Palmgren Karlsson *et al.*, 2002). In general, there appears to be no adverse effects on performance capacity of replacing part of the cereals in the diet with beet pulp. On the contrary, Palmgren-Karlsson *et al.*, (2002) found a lowered post-exercise muscle lactate and plasma lactate concentration, and higher muscle glycogen content, when oats was replaced with molassed beet pulp (Table 1).

Table 1. Post exercise plasma, muscle lactate concentration and glycogen content in four horses fed a forage diet with 3.5 kg oats or a diet where 1.5 kg of the oats was replaced with molassed sugar beet pulp (MSBP) (after Palmgren Karlsson et al., *2002).*

	Plasma lactate (mmol/l)	Muscle lactate (mmol/kg dry weight)	Muscle glycogen (mmol/kg dry weight)
Oats diet	8.2±0.6	38.5±3.8	394±34
MSBP diet	6.8±0.6[a]	22.7±3.8[a]	484±34[a]

[a] Significant difference between diets.

Forage for athletic horses

Through the well developed symbiosis with the hindgut microbiota, it should be possible to provide the energy needs of most horses (NRC, 2007) with a forage-only diet. However, the successful use of such a feeding strategy in practice in athletic horses has, as far as we know, not been reported until recently (Connysson *et al.*, 2006; Jansson and Lindberg, 2008). A prerequisite for this feeding practice to be applicable to athletic horses is to use forage with a high energy content, corresponding to the energy content of typical concentrates.

The major factor determining if forage can contribute to a great extent to the energy intake is the fibre content and the digestibility of the fibre fraction. Forages are composed of cell contents (protein, fat, soluble carbohydrates) and cell walls (cellulose, non-starch polysaccharides, lignin), which may vary in their relative proportions. The cell content is highly digestible (80-100%; Fonnesbeck, 1968; 1969), while the true digestibility of the cell wall is more limited (40-50%; Fonnesbeck, 1968; 1969). With increasing plant maturity the proportion of cell-contents decreases and that of cell-walls increases. Thus, the stage of maturity will have a profound effect on the energy and nutrient content of the forage (Ragnarsson and Lindberg, 2008). By offering high energy forage the diet proportions between forage and concentrate can be markedly affected, and made more favourable with regards to voluntary feed intake and digestive functions (Willard *et al.*, 1977; Radicke *et al.*, 1991), then with lower quality forage. There is little information on the importance of the botanical composition of the forage used for the response to exercise but one study, comparing alfalfa roughage with non-alfalfa roughage, showed no differences in plasma lactate, blood gas, respiratory and acid base measures (Southwood *et al.*, 1993a).

Jansson and Lindberg (2008) have shown that Standardbred trotters in racing condition fed a forage-only diet can train at a comparable level as those fed a typical forage: concentrate diet. The only disadvantage observed was a lower muscle glycogen content (-13%) compared to the forage: concentrate diet (unpublished data). This indicates that it could be possible to reduce the need for supplementing with concentrates.

Voluntary intake is a crucial factor in increasing the forage allowance to high performance horses. In this context, the palatability of the forage will be of great importance and this calls for methods to characterize different forage sources with respect to their palatability properties. Connysson *et al.* (2006), Jansson and Lindberg (2008) and Muhonen *et al.* (2009, experiment 1) reported good appetites in their studies on forage-only diets offered to Standardbred horses in training. However, in experiment 2 in Muhonen *et al.* (2009) the horses failed to maintain body weight and the diet had to be supplemented with a palatable concentrate (molassed sugarbeet pulp, corresponding to 20% of the estimated energy intake). The reason for the reduced appetite in that study was unclear but interestingly, a few weeks later the same horses were able to maintain body weight on another forage-only diet. Therefore, it is tempting to suggest that it was the chemical and/or microbial composition of the forage used in Muhonen *et al.* (2009, experiment 2) that affected the voluntary intake. However, when comparing data from the forages used in the studies mentioned above there is no obvious explanation for the reduced appetite (Table 2). Possibly, a slightly higher microbial activity in the forage used in Muhonen *et al.* (2009, experiment 2) may be one reason for the reduced appetite and voluntary feed intake. However, no health disturbances were observed.

In the study by Jansson and Lindberg (2008) the *in vitro* organic matter digestibility was 10% lower than in all the other studies (see Table 2) but in general the horses were able to maintain body weight and condition. However, the horses in that study showed a selective forage intake behavior on the forage-only diet (but not on the forage: concentrate diet) and the *in vitro* organic matter digestibility in the individual leftovers was consistently lower than in the forage sample on the forage-only diet. This implies that if athletic horses are offered excessive amounts of forage they might optimize their feed intake by selecting the most digestible parts and thereby achieving a higher energy intake compared to that estimated from a general feed analysis.

Practical recommendations

It can be concluded that high quality roughage should always be the foundation of the horse diet and that the palatability could be a limitation for a high intake. Although there is only limited information available on the effects of diet ingredients and macronutrient composition on performance capacity in horses, recently published data indicates that this is an area worth exploiting further. However,

Table 2. In vitro *digestible organic matter (IVDOM)*[1], *chemical and microbial composition*[2] *of grass forages (timothy and meadow fescue) and the horses (5-6 per study) appetite for the forage*[3].

Source[a]	IVDOM, %	DM, %	pH	Bu[b]	WSC[b]	Yeast	Mould	C	E	Appetite
1, diet RP	88	40-50	4.4	0.2	69	3.1	1.5	0	0	Good
1, diet HP	88	40-50	5.4	0	85	2.5	0	0	0	Good
2	78[c]	80	5.8	-	79	≤2.5	<2.0	<2.0	<1.0	Good
3, exp 1	89	82	-	-	157	0.4	1.1	0.5	1.5	Good
3, exp 1	90	45	5.3	<0.3	140	0	0	0.5	0	Good
3, exp 2	88	68	5.8	<0.2	132	3.7	2.2	<2.0	2.7	Not good
3, exp 2	89	41	4.8	<0.4	106	5.4	<2.0	<2.0	<2.0	Not good

[1] Lindgren (1979); [2] log colony forming units/g fresh matter, C=Clostridia, E=Enterobacteria;[3] good=voluntary intake high enough to maintain body weight within ±1%, not good=voluntary intake causing a body weight decrease of 0.6-2.9% within 5 days; [a] 1=Connysson *et al.* (2006), 2=Jansson and Lindberg, unpublished, 3=Muhonen *et al.* (2009); [b] Bu=butyric acid, g/kg dry matter; [c] IDOM of the leftovers were lower, indicating that the IVDOM of the feed ingested was higher than 78.

based on the literature reviewed above we suggest that the following criteria could be used as guidelines when composing safe and 'successful' diets to exercising horses with an energy demand twice the maintenance level; free access to forage with an *in vitro* digestibility of organic matter of 78% or higher (corresponds to ≥12.5 MJ digestible energy per kg DM; Lindberg and Ragnarsson, unpublished), the concentrate allowance should be less than 4.5 kg per day and starch intake should not exceed 1.1 g per kg body weight and meal. Moreover, supplementing the diet with fat appears to be a safe and efficient way to improve energy density, reduce starch intake and support performance.

References

Beyer, M., 1998. Colic. In: Advances in Equine Nutrition, ed. J. D. Pagan, Nottingham University Press, Manor Farm, Trumpton, UK, pp. 483-488.

Brooks, G.A., 1998. Mammalian fuel utilization during sustained exercise. Comparative Biochemistry and Physiology Part B 120: 89-170.

Burk, A.O. and C.A. Williams, 2008. Feeding management practices and supplement use in top-level event horses. Comparative Exercise Physiology 5: 85-93.

Connysson, M., S. Muhonen, J.E. Lindberg, B. Essén-Gustavsson, G. Nyman, K. Nostell and A. Jansson, 2006. Effects on exercise response, fluid and acid-base balance of protein intake from forage-only diets in Standardbred horses. Equine Veterinary Journal, Supplement 36: 648-653.

Crandell, K.G., J.D. Pagan, P. Harris and S.E. Duren, 1999. A comparison of grain, oil and beet pulp as energy sources for the exercised horse. Equine Veterinary Journal, Supplement 30: 485-489.

Fonnesbeck, P.V., 1968. Digestion of soluble and fibrous carbohydrates of forages by horses. Journal of Animal Science 27: 1336-1344.

Fonnesbeck, P.V., 1969. Partitioning the nutrients of forage for horses. Journal of Animal Science 28: 624-638.

Geor, R.J., 2006. The role of nutritional supplements and feeding strategies in equine athletic performance. Equine Comparative Exercise Physiology 3: 109-119.

Geor, R.J and L.J McCutcheon, 1998. Hydration effects on physiological strain of horses during exercise-heat stress. Journal of Applied Physiology 84(6): 2042-2051

Gillham, S.B., N.H. Dodman, L. Shuster, R.K. Kream and W. Rand, 1994. The effect of diet on cribbing behaviour and plasma β-endorphins in horses. Applied Animal Behaviour Science 41: 147-153.

Glade, M.J., 1983. Nutrition and performance of racing Thoroughbreds. Equine Veterinary Journal 15(1): 31-36.

Harris, P.A. and R.C. Harris, 2005. Ergogenic potential of nutritional strategies and substances in horses. Livestock Production Science 92: 147-165.

Jansson, A. and J.E. Lindberg, 2008. Effects of a forage-only diet on body weight and response to interval training on a track. In: Saastamoinen M.T. and W. Martin-Rosset (eds.), Nutrition of the exercising horse. EAAP publication no 125. Wageningen Academic Publishers, Wageningen, the Netherlands, p. 345-349.

Jansson, A., S. Nyman, A. Lindholm and J.E. Lindberg, 2002. Effect of exercise metabolism of varying dietary starch and sugar proportions. Equine Veterinary Journal, Supplement 34: 17-21.

Kavazis, A.N., J. Kivipelto, H.S. Choe, P.T. Colahan and E.A. Ott, 2004. Effects of ribose supplementation on selected metabolic measurements and performance in maximally exercising Thoroughbreds. Journal of Animal Science 82:619-625.

Kienzle, E., 1994. Small intestinal digestion of starch in the horse. Revue de Médecine Vétérinaire 145, 199-204.

Kusunose, R., 1992. Diurnal pattern of cribbing in stabled horses. Japan Journal of Equine Science 3: 173-176.

Lacombe, V.A., K.W. Hinchcliff, R.J. Geor and C.R. Baskin, 2001. Muscle glycogen depletion and subsequent replenishment affect anaerobic capacity of horses. Journal of Applied Physiology 91: 1782-1790.

Lindgren, E., 1979. The nutritional value of roughages determined *in vivo* and by laboratory methods. Department of Animal Nutrition and Management, Swedish University of Agricultural Sciences, Report 45, 63 pp.

MacLeay, J.M., S.A. Sorum, S.J. Valberg, W.E. Marsh and M.D. Sorum, 1999. Epidemiologic analysis of factors influencing exertional rhabdomyolysis in Thoroughbreds. American Journal of Veterinary Research 60 (12): 1562-1566.

Muhonen, S., J.E. Lindberg, J. Bertilsson and A. Jansson, 2009. Effects on fluid balance and exercise response in Standardbred horses feed silage, haylage and hay. Comparative Exercise Physiology 5(3-4): 133-142.

NRC, 2007. Nutrient requirement of horses. 6th ed., National Academies Press, Washington, D.C.

Mullen, P.A., R. Hopes and J. Sewell, 1979. The biochemistry, nutrition and racing performance of two-year-old thoroughbreds throughout their training and racing season. The Veterinary Record 104: 90-95.

O'Connor, C.I., L.M. Lawrence, A.C. St. Lawrence, K.M. Janicki, L.K. Warren and S. Hayes, 2004. The effect of dietary fish oil supplementation on exercising horses. Journal of Animal Science 82: 2978-2984.

Pagan, J.D., R.J. Geor, P.A. Harris, K. Hoekstra, S. Gardner, C. Hudson and A. Prince, 2002. Effects of fat adaptation on glucose kinetics and substrate oxidation during low-intensity exercise. Equine Veterinary Journal, Supplement 34: 33-38.

Palmgren Karlsson, C., A. Jansson, B. Essén-Gustavsson and J.E. Lindberg, 2002. Effect of molassed sugar beet pulp on nutrient utilisation and metabolic parameters during exercise. Equine Veterinary Journal, Supplement 34: 44-49.

Potter, G.D., F.F. Arnold, D.D Householder, D.H. Hansen and K.M. Brown. 1992a. Digestion of starch in the small or large intestine of the equine. Pferdeheilkunde 107-111.

Potter, G.D., S.L. Hughes, T.R Julen and D.D.L. Swinney, 1992b. A review of research on digestion and utilization of fat by the equine. Pferdeheilkunde, Sonderausgabe 119-123.

Radicke, S., E. Kienzle and H. Meyer, 1991. Preileal apparent digestibility of oats and corn starch and consequences for cecal metabolism. Proceedings of the 12th Equine Nutrition Symposium. Calgary, Canada. Pp. 43.

Ragnarsson, S. and J.E. Lindberg, 2008. Nutritional value of timothy haylage in Icelandic horses. Livestock Science 113: 202-208.

Redbo, I., P. Redbo-Torstensson, F.O. Ödberg, A. Hedendahl and J. Holm, 1998. Factors affecting behavioural disturbances in race-horses. Animal Science 66: 475-481.

Schuback, K., B. Essén-Gustavsson and S.G. Persson, 2000. Effect of creatine supplementation on muscle metabolic response to a maximal treadmill exercise test in Standardbred horses. Equine Veterinary Journal 32(6): 533-540.

Southwood, L.L., D.L. Evans, D.R. Hodgson, W.L. Bryden and R.J. Rose, 1993a. The effect of roughage source on exercise performance and metabolism in Thoroughbred horses. Cornell Veterinarian 83(3): 243-255.

Southwood, L.L., D.L. Evans, W.L. Bryden and R.J. Rose, 1993b. Nutrient intake of horses in Thoroughbred and Standardbred stables. Australian Veterinary Journal 70(5): 164-168.

Vervuert, I., M. Coenen and C. Bothe, 2003. Effect of oat processing on the glycaemic and insulin responses in horses. Journal of Animal Physiology and Animal Nutrition 87 (3-4): 96-104.

Vervuert, I., M. Coenen and C. Bothe, 2004a. Effect of corn processing on the glycaemic and insulinaemic responses in horses. Journal of Animal Physiology and Animal Nutrition 88 (9-10): 348-355.

Vervuert, I., M. Coenen and M. Bichman, 2004b. Comparison of the effect of fructose and glucose supplementation on metabolic responses in resting and exercising horses. Journal of Veterinary Medicine A 51:171-177.

Vervuert, I., C. Bothe and M. Coenen, 2007. Glycaemic and insulinaemic responses to mechanical and thermal processed barley in horses. Journal of Animal Physiology and Animal Nutrition 91 (5-6): 263-268.

Vervuert, I., K. Voigt, T. Hollands, D. Cuddeford and M. Coenen, 2009. Effect of feeding increasing quantities of starch on glycaemic and insulinaemic responses in healthy horses. Veterinary Journal 182 (1): 67-72.

Voiton, D.M., R. Navet, V.A. Lacombe, F. Sluse, B. Essen-Gustavsson, K.W. Hinchcliff, J.L.L. Rivero, D. Serteyn and S. Valberg, 2007. Muscle energetic in exercising horses. Equine Comparative Exercise Physiology 4:105-118.

Willard, J.G., J.C. Willard, S.A. Wolfram and J.P. Baker, 1977. Effect of diet on cecal pH and feeding behaviour of horses. Journal of Animal Science 45:87-93.

Waller, A.P., R.J. Geor, L.L. Spriet, G.J.F. Heigenhauser and M.I. Lindinger, 2009. Oral acetate supplementation after prollonged moderate intensity exercise enhances early muscle glycogen resynthesis in horses. Experimental Physiology 94: 888-898.

Willing, B., A. Vörös, S. Roos, C. Jones, A. Jansson and J.E. Lindberg, 2009. Changes in faecal bacteria associated with concentrate and forage-only diets fed to horses in training. Equine Veterinary Journal 41: 908-914.

Effect of growth rate on plasma leptin and bone markers, and metacarpal ultrasound measurements in the young Lusitano horse

M.J. Fradinho[1], M. Barbosa[2], M.J.C. Vila-Viçosa[3], A. Rosa[4], F. Perestrello[5], R.J.B. Bessa[1], R.M. Caldeira[1] and G. Ferreira-Dias[1]
[1]CIISA, Faculdade de Medicina Veterinária, TULisbon, 1300-477 Lisboa, Portugal
[2]Fundação Alter Real, Alter-do-Chão, Portugal
[3]Departamento de Medicina Veterinária, Universidade de Évora, Portugal
[4]Qta. da Lagoalva de Cima, Alpiarça, Portugal
[5]Companhia das Lezírias, Samora Correia, Portugal

Abstract

In the foal, the relationship between skeletal development and growth rate has been studied. Bone markers as osteocalcin (OC) and bone alkaline phosphatase (BAP) are known to reflect bone metabolism and, in humans and rodents, leptin was also involved in the regulation of bone mass. Quantitative ultrasonography could provide information on bone quality as it reflects mineral density and bone mechanical properties. The main objective of this study was to evaluate the relationship between growth rate and (1) bone biochemical markers; (2) leptin, and (3) ultrasound measurements of the third metacarpal bone in the young Lusitano horse, up to one year of age. Thirty three Lusitano foals from four stud farms were monitored every two months from birth to one year of age. Animals were weighed and blood samples were collected for analysis of BAP, OC and leptin. Measurements of speed of sound (SOS) on the mid section of the right third metacarpal bone (MC III) (dorsal and lateral aspects) were performed. Individual growth curves were adjusted and the derivatives as the instantaneous growth rate (IGR) were calculated for each sampling time. Pearson's correlation coefficients were used to examine the variables relationships. Effects of group and age were evaluated. Positive correlations were found between IGR and plasma leptin (0.26; $P<0.001$), OC (0.67; $P<0.0001$) and BAP (0.34; $P<0.0001$). The negative correlation between IGR and lateral SOS MCIII measurements (-0.47; $P<0.0001$), confirm its increase with age. Negative correlations were observed between SOS lateral measurements and OC (-0.38; $P<0.0001$) and BAP (-0.20; $P<0.05$). A correlation between OC and BAP was also observed (0.21; $P<0.01$). Differences on OC and BAP were detected between groups. This study suggests that bone markers and SOS measurements could be influenced by different growth rates in the Lusitano foal.

Keywords: growth rate; bone formation markers; leptin; quantitative ultrasonography; Lusitano foals

Introduction

Nowadays there is a growing interest on understanding the physiologic mechanisms of horse' bone metabolism in order to improve sports performance.

In the foal, the relationship between skeletal development and growth rate has been studied (e.g. Fleurence *et al.*, 2005). Bone biochemical markers such as osteocalcin (OC) and bone alkaline phosphatase (BAP) are known to reflect horse bone metabolism (e.g. Lepage *et al.*, 2001) and, in humans and rodents, the role of leptin as a possible modulator of bone mass was also emphasized (Cirmanová *et al.*, 2008).

Beside bone markers, quantitative ultrasonography (QUS) is a non-invasive technique, previously used in the horse, which could provide an indication of bone quality, as it reflects mineral density and mechanical properties of cortical bone (Carstanjen *et al.*, 2002).

The main objective of this study was to evaluate the relationship between growth rate and (1) bone biochemical markers; (2) leptin, and (3) ultrasound measurements of the third metacarpal bone in the young Lusitano horse, from birth to one year of age.

Material and methods

Thirty-three Lusitano foals from four stud farms (groups A, B, C, D) were monitored every two months from birth to one year old. Animals were weighed and blood samples collected for analysis of BAP, OC and leptin. Concentrations of BAP and OC were determined with two specific immunoassays (Quidel Corporation, USA) and leptin with a multi-species radioimmunoassay (Linco, USA). Measurements of speed of sound (SOS) on the mid section of the right third metacarpal bone (MC III) (dorsal and lateral aspects) were performed with a QUS device.

Until weaning all foals were kept on pasture with their dams: A and B mares were daily supplemented (compound feeds and preserved forages) according to pasture availability and farm practices, while C

Table 1. Correlation coefficients between analysed variables.

	IGR (kg/d) r *P* value n	SOS D (m/s) r *P* value n	SOS L (m/s) r *P* value n	Leptin (ng/ml) r *P* value n	OC (ng/ml) r *P* value n	BAP (U/L) r *P* value n
IGR (kg/d)						
r	1	-0.084	**-0.474**	**0.263**	**0.673**	**0.34**
P value		0.284	**<0.0001**	**0.0019**	**<0.0001**	**<0.0001**
n		164	**164**	**137**	**162**	**165**
SOS D (m/s)						
r		1	**0.228**	-0.044	-0.097	0.111
P value	-		**0.0035**	0.612	0.221	0.156
n			**163**	136	161	164
SOS L (m/s)						
r			1	-0.1	**-0.384**	-0.198
P value	-	-		0.246	**<0.0001**	**0.011**
n				136	**161**	**164**
Leptin (ng/ mL)						
r				1	0.164	0.048
P value	-	-	-		0.057	0.576
n					135	137
OC (ng/mL)						
r					1	**0.212**
P value	-	-	-	-		**0.007**
n						**162**
BAP (U/L)						
r						1
P value	-	-	-	-	-	
n						

IGR: instantaneous growth rate; SOS D: speed of sound – dorsal measurement; SOS L: speed of sound – lateral measurement; OC: Osteocalcin; BAP: bone alkaline phosphatise; the different numbers of observations on the correlation coefficients are due to missing values.

and D mares were rarely supplemented. After weaning (223±5.6 d), foals were group fed in paddocks during an adaptation period and returned to pasture afterwards, where they were daily supplemented with grass hay and compound feeds during the winter time.

Individual growth curves were adjusted and the derivatives as the instantaneous growth rate (IGR) were calculated for each animal in each sampling day. Statistical analysis was performed with SAS and Pearson's correlation coefficients were used to examine variables relationship. Effect of group was evaluated at 45 d (48.8±2.8), 105 d (103±2.8), 165 d (164.6±2.9), 225 d (226±2.9), 285 d (291.8±2.9) and 345 d (346.8±3.8) by ANOVA. In addition, a mixed model considering repeated measures on time was used to evaluate changes of different variables with age. The model considered the group, age and its interaction as fixed factors and foals within group as a random factor.

Results and discussion

For the considered period, positive correlations were found between IGR and plasma concentrations of leptin (0.26; $P<0.001$), OC (0.67; $P<0.0001$) and BAP (0.34; $P<0.0001$) (Table 1).

Differences between groups were observed in IGR on 45, 105, 165 and 225 d (Figure 1), which could potentially be a management effect, probably related with mares' feeding regimen. After that period (post-weaning) no differences between groups were found in terms of IGR.

The negative correlation between IGR and the SOS obtained at lateral measurements of the MCIII (-0.47; $P<0.0001$), confirms its increase with age (Figure 2), as already observed in previous work (Carstanjen *et al.*, 2003; Fradinho *et al.*, 2009). In the present study, besides the significant effect of age ($P<0.01$) on SOS dorsal measurements, there was also a significant effect of group ($P<0.01$).

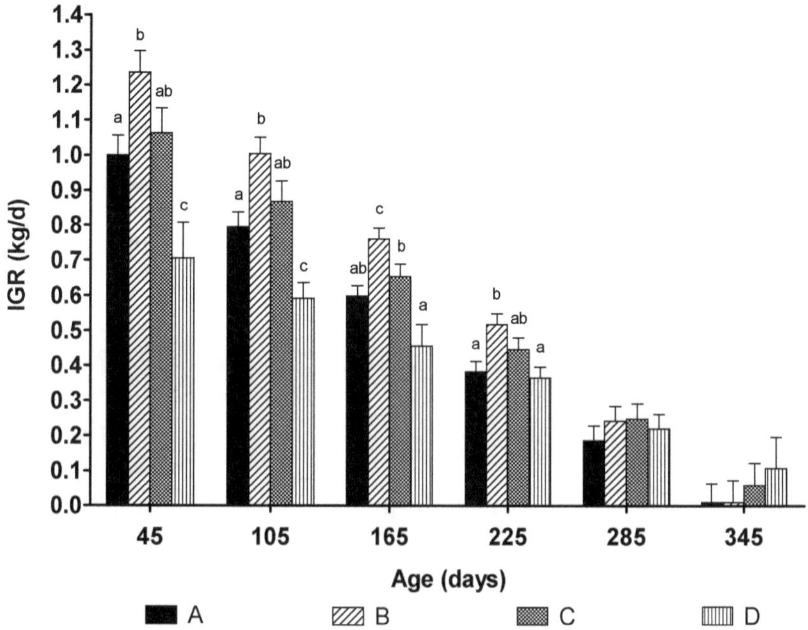

Figure 1. Instantaneous growth rate in the four groups of foals (LSMeans±SE). Case letters indicate differences between groups in each sample time (P<0.01).

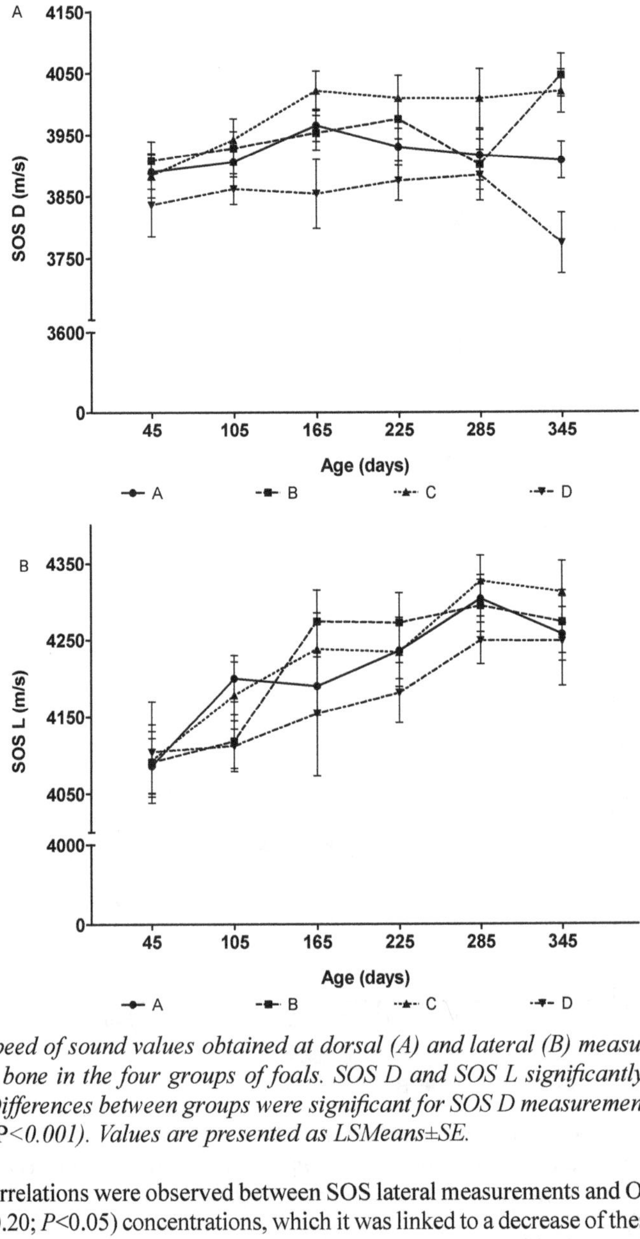

*Figure 2. Speed of sound values obtained at dorsal (A) and lateral (B) measurements on the third metacarpal bone in the four groups of foals. SOS D and SOS L significantly increases with age (*P*<0.01). Differences between groups were significant for SOS D measurements at 225 d (*P*<0.05) and 345 d (*P*<0.001). Values are presented as LSMeans±SE.*

Negative correlations were observed between SOS lateral measurements and OC (-0.38; *P*<0.0001) and BAP (-0.20; *P*<0.05) concentrations, which it was linked to a decrease of these bone markers with age (P<0.0001) (Figure 3). OC and BAP concentrations were positively correlated (0.21; *P*<0.01) and the observed age-related changes are consistent with previous studies in the foal (Price *et al.*, 2001; Fradinho *et al.*, 2006; Donabédian *et al.*, 2008). Differences on OC and BAP were detected between groups at each sampling time.

No differences were found for leptin concentrations between groups at each sampling time (Table 2). Observed values are within the range reported for two months old Quarter-Horse foals (Berg *et al.*, 2007) and slightly above the values referred for Lipizzan foals (Cebulj-Kadunc and Cestnik, 2005).

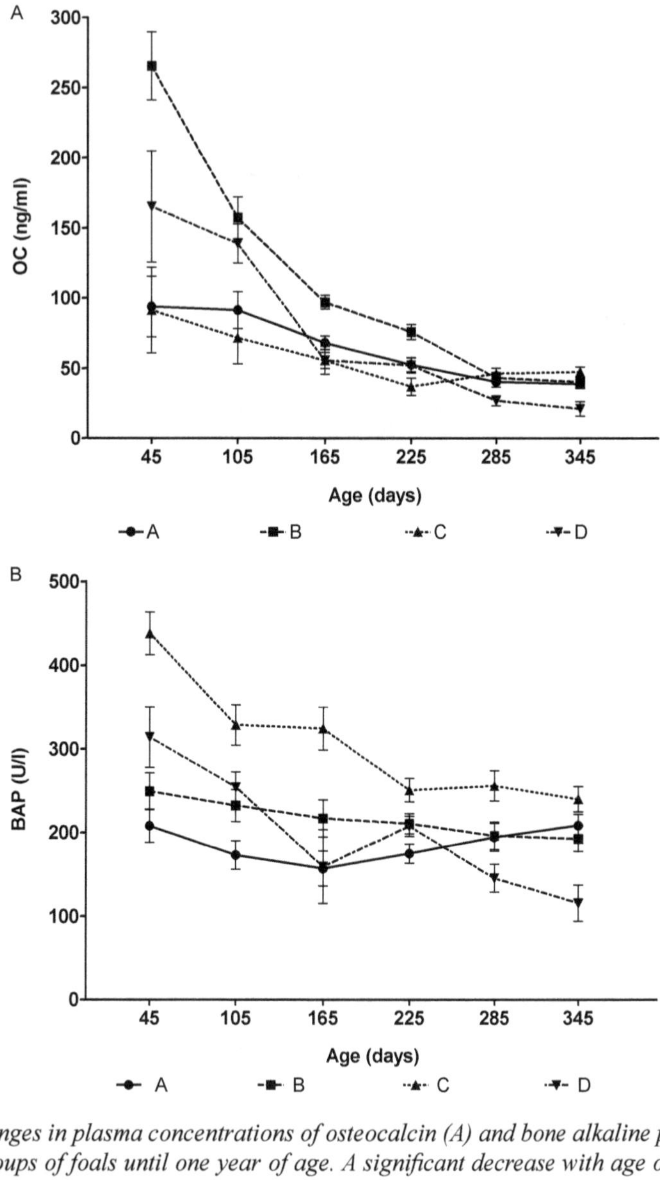

Figure 3. Changes in plasma concentrations of osteocalcin (A) and bone alkaline phosphatase (B) in the four groups of foals until one year of age. A significant decrease with age occurred in both bone markers (P<0.0001). Differences between groups were significant for OC and BAP values in each sampling time (P<0.01).
Values are presented as LSMeans±SE.

This study suggests that bone markers and SOS measurements could be influenced by different growth rates in the Lusitano foal, which could be probably related to the feeding management options.

Acknowledgements

Work supported by grants POCI 2010, N° GG/GGP/ME611-0166/05 and SFRH/BD/29890/2006 from FCT, Lisbon, Portugal.

Table 2. Plasma leptin concentrations (ng/ml) in the four groups of foals throughout the study.

Groups	Age (days)					
	45	105	165	225	285	345
A	2.22±0.21	2.32±0.14	1.77±0.16	1.33±0.14	1.61±0.24	1.93±0.20
B	1.84±0.27	1.87±0.16	1.99±0.17	1.48±0.15	1.17±0.28	1.57±0.23
C	1.09±0.65	1.25±0.48	1.34±0.20	1.78±0.16	2.22±0.33	2.27±0.25
D	2.31±0.38	1.90±0.15	1.42±0.31	1.24±0.15	1.55±0.29	-

Values are presented as LSMeans±SE.

References

Berg, E.L., D.L. McNamara and D.H. Keisler, 2007. Endocrine profiles of periparturient mares and their foals. J. Anim. Sci. 85, 1660-1668.

Carstanjen, B., O. Lepage, J. Detilleux, F. Duboeuf and H. Amory, 2002. Use of quantitative ultrasonography for noninvasive assessment of bone in horses. Am. J. Vet. Res. 63(10), 1464-1469.

Carstanjen, B., O. Lepage, O. Hars, P. Langlois, F. Duboeuf and H. Amory, 2003. Speed of sound measurements of the third metacarpal bone in young exercising thoroughbred racehorses. Bone. 33, 411-418.

Cebulj-Kadunc, N. and V. Chestnik, 2005. Circulating leptin concentrations in Lipizzan horses and Jezersko-Solchava sheep. Slov Vet Res. 42(1/2), 11-14.

Cirmanová, V., M. Bayer, L. Stárka and K. Zajicková, 2008. The effect of leptin on bone – An Evolving concept of action. Physiol. Res. 57 (Suppl. 1), S143-S151.

Donabédian, M., G. Fleurence, G. Perona, C. Robert, O. Lepage, C. Trillaud-Geyl, S. Leger, A. Ricard, D. Bergero and W. Martin-Rosset, 2006. Effect of fast vs. moderate growth rate related to nutrient intake on developmental orthopaedic disease in the horse. Anim. Res. 55, 471-486.

Donabédian, M., P.R. Van Weeren, G. Perona, G. Fleurence, C. Robert, S. Léger, D. Bergero, O. Lepage and W. Martin-Rosset, 2008. Early changes in biomarkers of skeletal metabolism and their association to the occurrence of osteochondrosis (OC) in the horse. Equine vet. J. 40(3), 253-259.

Fleurence, G., C. Trillau-Geyl, M. Donabedian, G. Perona, G. Bigot, G. Arnaud, H. Dubroeucq and W. Martin-Rosset, 2005. Effect of body weight gain on skeletal growth in the sport horses. *In*: Nineteenth Symposium Proceedings of the Equine Science Society. May 31-June 3. Tucson, Arizona, 129-134.

Fradinho, M.J., G. Ferreira-Dias, L. Mateus, M.F. Santos-Silva, R. Agrícola, M. Barbosa and J.M. Abreu, 2006. The influence of mineral supplementation on skeleton formation and growth in Lusitano foals. Livest. Sci; 104, 173-181.

Fradinho, M.J., M.J. Correia, M.J.C. Vila-Viçosa, A. Farrim, P.A. Harris, R.J.B. Bessa, R.M. Caldeira and G. Ferreira-Dias, 2009. Non-invasive bone assessment by quantitative ultrasonography in Lusitano suckling foals: A preliminary study. In: Lindner, A. (ed.), Applied equine nutrition and training. Wageningen Academic Publishers, Wageningen, the Netherlands, p. 191-196.

Lepage, O., B. Carstanjen and D. Uebelhart, 2001. Non-invasive assessment of equine bone: an update. Vet. J. 156, 169-175.

Price, J.S., B.F. Jackson, J.A. Gray, P.A. Harris, I.M. Wright, D. Pfeiffer, S.P. Robins, R. Eastell and S.W. Ricketts, 2001. Biochemical markers of bone metabolism in growing thoroughbreds: a longitudinal study. Res Vet Sci. 71, 1-9.

Relationship between body weight variations and biological parameters during the work season in show horses

C. Amato, L. Martin, B. Siliart, P. Nguyen and H. Dumon
Nutrition and Endocrinology Unit, ONIRIS, USP 99025305, Atlanpole-La Chantrerie, BP 40706, 44307 Nantes, France

Abstract

Exercise-induced fatigue is associated with various changes in physiological parameters. The aim of the present study was to determine the relationship between some biological parameters and body weight loss, energy intake and workload. Twenty-one healthy show horses of Andalusian or Lusitanian breeds were included in the study. All the horses were weighed twice monthly and body condition score was determined monthly using INRA's method (scale 0 to 5). Blood samples were collected monthly at rest for standard biochemical assays. Workload was determined according of heart rate and correlations between zootechnical parameters and biological parameters were calculated. Variations in the biological parameters were determined with a linear mixed effects model ($P<0.05$). A significant decrease in body weight occurred during follow-up despite the increased energy intake. The variations in body weight, energy intake and workload were significantly correlated with creatinine, hematocrit, fibrinogen, aspartate aminotransferase, gammaglutamyl transferase, glucose, lactate, cholesterol, HDL-cholesterol, sodium, potassium, magnesium, chloride, calcium and phosphates at rest parameters. A decrease in body weight despite an increase in energy intake may be useful for detecting chronic fatigue in show horses and this may be due to excessive catabolic effect of some hormones and cytokines stimulated by workload and stress.

Keywords: biological parameters, body weight variation, energy intake

Introduction

Horses are able to perform high levels of physical exercise, but repetitive exercise induces a multitude of physiologic adaptations in the horse, which reduce the effect of damage due to physiologic stressors associated with exercise (Hinchcliff and Geor, 2004). Overtraining has been described as a state of prolonged fatigue caused primarily by an imbalance between training and recovery (Kuipers, 1998). Excessive training loads with inadequate recovery periods lead to a reduction in performance capacity. Even with complete rest, recovery from the overtrained condition can take weeks to months (Bayly, 2002; Golland, 2003). Early studies in horses identified overtraining as a syndrome of poor performance and physiological or hormonal signs (Golland, 1999; Hamlin, 2002; McGowan, 2008; Tyler-McGowan, 1999). These studies revealed that overtraining in horses was associated with one or more physiological signs, including a decrease in body weight (Tyler, 1996). The aim of the present study was to investigate the variations in body weight, body condition and energy intake during a season of work along with some biochemical parameters, to determine their relationship according to workload and intensity in historic re-enactment shows.

Material and methods

Twenty-one clinically healthy show horses (14 geldings and 7 stallions, from 7 to 19 years old, mean BW= 482 kg, mean BCS= 3) of Andalusian or Lusitanian breeds were included in the study. The follow-up started in October 2008 and ended in September 2009. General health and feeding schedules were recorded. All horses were weighed twice a month and each received a suitably ration adjusted according to the measured body weight variations (ΔBW) and Body Condition Score (BCS). The ration was based on hay, concentrate and bran mash and horses had free access to water. Energy intake was discussed according to INRA requirements (INRA, 1990). Body Condition Score was evaluated monthly, on a scale of 1 to 5 using both palpation of fat cover with hand and a visual appraisal of the elected areas (Martin-Rosset, 2008).

Blood samples were collected by jugular vein puncture into Vacutainer® tubes monthly, before exercise. One tube (lithium heparin) was used for biochemical analyses: albumin, creatinine, urea, total protein (TP), creatine phosphokinase (CPK), lactate dehydrogenase (LDH), aspartate aminotransferase (ASAT), alanine aminotransferase (ALAT), alkaline phosphatase (ALP), gammaglutamyl transferase (GGT), glucose, cholesterol, HDL-cholesterol, triglycerides (TG), sodium (Na), potassium (K), magnesium (Mg), chloride (Cl), calcium (Ca) and phosphates (P). A second tube (EDTA) was used for the hematocrit (Hct) and fibrinogen assay. Blood lactate and blood glucose were determined at rest and during work by Accutrend® Lactate and Accu-Chek® Active, analyser (Roche®).

Exercise intensity was assessed using a combined global positioning system (FRWD W series®) and a heart rate monitoring system (Polar Equine Transmitter®). To quantify exercise intensity, an in-house scale was established then validated by heart rate variations recorded during work. The shows were subdivided into various scenes or exercise bout, and for each scene according to heart rate and duration was assigned a score from 0 to 2 (score of intensity). Moreover, for each horse, the number of the executed shows was numbered weekly. Therefore the intensity of the workload was calculated monthly by multiplying the number of the shows with the score of intensity. The intensity of the cumulative workload was then calculated and categorized into six groups from 0 (I_0 period without show) to >600 (I_5 heaviest workload). Workload intensity $=\sum (I_n \times j_n)$ with I_n= show intensity, j_n= number of shows.

The correlation between zootechnical parameters and biological parameters was calculated by Pearson's test. Variations in the biological parameters according to work intensity were determined using a linear mixed effects model, $P<0.05$ was considered significant. All analyses were performed with statistical software XLstat-Pro 2010 (Addinsoft® SARL, USA).

Results and discussion

Workload increased gradually from April/May (3.3 shows/week) to August (7.6 shows/week) then decreased in September (2.4 shows/week). These horses performed historical re-enactment shows (medieval battle and vaulting on a lunge). In these shows, the stress is a major compound of the exercise. In all scenes, there are many short and promptness exercises disrupted by short but stressing waiting periods. In the present conditions, heart rate is the best indicator of the demands endures by the horses and so is the base element of the workload estimation. The horses performed two different historic shows: first show: mean distance = 3,700±1,200 m, max speed = 5.9±1.2 m/s, max heart rate = 201±29 bpm and max lactate = 9.2 mmol/l; second show: mean distance = 8,000±1,500 m, max speed = 12.7±1.4 m/s, max heart rate = 222±10 bpm and max lactate = 18.5 mmol/l.

A significant decrease in body weight occurred during the follow-up despite a significant increased energy intake (Table 1).

Several parameters were affected by the workload (Table 2 and 3). Plasma Na, Cl and P decreased significantly during the work season but the concentrations remained within the reference ranges and horses never showed dehydration (Table 3). Albumin g/l (mean=34.8±2), LDH U/l (mean=602±170), ASAT U/l (mean=395±163), ALAT U/l (mean=14±82), total protein g/l (mean=66±4), triglycerides g/l (mean=0.25±0.08), K mmol/l (mean=4.1±0.6), and Ca mmol/l (mean=3.1±0.1) were not affected by workload.

Table 1. Variations of zootechnical parameters according to exercise intensity (mean ± SEM, linear mixed effect).

	I_0	I_1	I_2	I_3	I_4	I_5	P^1 cumulative workload
% Δ BW	0.0±0.0	1.3±0.4**	0.6±0.2	-0.5±0.2*	-0.6±0.2*	0.0±0.2	<0.001
Kg Δ BW	0.0±0.0	6.1±1.9**	2.9±1	-2.4±1*	-2.8±0.9*	0.0±1	<0.001
Energy ration (UFC^2)	5.3±0.12*	6.3±0.12*	6.3±0.16	6.6±0.13*	6.7±0.15	6.7±0.15	<0.0001

Categories of intensity cumulative workload: Workload intensity $=\sum (I_n \times j_n)$; I_n= show intensity, j_n= show numbers. The cumulative intensity workload was categorized into six categories noted from I_0: period without show; I_1: [from 5, to 30]; I_2: [from 31, to 120]; I_3: [from 121, to 300]; I_4: [from 301, to 600]; I_5: >600; level of significance for effect of work in different categories, versus I_0 (period without show): * $P<0.05$, ** $P<0.001$, *** $P<0.0001$; P^1 cumulative workload: levels of significance for effect cumulative workload; UFC^2= Unite Fourragère Cheval, e.g. Horse Feed Unit.

Table 2. Variations of biological parameters according to exercise intensity (mean ± SEM, linear mixed effect).

	I_0	I_1	I_2	I_3	I_4	I_5	P^1 cumulative workload
Hct (%)		38.5±0.9*	37.4±0.9	36.5±0.6	35.7±0.6*	35.0±0.6	0.001
Creatinine (mg/l)	13.3±0.4	12.6±0.2	13.5±0.3*	13.3±0.3	12.7±0.3	12.1±0.3**	0.021
Fibrinogen (g/l)		0.1±0.03	0.09±0.03	0.12±0.01	0.08±0.02	0.05±0.01	0.020
CPK (U/l)	146.6±8.1	174.4±10.2	198.0±19.3	167.0±9.8	283.8±94.4	244.4±13.9	0.005
GGT (U/l)		29.4±5.1*	32.2±6.3*	20.5±2.2	18.3±1.2	17.9±0.9*	0.002
ALP (U/l)	285.4±17.8	275.5±13.5	296.5±16.4	296.5±16.5	303.0±16.6	328.7±14.5*	<0.0001
Urea (g/l)	0.27±0.01**	0.30±0.01	0.33±0.01*	0.29±0.01	0.33±0.0*	0.32±0.01	0.001
Cholesterol (g/l)	0.81±0.03*	0.88±0.03	0.91±0.03	0.89±0.03	0.93±0.03	0.93±0.03	0.002
HDL-Cholesterol (g/l)	0.45±0.02	0.49±0.02	0.52±0.01*	0.52±0.01	0.54±0.01*	0.54±0.02	<0.0001

Categories of intensity cumulative workload: workload intensity $=\sum (I_n \times j_n)$; I_n= show intensity, j_n= show numbers. The cumulative intensity workload was categorized into six categories noted from I_0: period without show; I_1: [from 5, to 30]; I_2: [from 31, to 120]; I_3: [from 121, to 300]; I_4: [from 301, to 600]; I_5: >600; level of significance for effect of work in different categories, versus to I_0 (period without show): * $P<0.05$, ** $P<0.001$, *** $P<0.0001$; P^1 cumulative workload: level of significance for effect cumulative workload.

The Pearson analysis showed that the ΔBW (r^2= -0.4, $P\leq0.0001$) and energy intake (r^2= 0.4, $P\leq0.0001$) were significantly and positively correlated with monthly workload. ΔBW (r^2= -0.3, P=0.003), BCS (r^2= -0.2, P=0.008) and energy intake (r^2= 0.2, P=0.004) were significantly correlated with the cumulative workload. Moreover the variations in body weight and workload were significantly correlated with several changes in routine biochemical variables: ΔBW were positively correlated with ASAT (r^2= 0.2, P=0.03) and Na (r^2= 0.2, P=0.02) but negatively correlated with cholesterol (r^2= -0.3, P=0.001), HDL-cholesterol (r^2= -0.3, P=0.000) and calcium (r^2= -0.2, P=0.01). The

Table 3. Variations of biological parameters according to exercise intensity (Mean ± SEM, linear mixed effect).

	I_0	I_1	I_2	I_3	I_4	I_5	P^1 cumulative workload
Na (mmol/l)	147.6±0.8	150.5±1.6***	144.1±0.5*	146.8±0.3	145.2±0.4	144.3±0.3*	<0.0001
Mg (mmol/l)	0.8±0.01	0.8±0.02	0.7±0.01	0.8±0.01	0.8±0.01	0.8±0.01	0.006
Cl (mmol/l)	107.8±0.6*	108.3±0.8**	105.2±0.3*	106.9±0.3	106.1±0.4	105.7±0.3*	<0.0001
P (mmol/l)	0.9±0.03	1.0±0.05*	1.0±0.04	0.8±0.03*	1.0±0.04	0.8±0.03**	<0.0001
Lactate² (mmol/l)	0.8±0.07**	1.2±0.05*	1.1±0.06	1.2±0.1	1.1±0.04	1.0±0.05*	0.007
Lactate³ (mmol/l)	1.0±0.04**	4.5±0.6	5.6±0.8	8.4±1.5*	6.1±0.7	6.0±0.9	0.002
Glucose² (mmol/l)	5.4±0.08	5.0±0.1	5.1±0.2	5.1±0.1	5.1±0.1	4.9±0.2	0.035
Glucose³ (mmol/l)	4.7±0.1	4.7±0.1*	5.1±0.2	5.5±1.03*	5.1±0.1	5.2±0.2	0.006

Categories of intensity cumulative workload: Workload intensity $=\sum (I_n \times j_n)$; I_n= show intensity, j_n= show numbers. The cumulative intensity workload was categorized into six categories noted from I_0: period without show; I_1: [from 5, to 30]; I_2: [from 31, to 120]; I_3: [from 121, to 300]; I_4: [from 301, to 600]; I_5: >600; level of significance for effect of work in different categories, versus I_0 (period without show): * $P<0.05$, ** $P<0.001$, *** $P<0.0001$; P^1 cumulative workload: level of significance for effect cumulative workload; ² blood lactate and glucose dosage at rest; ³ blood lactate and glucose dosage during work.

cumulative workload was positively correlated with HDL-cholesterol (r^2= 0.3, P=0.03) and ALP (r^2= 0.2, P=0.006) and negatively with hematocrit (r^2= -0.22, P=0.01), creatinine (r^2= -0.3, P=0.01), GGT (r^2= -0.2, P=0.008), Na (r^2= -0.3, P=0.000), Cl (r^2= -0.3, P=0.003), P (r^2= -0.3, P=0.002) and lactate at rest (r^2= -0.3, P=0.003). Albumin, total protein, urea, triglycerides, K did not show any correlation with body weight variations or cumulative workload.

Although energy intake was adjusted according to workload (1.2 to 1.5 times maintenance energy requirements: 4.3 UFC for geldings and 4.7 UFC for stallions) horses exhibited a significant but moderate weight loss, instead the BCS remained relatively constant. It was hypothesized that this loss in body weight might be due to fatigue and to the excessive catabolic effect of some hormones and cytokines stimulated by workload and stress. The increase of CPK, initially, could be the result of an increase fitness level, whereas, the increase observed during intense exercise period could be the result of a muscular suffering probably related with excessive training loads. The observed decrease in Na and Cl during the test period might be due to sweat production and the accompanying loss of electrolytes which was higher than the water loss and consequent reduction of these ions in the plasma, without any signs of dehydration. However this decrease was progressive and correlated with cumulative workload and might, therefore be a marker of exercise intolerance (McKeever, 2008).

References

Bayly, W.M., 2002. Overtraining in the racehorse. Ippologia, 13, 33-37.

Golland, L.C., D.L. Evans, G.M. Stone, C.M. Tyler-McGowan, D.R. Hodgson and R.J. Rose, 1999. Plasma cortisol and β-endorphin concentrations in trained and over-trained Standardbred racehorses. Eur. J. Physiol. 439, 11-17.

Golland, L.C., D.L. Evans, C.M. McGowan, D.R. Hodgson and R.J. Rose, 2003. The effects of overtraining on blood volumes in Standardbred racehorses. Vet. J. 165, 228-223.

Hamlin, M.J., J.P. Shearman and W.G. Hopkins, 2002. Changes in physiological parameters in overtrained Standardbred racehorses. Eq. Vet J. 34, 383-388.

Hinchcliff, K.W. and R.J. Geor, 2004. Integrative physiology of exercise. In: Hinchcliff, K.W., A.J. Kaneps and R.J. Geor (eds.), Equine sports medicine and surgery. Sunders company, Edinburgh, London, New York, Oxford, Philadelphia, St Louis, Sydney, Toronto, 3-8.

Kuipers, H., 1998. Training and overtraining – an introduction. Med. Sci. Sports Exer. 30, 1137-1139.

Martin-Rosset, W. (ed.), 1990. Alimentation des chevaux. INRA Editions, Versailles, France, 232 pp.

Martin-Rosset, W., J. Vernet, H. Dubroeucq, G. Arnaud, A. Picard and M. Vermorel, 2008. Variation of fatness and energy content of body with body condition score in sport horses and its prediction. In: M. Saastamoinen and W. Martin-Rosset (eds.), Nutrition of the exercising horse. EAAP publication no 125. Wageningen Academic Publishers, Wageningen, the Netherlands, p. 167-176.

McGowan, C.M., L.C. Golland, D.L. Evans, D.R. Hodgson and R.J. Rose, 2002. Effects of prolonged training, overtraining and detraining on skeletal muscle metabolites and enzymes. Eq. Vet. J. 34 (suppl), 257-263.

McGowan, C.M. and D.J. Whitworth, 2008. Overtraining syndrome in horses. Comp. Ex. Physiol. 5(2), 57-65.

McKeever, K.H., 2008. Body fluid and electrolytes : responses to exercise and training. In: Hinchcliff, K.W., R.J. Geor and A.J. Kaneps (eds.), Equine exercise physiology. Sunders Elsevier, Edinburgh, London, New York, Oxford, Philadelphia, St Louis, Sydney, Toronto, pp 328-346.

Tyler, C.M., L.C. Golland, D.L. Evans, D.R. Hodgson and R.J. Rose, 1996. Changes in maximum oxygen uptake during prolonged training, overtraining and detraining in horses. J. Appl. Physiol. 81, 2244-2249.

Tyler-McGowan, C.M., L.C. Golland, D.L. Evans, D.R. Hodgson and R.J. Rose, 1999. Haematological and biochemical responses to training and overtraining. Eq. Vet. J. 30 (suppl), 621-625.

Effect of a new alive lactic bacteria as a probiotic on organic matter and cell-walls digestibility in competing endurance horses

A.G. Goachet[1], C. Berger[2] and V. Julliand[1]
[1]AgroSup Dijon, 26 bd Dr Petitjean, 21000 Dijon, France
[2]Danisco, 75000 Paris, France

Abstract

This study aimed at measuring the impact of an alive lactic bacterial probiotic on organic matter and cell-walls digestibility in endurance horses, before, during and after a 130 km endurance event. Six endurance horses, in regular endurance training program and competing in two 130 km endurance events, were used in a 3x2 Latin square design and received two dietary treatments : control and TSH-MS01 (5.10^9 CFU/day of each bacterial strain: *Propionibacteria* and *Lactobacillus plantarum)*. The supplementation was given for five weeks: from d-21 before to d+13 after each 130 km event. Digestibility of organic matter (OM), neutral ditergent fibre (NDF), acid detergent fibre (ADF), (NDF-ADF) and (ADF-ADL) was determinated by a 4-day partial fecal collection (2 grab samples/ day at 08:30 and 17 :30) using ADL as an internal marker. Five pools were constituted from fecal samples taken from d-7 to d-4 (Pool 1), d-1 to d+2 (Pool 2), d0 to d+3 (Pool 3), d+1 to d+4 (Pool 4) and d+10 to d+13 (Pool 5). When the data obtained from pools 1, 2 and 5 were compiled, digestibility coefficients were higher when horses were supplemented. In addition, OM and ADF digestibility for the period around the event (from d-1 to d+2) was higher when this bacterial type probiotic was fed. These results suggest that these alive lactic bacteria enhanced OM and cell-walls digestibility in endurance horses. The better fibre digestibility is of particular interest as it suggests a greater energy supply, during training and competition.

Keywords: probiotic, lactic-acid bacteria, organic matter digestibility, cell wall digestibility

Introduction

Endurance horses diet commonly contained 80% forages (Crandell, 2002), which is higher in comparison to common diet of other athletic horses such as racing and show-jumping horses (Ellis and Saastamoinen, 2008; Martin *et al.*, 2008). In endurance horses, it can be assumed that energy from the dietary feeds may be essentially provided by the microbial degradation of cell-walls in the hindgut. Therefore, it is crucial to optimize the forage digestibility for increasing energy from fibre degradation during training and competition.

Cell-walls digestibility was reported to increase with yeast type probiotic by improving the fibrolytic microbial activity in the hindgut of horses (Jouany *et al.*, 2008; 2009). Such an impact has not been demonstrated with bacterial type probiotic (Swyers *et al.*, 2008). This study aimed at measuring the impact of an alive lactic bacterial probiotic on organic matter and cell-walls digestibility in endurance horses, before, during and after a 130 km endurance.

Material and methods

Six endurance horses (pure-bred Arabians, aged 10.8±2.1 years, with a body weight (BW) of 422.5±20.7 kg and a Body Condition Score (BCS) of 3.2±0.2 (on a scale of 0 to 5) at the beginning of the experiment) were used in 3x2 Latin square design. Two treatments were tested : control and TSH-MS01.

Horses underwent a regular endurance training program for 8 consecutive months and competed in two 130 km-endurance events after 14±2 and 24±4 weeks of conditioning. For each endurance event, horses were transported to the event locations two days before the competition, and transported back to the training center the day after the race.

Horses were fed a basal diet of 85% meadow hay (31% cellulose, 66% NDF, 36% ADF) and 15% concentrate pelleted feed (commercial horse feed containing 30% wheat bran, 14% barley, 10% alfalfa, providing 12.5% CP, 3.5% fat, 15% cellulose, 40% NDF and 17% ADF) at 2.6 kg DM / 100 kg BW. Hay and concentrate were offered in two equal meals, at 10:00 and 16:00 and at 8:00 and 17:30 respectively. During the two days spent on the event location prior to the race, concentrate and hay were offered according to the same schedule and amounts as previously described, excepted during the night before the race, where 10 kg hay was offered. On the race day, any concentrate meal was offered right before departure. At each vet-gates, 1.2 kg concentrate feed and 3 kg meadow hay were offered. At the end of the race, horses received no concentrate feed but 10 kg hay. During the measurement periods, i.e. from 7 days to 13 days before and after each race, all the feeds offered and refused were weighed and recorded for each horse.

2 g of TSH-MS01 (5.10^9 CFU/day of each bacterial strain : *Propionibacteria* and *Lactobacillus plantarum*) was offered mixed to 50 g of cereal blend with the 8:00 concentrate meal for five weeks : from d-21 before to d+13 after each 130 km event. During the endurance event, 2 g of TSH-MS01 mixed to 50 g of cereal blend was given with each concentrate meal at the vet-gates.

Total tract apparent digestibility of organic matter (OMd), neutral detergent fiber (NDFd), acid detergent fiber (ADFd), NDF-ADF as considered as an estimation of hemicelluloses (NDF-ADFd) and ADF-ADL as considered as an estimation of celluloses (ADF-ADLd) was determinated by a 4-day partial fecal collection using ADL as an internal marker (Goachet *et al.*, 2009). Samples of 300 g of fresh faeces were taken from each horse by rectal sampling at 08:30 and 17:30 every day during 21 days : from 7 days to 13 days before and after each race. From d-7 to d-3, samples were taken at the stable before travelling to the race place, when horses were in training. From d-2 to d+1, samples were taken at the event location. From d+2 to d+13, samples were taken at the stable, during the resting period allowed after each endurance event ; during this period, horses were not exercised. Faecal as well as feed samples were dried in an air-forced oven at 75 °C to constant weigh, for DM determination, and ground to pass through a 1 mm screen. For each horse and each period, five pools of feces were then constituted from samples taken from d-7 to d-4 (Pool 1), d-1 to d+2 (Pool 2), d0 to d+3 (Pool 3), d+1 to d+4 (Pool 4) and d+10 to d+13 (Pool 5). Feed and faecal samples were analysed for OM by incineration at 550 °C for 5 hours (71/250/CEE), and sequential analysis of NDF, Acid ADF and ADL (Van Soest *et al.*, 1991).

Results have been statistically analysed by the GLM procedure of SASv8 in a model including repeated measures. The model included the effects of treatment, animal and period. Differences were considered statistically significant at $P<0.05$.

Results

Horses consumed easily the TSH-MS01 supplementation during all the experiment. When analyzed on each pool, digestibility coefficients of OM and ADF were higher in Pool 2 in horses fed TSH-MS01 treatment (Table 1).

When the data obtained from pools 1, 2 and 5 (which have no sampling days in common) were compiled, digestibility coefficients were higher when horses were supplemented (Table 2).

Table 1. Mean ± SD apparent total tract digestibility of OM and cell-walls in horses fed control or TSH-MS01 supplementation, at each time of measurement (n=6).

		Pool 1 (d-7 to d-4)	Pool 2 (d-1 to d+2)	Pool 3 (d0 to d+3)	Pool 4 (d+1 to d+4)	Pool 5 (d+10 to d+13)
OMd	Control	53.5±4.8	51.7±2.9 [a]	51.6±3.7	51.8±4.5	53.4±3.3
	TSH-MS01	55.7±1.8	56.6±3.2 [b] (*P*=0.02)	55.5±4.7	55.2±2.4	56.5±3.1
NDFd	Control	42.7±6.7	40.9±3.0	40.8±3.2	41.5±3.6	43.5±5.1
	TSH-MS01	45.6±2.7	45.8±4.3	44.5±5.9	44.7±3.3	47.5±4.7
ADFd	Control	38.0±6.8	36.3±3.6 [a]	36.5±3.0	37.6±3.7	37.9±4.9
	TSH-MS01	41.2±2.4	39.8±3.7 [b] (*P*=0.03)	39.4±4.7	39.5±3.4	42.1±3.7
(NDF-ADF)d	Control	48.1±7.1	46.0±2.6	45.7±3.8	46.0±4.2	49.8±5.4
	TSH-MS01	50.7±3.2	52.4±5.1	50.2±7.3	50.6±3.4	53.6±6.1
(ADF-ADL)d	Control	43.5±7.7	41.6±4.2	41.8±3.5	43.0±4.3	43.3±4.2
	TSH-MS01	47.0±2.9	45.7±4.3	45.3±5.5	45.3±3.9	48.2±4.2

[a,b] Means in column with different superscripts are significantly different ($P<0.05$); OMd: digestibility of organic matter, NDFd: neutral detergent fiber, ADFd: acid detergent fiber.

Table 2. Mean ± SD apparent total tract digestibility of OM and cell-walls in horses fed control or TSH-MS01 supplementation, before and after endurance events (n=6).

	Control	TSH-MS01	*P* value
OMd	52.9±3.7[a]	56.3±2.7[b]	0.001
NDFd	42.4±5.0[a]	46.3±3.9[b]	0.003
ADFd	37.4±5.1[a]	41.0±3.3[b]	0.001
(NDF-ADF)d	48.0±5.3[a]	52.3±4.9[b]	0.008
(ADF-ADL)d	42.8±5.8[a]	47.0±3.8[b]	0.001

[a,b] Means in rows with different superscripts are significantly different ($P<0.05$); OMd: digestibility of organic matter, NDFd: neutral detergent fiber, ADFd: acid detergent fiber.

Discussion and conclusion

Our data suggest that TSH-MS01 enhanced OM and cell-walls digestibility in endurance horses, who consumed high amounts of forage in their ration. In addition, OM and ADF digestibility for the period around the event (from d-1 to d+2) was higher when this bacterial type probiotic was fed. This impact could be explained by an increase in fibrolytic micro-organisms or in their activity as suggested by Faubladier *et al.*, (2010) and previously shown with yeast (Médina *et al.*, 2002; Santos *et al.*, 2008). Further investigations are needed to confirm the mechanisms of action of lactic bacteria. Also more studies are necessary to corroborate our first data, in order to register this new probiotic in agreement with EU directives. The better fibre digestibility is of particular interest as it suggests a greater energy supply for endurance horses, during training and competition.

References

Crandell, K., 2005. Trends in feeding the American endurance horse. In: Pagan, J. and R.J. Geor (eds.), Advances in Equine Nutrition III, Versailles, Kentucky, USA, 135-138.

Ellis, A. and M. Saastamoinen, 2008. Feeding growing race horses at work. In: Saastamoinen, M.T. and W. Martin-Rosset (eds.), Nutrition of the exercising horse. EAAP publication no 125, Wageningen Academic Publishers, Wageningen, the Netherlands, p. 311-331.

Faubladier, C., E. Jacotot, C. Berger and V. Julliand, 2010. Effects of a live or heat-treated lactic-acid bacteria versus a placebo on faecal microbial communities and activities in horses. In: Ellis, A.D., A.C. Longland, M. Coenen and N. Miragli (eds.), The impact of nutrition on the health and welfare of horses. EAAP publication no. 128, Wageningen Academic Publishers, Wageningen, the Netherlands, p. 288-292.

Goachet, A.G., C. Philippeau, M. Varloud and V. Julliand, 2009. Adaptations to standard approaches for measuring gastro-intestinal retention time and apparent digestibility in horses in training. Anim. Feed Sci. Technol., 152, 141-151.

Jouany, J.P., J. Gobert, B. Medina, G. Bertin and V. Julliand, 2008. Effect of live yeast culture supplementation on apparent digestibility and rate of passage in horses fed a high-fiber or high-starch diet. J. Anim. Sci., 86, 339-347.

Jouany, J.P., J. Gobert, B. Medina, G. Bertin and V. Julliand, 2009. Effect of live yeast culture supplementation on hindgut microbial communities and their polysaccharidase and glycoside hydrolase activities in horses fed a high-fiber or high-starch diet. J. Anim. Sci., 87, 2844-2852.

Martin, L., O. Geffroy, A. Bonneau, C. Barré, P. Nguyen and H. Dumon, 2008. Nutrient intake in high performance show-jumping horses in France. In: Saastamoinen, M.T. and W. Martin-Rosset (eds.), Nutrition of the exercising horse. EAAP publication no 125, Wageningen Academic Publishers, Wageningen, the Netherlands, p 333-339.

Médina, B., E. Girard, E, Jacotot and V. Julliand, 2002. Effect of a preparation of saccharomyces cerevisiae on microbial profiles and fermentation patterns in the large intestine of horses fed a high fiber or a high starch diet. J. Anim. Sci., 80, 2600-2609.

Santos, A.S., C.M.V. Guedes, L.M.M. Ferreira, L. Evangelista, P.H. Gaspar, G. Bertin, M.A.M. Rodrigues, 2008. The effect of Saccharomyces cerevisiae CBS 493.94 level on *in vitro* fermentation using faecal inoculum from horses fed diets with different fibre concentrations. In: Saastamoinen, M.T. and W. Martin-Rosset (eds.), Nutrition of the exercising horse. EAAP publication no 125, Wageningen Academic Publishers, Wageningen, the Netherlands, p. 361-363.

Swyers, K.L., A.O. Burk, T.G. Hartsock, E.M. Ungerfeld and J.L. Shelton, 2008. Effects of direct-fed microbial supplementation on digestibility and fermentation end-products in horses fed low- and high-starch concentrates. J. Anim. Sci., 86, 2596-2608.

Van Soest, P.J., J.B. Robertson and B.A. Lewis, 1991. Methods for dietary fibre, neutral detergent fibre, and non-starch polysaccharides in relation to animal nutrition. J. Dairy Sci., 74, 3583-3597.

Accelerometer based system for the monitoring of exercise in horses

K. Kubus[1], R.P. Wilson[2], I. Vervuert[1] and M. Coenen[1]
[1]Institute of Animal Nutrition, Nutrition Diseases & Dietetics, Gustav-Kühn-Straße 8, 04159 Leipzig, Germany
[2]Institute of Environmental Sustainability, School of the Environment and Society, University of Wales Swansea, United Kingdom

Introduction

Accelerometer based systems are of interest in equine locomotion studies (Keegan *et al.*,2002). Additionally, accelerometric measurements have been used successfully in wild animals (Wilson *et al.*, 2006) to obtain data on energy expenditure by transforming the raw kinetic data into caloric dimensions. The aim of the present study was to examine whether the accelerometer can monitor exercise intensity and reflect energy expenditure in horses

Material and methods

Heart rate (HR, Polar®) was used as an indicator for energy expenditure (Coenen, 2008). A simple accelerometer device ('Goldfinger') recording movement in a three-dimensional scale at 32 Hertz was fixed at the equine neck. Four horses (506-602 kg BW) were exercised in hand (5 min walk, 3 min trot, 3 min walk, 6 repetitions) or free exercised in a fenced oval track (5 min walk, 10 min trot, 10 min walk, 6 rep.). Five horses (330-535 kg BW) were ridden in all gaits (5 min walk, 10 min trot, 5 min gallop, 10 min walk, 5 rep.). The data of the accelerometer units (AU, corresponding to gravity) and HR were analyzed by Statistica 7.1®.

Results and discussion

A simple linear regression model described the relationship between AU and HR; however the scatter indicated that AU´s are not as precise as required to give a prediction of HR. The inclusion of gait (coded by 1-3 for walk, trot, gallop) as an independent variable yielded improved models (handheld exercise: HR=28.0+6.6x+34z, r^2=0.81; free exercise HR=33.4+23.1x+20.9z, r^2=0.8; ridden, HR=24.1+25.3x+19.5z, r^2=0.72, x =AU, z =gait code). This showed that the accelerometer was insensitive to mirror changes in HR during a non-uniform profile of exercise. Important was, that after lowering speed HR dropped by a delay while AU´s did so immediately. Furthermore, the plasticity of HR may mask a closer link of AU´s to oxygen consumption.

References

Coenen, M., 2008. The suitability of heart rate in the prediction of oxygen consumption, energy expenditure and energy requirement for the exercising horse. In: Saastamoinen, M.T. and W. Martin-Rosset (eds.), Nutrition of the exercising horse. EAAP publication no 125, Wageningen Academic Publishers, Wageningen, the Netherlands, p. 139-146.

Keegan, K.G., Y. Yonezawa, P.E. Pai and D.A. Wilson, 2002. Accelerometer-based system for the detection of lameness in horses. Biomed. Sci. Instrum. 38, 107-112.

Wilson, R.P.,C.R. White, F. Quintana, L.G. Halsey, N. Liebsch, G.R. Martin and P.J. Butler, 2006. Moving towards acceleration for estimates of activity-specific metabolic rate in free-living animals: the case of the cormorant. J. Anim. Ecol. 75, 1081-1090.

Keyword index

Author index

Sponsors

CAVALOR®

Alltech®

HORSE HEALTH NUTRITION
5th Edition
April 15-16 2011
European Equine Health & Nutrition Congress

SARACEN
HORSE FEEDS

Protexin®
EQUINE PREMIUM

SPILLERS®

LITOVET®
EQUINE JOINT HEALTH
Active Ingredient
GOPO®
from 100% natural ROSE-HIP

TRM

WALTHAM®
the science behind SPILLERS® & WINERGY®

FIT FOR LIFE WITH NEW
LITOVET® JOINT HEALTH
THE UK'S NEW UNIQUE JOINT SUPPLEMENT SUPPORTED BY PUBLISHED RESEARCH
"The difference is remarkable"
James Fanshawe, Racehorse Trainer, Newmarket
CONTAINS THE UNIQUE PATENTED ACTIVE INGREDIENT GOPO®
LitoVet® a rich standardised source of GOPO® (glycoside of mono and diglycerol), a key component of the specially cultivated Rosa Canina variety of rose-hips. LitoVet® is also naturally high in antioxidants especially bio-available Vitamin C essential for the formation and maintenance of connective tissues – including joint cartilage, tendons, ligaments and bone. Together these active ingredients support the body's natural recovery processes following exercise.
SUPPORTED BY PUBLISHED RESEARCH
Vigorous research has established that LitoVet® can help maintain soundness, suppleness and mobility in equine joints and joint tissues, helping to offset the rigours of training or competition. This can reduce the likelihood of loss of performance, due to over-reaching or over-training.
Rose-hip extract that does not contain GOPO® has not been shown to help maintain joint mobility.
VETERINARY RECOMMENDED
100% NATURAL
SUITABLE FOR USE WITH ALL COMPETITION HORSES AND PONIES
WINNER OF THE BETA (British Equestrian Trade Association) INNOVATION AWARD 2010
LITOVET®
EQUINE JOINT HEALTH
Active Ingredient
GOPO®
from 100% natural
ROSE-HIP
Available from all leading stockists or call G R Lane Health Products Ltd. on the number below.
G R Lane Health Products Ltd., Sisson Road, Gloucester, GL2 0GR Tel: 01452 524012
email info@litovet.co.uk www.litovet.co.uk

IT'S TIME TO...
Only 63p per day*
SARACEN
SHAPE-UP™
CONCENTRATED LOW INTAKE FEED
FORMS PART OF A CALORIE CONTROLLED DIET
INNOVATORS IN Kentucky Equine Research, Inc. TEAM MEMBER NUTRITION
SUPER FIBRE TECHNOLOGY
BIOPLEX MINERALS
E-MAX ESTER
OMEGA 3
ACID BUFFER
UPTO 25 SERVINGS
AN ESSENTIAL PART OF YOUR 5 A DAY
1 EXERCISE
2 WATER
3 FORAGE
4 FEED
5 HEALTH CHECK
www.fitfreddy.org
HE PERFECT STABLEMATE
O KEEPING YOUR HORSE
OR PONY IN SHAPE.
Saracen Slim-Chaff™ is the ideal low sugar chaff to accompany Shape-Up™
SLIM-CHAFF
Your horse relies on you to keep him happy, fit and healthy through regular exercise, health checks and a balanced, nutritional diet. He counts on you to make sure his body weight is not affecting his health, leading to problems like cardiovascular and skeletal system disorders or the risk of laminitis and hyperlipemia, which can all have severe consequences. To help you keep your horse's weight stable, Saracen have created Shape-Up™. This nutrient dense, low sugar, low starch feed complements forage and helps form part of a carefully calorie controlled diet – helping to maintain a healthy and normal metabolism. So, for a horse as fit as a fiddle, keep him in tune with Saracen Shape-Up™.
For those who know what they're on.
SARACEN HORSE FEEDS
don't know what you're on, visit our online horse feed service: www.saracenhorsefeeds.com or call 01622 718487 for more information.
*based on RRP of £15.80 and pony being fed 0.8kg per day.

LIFEFORCE™
FORMULA
Designed to bring out the very best
in the equine athlete by ensuring
overall health of the horse
From the
Alltech®
advantage series
PROUD SPONSOR OF
Alltech FEI™
World Equestrian Games™
Kentucky 2010
www.alltech.com/games

www.ingramcontent.com/pod-product-compliance
Lightning Source LLC
LaVergne TN
LVHW082053011026
849162LV00020B/1637
* 9 7 8 9 0 8 6 8 6 1 5 5 2 *